Petrogenesis of Metamorphic Rocks

Springer
Berlin
Heidelberg
New York
Barcelona
Hong Kong
London
Milan
Paris
Tokyo

Kurt Bucher · Martin Frey (†)

Petrogenesis of Metamorphic Rocks

7th Completely Revised and Updated Edition

With 99 Figures and 25 Tables

 Springer

Professor Dr. Kurt Bucher
University of Freiburg
Institute of Mineralogy, Petrology and Geochemistry
Albertstr. 23b
79104 Freiburg i. Br.
Germany
E-mail: bucher@uni-freiburg.de
Web site: http://www.minpet.uni-freiburg.de

Professor Dr. Martin Frey (†)

ISBN 3-540-43130-6 Springer-Verlag Berlin Heidelberg New York
ISBN 3-540-57567-7 6th ed. Springer-Verlag Berlin Heidelberg New York

Library of Congress Cataloging-in-Publication Data
Bucher, Kurt, 1946–
    Petrogenesis of metamorphic rocks / Kurt Bucher, Martin Frey. – 7th, completely
revised and updated ed.
        p. cm.
    Includes bibliographical references and index.
    ISBN 3540431306 (alk. paper)
    1. Rocks, Metamorphic. I. Frey, Martin, 1940– II. Title.
QE475.A2 B84 2002
552'.4–dc21   2002070560

Springer-Verlag Berlin Heidelberg New York
a member of BertelsmannSpringer Science+Business Media GmbH

http://www.springer.de

© Springer-Verlag Berlin Heidelberg 2002
Printed in Germany

Production: PRO EDIT GmbH, Heidelberg, Germany
Cover-Design: Erich Kirchner, Heidelberg, Germany
Typesetting: K+V Fotosatz, Beerfelden, Germany

Printed on acid-free paper   SPIN 10834875   32/3130/Di   5 4 3 2 1 0

Preface to Sixth Edition

Metamorphic rocks are one of the three classes of rocks. Seen on a global scale they constitute the dominant material of the Earth. The understanding of the petrogenesis and significance of metamorphic rocks is, therefore, a fundamental topic of geological education. There are, of course, many different possible ways to lecture this theme. This book addresses rock metamorphism from a relatively pragmatic viewpoint. It has been written for the senior undergraduate or graduate student who needs practical knowledge of how to interpret various groups of minerals found in metamorphic rocks. The book is also of interest for the non-specialist and non-petrologist professional who is interested in learning more about the geological messages metamorphic mineral assemblages are sending as well as pressure and temperature conditions of formation.

The book is organized into two parts. The first part introduces the different types of metamorphism, defines some names, terms and graphs used to describe metamorphic rocks, and discusses principal aspects of metamorphic processes. Part I introduces the causes of metamorphism on various scales in time and space and some principles of chemical reactions in rocks that accompany metamorphism, but without treating these principles in detail and presenting the thermodynamic basis for quantitative analysis of reactions and their equilibria in metamorphism. Part I also presents concepts of metamorphic grade or intensity of metamorphism, such as the metamorphic-facies concept. Also, a brief presentation of techniques and methods for estimating pressure and temperature conditions at which metamorphic rocks once were formed a long time ago can be found in Part I (geothermobarometry).

Part II deals systematically with prograde metamorphism of six different classes of bulk rock compositions. The categories of rock compositions treated cover most of the sedimentary, igneous and metamorphic precursor material that is commonly encountered. Some more exotic rock compositions, for example, ironstones, evaporites, laterites and manganese-rich rocks, are not covered by the book. Part II can, in principle, be used alone provided that the reader is familiar with common types of petrologic phase diagrams. Part II makes extensive use of computed phase diagrams. The routine computation of phase diagrams is one of the major developments in petrology of the last few years. It was made possible by a continuous and tremendous effort by scientists in experimental laboratories who have vastly expanded the database of material of geological interest through phase equilibria and thermochemical studies. Advanced mathematical and thermodynamic analysis of the experimental data resulted in the compilation of thermodynamic data of phase components of geologic substances and in the development of solution models for solid mineral solutions and gas mixtures at high pressures and temperatures. Finally, several computer software packages were developed during recent years that make the computation of phase diagrams convenient and easy. Computation of phase diagrams is becoming increasingly widespread among geologists, and as the reader will see, this book in its present form could not been written 5 years ago.

In general, the book focuses on the interpretation and significance of mineral assemblages and chemical reactions in metamorphic rocks. Several important topics concerning metamorphism and metamorphic rocks are not covered by this book. These topics include metamorphic micro-structures and their interpretation, deformation and metamorphism, geodynamic aspects of metamorphism and the thermodynamic basis for the quantitative treatment of chemical reactions in rocks.

The references are cited at the end of each chapter, the lists of references include a selection of important papers and books on the subject covered by the respective chapter that are highly recommended for further reading. The bibliography includes classical papers in the field as well as more recent contributions that may help to provide access to research in specialized fields via the references cited in the listed research (e.g., granulite-facies metamorphism of pelitic rocks, etc.).

Many of the phase diagram figures show numbered reaction equilibria that are formulated and listed in tables. Reaction numbers are also frequently used in the text. We recommend copying the tables with the reaction stoichiometries and keeping them on hand, which will help eliminate the need to flip back and forth between text and reaction tables.

We also strongly recommend regular reading of current issues of scientific journals that arrive at your library. In the field of metamorphic petrology, the Journal of Metamorphic Geology is essential reading and a few of the other particularly relevant journals are *Journal of Petrology, Contributions to Mineralogy and Petrology, American Mineralogist, European Journal of Mineralogy, Mineralogical Magazine, Lithos, Chemical Geology* and *Earth and Planetary Science Letters.*

Freiburg and Basel (March 1994)   Kurt Bucher and Martin Frey

Preface to Seventh Edition

The new edition of *Petrogenesis of Metamorphic Rocks* has several completely revised chapters and all chapters have updated references and redrawn figures. The chapters on metamorphic grade and metamorphism of ultramafic rocks have been rewritten. Also, the introduction and definitions have experienced some major changes. An important new addition is the reference made to important websites for metamorphic petrology tutorials, software, mail bases, and more. The links to these sites may fail to work over time. We will try to maintain and update metamorphism-related links at our own website: http://www.minpet.uni-freiburg.de.

We encourage you to regularly read (or glance through) the current issues of scientific journals that arrive at your library. In the field of metamorphic petrology, the *Journal of Metamorphic Geology* is essential reading and a few of the other particularly relevant journals are *Journal of Petrology, Contributions to Mineralogy and Petrology, Geofluids, American Mineralogist, European Journal of Mineralogy, Lithos, Chemical Geology* and *Earth and Planetary Science Letters*.

The work on the second edition of *Petrogenesis of Metamorphic Rocks* was deeply affected by the unexpected and tragic death of Martin Frey. He died in an accident during a hike in the Swiss Alps on September 10, 2000, a few months after celebrating his 60th birthday. It was not far from his beloved Lukmanier field area that he fell over a cliff during a steep descent and died from severe injuries. It was a clear Sunday in early fall and the message that reached us at Lukmanier Pass Hospiz during a mapping field course deeply shocked all of us. Martin was a very open and generous colleague and it was a great pleasure for me to first work in his Institute in Basel as Assistant Professor and later, during my stays in Oslo and Freiburg, to write and edit *Petrogenesis of Metamorphic Rocks* together with him. His e-mails and chats on the phone were always cheerful highlights of my working days. I miss him very much.

Freiburg, December 2001   Kurt Bucher

Acknowledgments

The revision of this book benefited from software help provided by Jamie Connolly (ETH Zürich) and Christian de Capitani (Basel). The patience of the Freiburg students in following my convoluted lectures and their helpful suggestions for improvement of the text are gratefully acknowledged. Very useful suggestions, comments and corrections were also supplied by Bernard Evans (and his students), B. Storre and H. Kawinski. Also, various discussions on the "Geometamorphism Mailbase" were useful for the improvement of the book. One round of discussion was, in fact, directly related to Fig. 2.1 of the sixth edition. Martin Engi provided the P–T contour map of the Central Alps. Martin Engi and Dave Pattison are thanked for stimulating dialogues on amphibolites and granulites. Christian de Capitani instructed me in the latest developments regarding the "pseudosection virus" that is rapidly infecting the geometamorphism community.

Contents

Part II

Part I

CHAPTER 1

Definition, Conditions and Types of Metamorphism

The process of rock **metamorphism** changes the mineralogical and chemical composition, as well as the structure of rocks. Metamorphism is typically associated with elevated temperature and pressure; thus it affects rocks within the earth's crust and mantle. The process is driven by changing physical and/ or chemical conditions in response to large-scale geological dynamics. Consequently, it is inherent in the term, that metamorphism is always related to a precursor state where the rocks had other mineralogical and structural attributes. Metamorphism, metamorphic processes and mineral transformations in rocks at elevated temperatures and pressures are fundamentally associated with chemical reactions in rocks. Metamorphism does not include, by definition, similar processes that occur near the earth's surface such as weathering, cementation and diagenesis. The transition to igneous processes is gradual; metamorphism may include partial melting. The term metasomatism is used if modification of the bulk composition of rocks is the dominant metamorphic process. **Metamorphic rocks** are rocks that developed their mineralogical and structural characteristics by metamorphic processes.

1.1
Conditions of Metamorphism

Large-scale geologic events such as global plate movements, subduction of oceanic lithosphere, continent-continent collision and ocean floor spreading all have the consequence of moving rocks and transporting heat. Consequently, changes in pressure (depth) and temperature are the most important variables in rock metamorphism. For example, consider a layer of sediment on the ocean floor that is covered with more sediment layers through geologic time and finally subducted at a destructive plate margin. The mineralogical, chemical and structural transformation experienced by the sediment can be related to a gradually increasing temperature with time. The question may be asked: at what temperature does metamorphism begin?

1.1.1
Low-Temperature Limit of Metamorphism

Temperatures at which metamorphism sets in are strongly dependent on the material under investigation. Transformation of evaporites, of vitreous material and of organic material, for example, begins to take place at considerably lower temperatures than chemical reactions in most silicate and carbonate rocks. This volume is not concerned with the metamorphism of organic material, i.e., coalification (maturation). For a review of this subject, the reader is referred to Teichmüller (1987). Correlations between the rank of coalification and mineralogical changes in slightly metamorphosed sediments and volcanic rocks have been reviewed by Kisch (1987).

In many rocks, mineral transformations begin shortly after sedimentation and proceed continuously with increasing burial. Whether such reactions are called "diagenetic" or "metamorphic" is largely arbitrary. Further examples of processes that are intimately related to metamorphism include low-temperature alteration of volcanic rocks, precipitation of mineral coatings in fractures and low-temperature rock-water reactions that produce mineral-filled veins and fissures. Hence, metamorphic processes occur in a temperature continuum from surface temperature upward. However, the low-temperature limit of metamorphism has been arbitrarily set at $150 \pm 50\,°C$ in this volume and most of our phase-diagrams show phase equilibria above $200\,°C$.

Some minerals are considered distinctly metamorphic, i.e., they form at elevated temperature and are not found in diagenetically transformed sediments. In a regime of increasing temperature, the first occurrence of such minerals in sediments thus marks the onset of metamorphism. Indicators of the beginning of metamorphism include: carpholite, pyrophyllite, Na-amphibole, lawsonite, paragonite, prehnite, pumpellyite and stilpnomelane. Note, however, that these minerals may also be found as detrital grains in unmetamorphosed sediments. In this case, textural evidence from a thin-section will distinguish between a neoformation and a detrital origin. For a more detailed discussion of problems dealing with the low-temperature limit of metamorphism and its delimitation from diagenesis, the reader is referred to Frey and Kisch (1987) and Robinson and Merriman (1999).

1.1.2
High-Temperature Limit of Metamorphism

Ultimately, at high temperature, rocks will start to melt and dealing with silicate melts is the subject of igneous petrology. However, partial melting has always, both, a metamorphic and an igneous aspect. Crustal rocks that are characteristically produced by partial melting, so called migmatites, are made up from a residual metamorphic rock and an igneous rock component. Nevertheless, melting temperatures of rocks define the high-temperature limit of metamorphism. Melting temperatures are strongly dependent on pressure, rock composition and the amount of water present. For example, at 5 kbar and in the presence of an aqueous fluid, granitic rocks begin to melt at a temperature of about $660\,°C$ while basaltic rocks need a much higher temperature of about $800\,°C$ (Fig. 1.1). If H_2O is absent, melting temperatures are much higher. Granitic gneisses will not melt below about $1000\,°C$; mafic rocks such as basalt require $>1120\,°C$ to melt. The highest temperatures reported from crustal metamorphic rocks are $1000–1100\,°C$ (e.g., Lamb et al. 1986, Fig. 3; Ellis 1980; Harley and Motoyoshi 2000; Hokada 2001). These

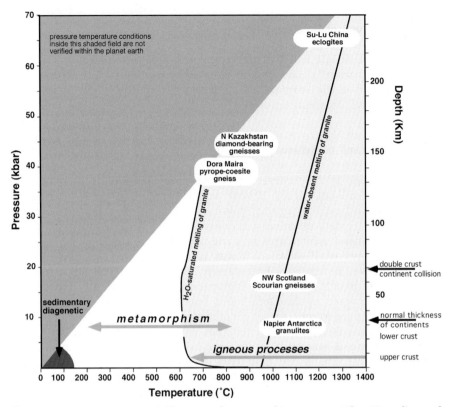

Fig. 1.1. Pressure–temperature (pT) range of metamorphic processes. The pT gradients of four typical geodynamic settings are shown. The boundary between diagenesis and metamorphism is gradational, but a T-value of ~150 °C is shown for convenience. Note that the metamorphic field has no upper pT limit on this diagram, and that there is a large overlap for metamorphic and magmatic conditions. Extreme pT conditions are shown for the following areas: (1) sapphirine quartzite from the Napier Complex, Antarctica (Harley and Motoyoshi 2000); (2) Scourian granulites from NW Scotland (Lamb et al. 1986, Fig. 3); (3) pyrope-coesite rocks from Dora Maira, Western Alps (Schertl et al. 1991); (4) diamond-bearing metamorphic rocks from the Kokchetav Massif, northern Kazakhstan (Shatsky et al. 1995); (5) garnet peridotite from the Su-Lu unit, east central China (Yang et al. 1993). The *wet granite melting curve* is after Stern and Wyllie (1981), and the *dry granite melting curve* is after Newton (1987, Fig. 1)

temperatures were determined by the indirect methods of thermobarometry that will be explained in Chapter 4.7. The temperature in the lower continental crust in geologically active areas may reach about 750–850 °C and the rocks produced by these extreme conditions are called granulites. This is also the typical upper temperature limit of crustal metamorphism. However, metamorphism is not restricted to the earth's crust. A given volume of rock in the convecting mantle continuously undergoes metamorphic processes such as recrystallization and various phase transformations in the solid state at temperatures in excess of 1500 °C.

1.1.3
Low-Pressure Limit of Metamorphism

Ascending hot silicate magmas are typical and globally important incidents in geologically active areas. The heat released by cooling silicate melts causes metamorphism in the surrounding country rocks in so-called contact aureoles (see below). This may occur at near-surface pressures of a few bars.

1.1.4
High-Pressure Limit of Metamorphism

For a long time it was believed that maximum pressures in metamorphic crustal rocks did not exceed 10 kbar, which corresponds to the lithostatic pressure at the base of a normal continental crust with a thickness of 30–40 km. As better data on mineral stability relations became available, it was found that mineral assemblages in some metamorphic mafic rocks often recorded pressures of 15–20 kbar. It was recognized early that these rocks, called eclogites, represented high-density and consequently high-pressure equivalents of basalts. Spectacular high-pressure rocks of clearly crustal origin were discovered in the Dora-Maira massif of the Alps (Chopin 1984). Gneisses with pure pyrope garnet containing coesite inclusions indicate pressures of at least 30 kbar. Today, eclogites with coesite inclusions in garnet and even diamond-bearing eclogites have been reported (Smith and Lappin 1989; Okay 1993; Zhang and Liou 1994; Coleman and Wang 1995; Schreyer 1995; Ernst 1999). Some of these rocks require at least 60 kbar to form. Rock transformations at such high pressures are called ultra-high-pressure (UHP) metamorphism. Similar super-high-pressure rocks have been discovered in many other places in the world (e. g., Reinecke 1991; Wain et al. 2000) since. It is clear that such super-high pressures must be related to subduction of crustal rocks to very great depths of >100 km.

However, metamorphism is not confined to the earth's crust or to subducted crustal rocks. Mantle rocks such as garnet peridotites (or garnet-olivine-pyroxene-granofelses) from ophiolite complexes or from xenoliths in kimberlite record pressures of >30–40 kbar (e.g., Yang et al. 1993). Consequently, it is possible to find and study rocks that formed 100–200 km below the earth's surface corresponding to pressures >30–60 kbar (3–6 GPa).

Temperature	Pressure
Low T-limit 0 °C for processes in near-surface environments, rock-water reactions Conventionally the term metamorphism implies T >150–200 °C	**Low P-limit** 1 bar at contact to lava flows at the surface
Upper T-limit In crustal rocks: 750–850 °C (max. recorded T~1050 °C) In many regional scale metamorphic areas T does not exceed ~650–700 °C	**High P-limit** Presently, some rocks collected at the earth's surface are known to have once formed at 100–200 km depth, = 30–60 kbar

1.2
Types of Metamorphism

On the basis of geological setting, we distinguish between metamorphism of local and regional extent:

Regional extent	Local extent
Orogenic metamorphism Subduction metamorphism Collision metamorphism Ocean-floor metamorphism Burial metamorphism	Contact metamorphism Cataclastic metamorphism Impact metamorphism Hydrothermal metamorphism

Such a subdivision is certainly useful, but it should be kept in mind that there exist commonly transitional forms between these genetic categories of metamorphism.

1.2.1
Orogenic Metamorphism

Orogenic metamorphism (Miyashiro 1973, p. 24) is caused by mountain building or orogenic process. It acts upon large volumes of rocks and the metamorphic effects of a specific mountain building episode are found over large, regional-scale dimensions. The term **regional metamorphism** can be used as a word with the same meaning. Metamorphism associated with a specific mountain building episode may show characteristic features. For example, Alpine metamorphism in the Alps is collectively used for all metamorphic processes associated with the formation of the Alps in late Cretaceous and Tertiary times. It affected rocks that experienced metamorphic transformations for the first time, i.e., Mesozoic sediments and volcanics. It also once more transformed rocks that already had been metamorphosed during earlier orogenic periods, such as Variscan gneisses, Caledonian metasediments and even material with Precambrian mineral assemblages. Rocks subjected to orogenic metamorphism usually extend over large belts, hundreds or thousands of kilometers long and tens or hundreds of kilometers wide, but may also be found within major strike-slip zones (Dempster and Bluck 1995).

In many higher-temperature parts of orogenic metamorphic belts, syn- or late-tectonic granites are abundant. Orogenic metamorphism and granitic plutons are often intimately associated. In the middle and upper crust, rising granitic plutons carry heat and thereby contribute to an increase in temperature on a regional scale leading to typical high-temperature low-pressure terrains. In the lower to middle crust, the granitic rocks were generated by partial melting as a consequence of high-grade metamorphism.

Orogenic metamorphism is commonly characterized by two fundamentally different kinds of regional-scale metamorphic transformations that follow each other in time. An early high-pressure, low-temperature type metamorphism is related to a subduction zone process, a later younger regional metamorphism following a moderate P–T gradient is related to continental collision. This dualism of orogenic metamorphism is related to major intervals of a Wilson cycle of mountain building.

Typically, metamorphic recrystallization in orogenic belts is accompanied by deformation. Such metamorphic rocks exhibit a penetrative fabric with preferred orientation of mineral grains. Examples are phyllites, schists and gneisses. Orogenic metamorphism is a long-lasting process of millions or tens of millions of years duration. It includes a number of distinct episodes of crystallization and deformation. Individual deformation phases appear to have definite characteristics, like attitude and direction of schistosity, folds, and lineations. Therefore, several phases of deformation can perhaps be recognized in the field and they often can be put into a time sequence. Microscope observations can unravel the relationships between structural features and mineral growth, and establish the time relations of deformation and metamorphism (e.g., Vernon 1976, p. 224).

Some features of orogenic metamorphism are summarized in Table 1.1. Rocks produced by regional orogenic and local contact metamorphism differ significantly in their fabric, being schistose as opposed to non-schistose, respectively. Furthermore, orogenic metamorphism takes place at higher pressures, while the temperature regime is often similar in both types.

1.2.2
Ocean-Floor Metamorphism

This type of regional metamorphism was introduced by Miyashiro et al. (1971) for transformations in the oceanic crust in the vicinity of mid-ocean ridges. The metamorphic rocks thus produced are moved laterally by ocean-floor spreading, covering large areas of the oceanic floor. The ocean-floor metamorphic rocks are mostly of basic and ultrabasic composition (basalts, peridotites). As most of these rocks are non-schistose, ocean-floor metamorphism resembles continental burial metamorphism (see below). A further

Table 1.1. Typical features of important types of metamorphism

Type of metamorphism	Orogenic (subduction type)	Orogenic (collision type)	Ocean-floor	Contact
Geologic setting	In orogenic belts, extending for several 1000 km². Early phase of orogenic metamorphism	In orogenic belts, extending for several 1000 km². Late phase of orogenic metamorphism	In oceanic crust and upper mantle, extending for several 1000 km²	Proximity to contacts to shallow-level igneous intrusions; contact aureole of a few m up to a some km width
Static/dynamic regime	Dynamic; generally associated with thrusting, slicing	Dynamic, generally associated with polyphase deformation, foliation and folding	± Static, extensive fracturing and veining, no foliation. Associated with extension and seafloor spreading	Static, no foliation
Temperature	150–700 °C (>700 °C in deep subduction)	150–850 °C (max. T ~1050 °C)	150–500 °C (>500 °C close to magmas)	150–600 °C (>600 °C at gabbro contacts)
Lithostatic pressure	2–30 kbar for crustal rocks	2–10 kbar (in some collision belts up to 14 kbar, "double crust")	<3 kbar	From a few hundred bars to 3 kbar
Temperature gradients	5–12 °C/km (vertical) gradient depends on subduction velocity	12–60 °C/km (vertical) gradients depends on associated igneous activity	50–500 °C/km (vertical or horizontal)	100 °C/km or higher (horizontal)
Processes	Associated with subduction of oceanic lithosphere (ophiolites) and partly also continental rocks	Continent-continent collision, lithospheric thickening, compression and heating	Heat supply by ascending asthenosphere and intruding mafic magmas at mid-ocean ridges, combined with circulation of seawater through fractured hot rocks in an extensional regime	Heat supply by igneous intrusions, commonly also associated with extensive metasomatism caused by convective hydrothermal circulation
Typical metamorphic rocks	Blueschist, eclogite, serpentinite	Slate, phyllite, schist, gneiss, migmatite, marble, quartzite, greenschist, amphibolite, granulite	Metabasalt, greenstone, metagabbro, serpentinite; primary structure usually well preserved	Hornfels, marble, calcsilicate granofels, skarn

characteristic of ocean-floor metamorphic rocks is their extensive veining, produced by the convective circulation of large amounts of heated seawater. This process leads to chemical interaction between rocks and seawater; in this respect, ocean-floor metamorphism resembles hydrothermal metamorphism (see below). Some features of ocean-floor metamorphism are stated briefly in Table 1.1.

1.2.3
Other Types of Regional Metamorphism

Burial metamorphism was introduced by Coombs (1961) for low-temperature regional metamorphism affecting sediments and interlayered volcanic rocks in a geosyncline without any influence of orogenesis or magmatic intrusion. Lack of schistosity is an essential characteristic of resultant rocks. This means that original rock fabrics are largely preserved. Mineralogical changes commonly are incomplete, so that the newly generated mineral assemblage is intimately associated with relict mineral grains inherited from the original rock. Burial metamorphism in fact merges into and cannot be sharply distinguished from deep-seated diagenesis. Metamorphic changes are often not distinguishable in hand specimens; only in thin sections can they be clearly recognized.

Well-known examples of burial metamorphism include those described from southern New Zealand (e.g., Boles and Coombs 1975; Kawachi 1975), eastern Australia (e.g., Smith 1969), Japan (e.g., Seki 1973), and Chile (e.g., Levi et al. 1982). Other examples of burial metamorphism in metapelitic and metabasic rocks have also been described by Merriman and Frey (1999) and Robinson and Bevins (1999), respectively.

Diastathermal metamorphism is a term proposed by Robinson (1987) for burial metamorphism in extensional tectonic settings showing enhanced heat flow. An example of this type of regional metamorphism was described from the Welsh Basin (Bevins and Robinson 1988; see also Merriman and Frey 1999; Robinson and Bevins 1999).

1.2.4
Contact Metamorphism

Contact metamorphism takes place in rocks in the vicinity of hot intrusive or extrusive igneous bodies. Metamorphic changes are caused by the heat given off by the cooling igneous body and also by gases and fluids released by the crystallizing magma. The zone of contact metamorphism is termed a **contact aureole**; the rocks affected are the **country rocks** of the intrusion. The width of contact metamorphic aureoles varies typically in the range of several meters to a few kilometers. Local deformation connected to the emplacement of the igneous mass can be observed in some aureoles.

Specifically, the width of aureoles depends on the volume, nature and intrusion depth of a magmatic body as well as the properties of the country rocks, especially their fluid content and permeability. A larger volume of magma carries more heat with it than a smaller one, and the temperature increase in the bordering country rock will last long enough to cause mineral reactions. The rocks adjacent to small dikes, sills or lava flows are barely metamorphosed, whereas larger bodies of igneous rocks give rise to a contact aureole of metamorphic rocks.

The temperature of the various types of magma differs widely. Typical solidus temperatures of gabbroic (mafic, basaltic) magmas are well over 1000 °C. In contrast, many granitic plutons formed from water-rich magmas melt at temperatures close to 650–700 °C. Large granitoid plutons formed from water-deficient magmas (e.g., charnockite, mangerite) have solidus temperatures close to 900–950 °C. However, because water-rich granites are the most common plutonic bodies in the continental crust, contact aureoles around granites are the most common type of contact metamorphism as well.

The intrusion depth of a magmatic body determines the thermal gradient and heat flow across the hot intrusive contact to the country rock. Exceptionally high thermal gradients are generally confined to the upper 10 km of the earth's crust because, at deeper levels, the country rocks are already rather hot and, hence, noticeable thermal aureoles are not produced.

The effects of contact metamorphism are most obvious where non-metamorphic sedimentary rocks, especially shales and limestones, are in contact with large magmatic bodies. On the other hand, country rocks that experienced medium- or high-temperature regional metamorphism prior to the intrusion show limited effects of contact metamorphic overprinting because the older mineral assemblages persist at contact metamorphic conditions.

Some features of contact metamorphism are stated briefly in Table 1.1; for more details, see Kerrick (1991). Contact metamorphic rocks are generally fine-grained and lack schistosity. The most typical example is called hornfels (see Chap. 2 for nomenclature); however, foliated rocks such as spotted slates and schists are occasionally present.

Pyrometamorphism is a special kind of contact metamorphism. It shows the effects of particularly high temperatures at the contact of a rock with magma under volcanic or quasi-volcanic conditions, e.g., in xenoliths. Partial melting is common, and, in this respect, pyrometamorphism may be regarded as being intermediate between metamorphism and igneous processes.

1.2.5
Other Types of Small-Scale Metamorphism

Cataclastic metamorphism is confined to the vicinity of faults and overthrusts, and involves purely mechanical forces that cause crushing and granulation of the rock fabric. Experiments show that cataclastic metamorphism is favored by high strain rates under high shear stress at relatively low temperatures. The resulting cataclastic rocks are non-foliated and are known as fault breccia, fault gauge, or pseudotachylite. A pseudotachylite consists of an aphinitic groundmass that looks like black basaltic glass (tachylite). Note that mylonites are no longer regarded as cataclastic rocks, because some grain growth by syntectonic recrystallization and neoblastesis is involved (see, e.g., Wise et al. 1984).

Dislocation and dynamic metamorphism are sometimes used as synonyms for cataclastic metamorphism; however, these terms were initially coined to represent what is now called regional metamorphism. In order to avoid misunderstanding, the name "cataclastic metamorphism" is preferred.

Impact metamorphism is a type of shock metamorphism (Dietz 1961) in which the shock waves and the observed changes in rocks and minerals result from the hypervelocity impact of a meteorite. The duration is very short, i.e., a few microseconds. Mineralogical characteristics involve the presence of shocked quartz and the neoformation of coesite and stishovite as well as minute diamonds. A recent review of impact metamorphism is provided by Grieve (1987) and the topic is not considered further in this volume.

Hydrothermal metamorphism was originally introduced as a term by Coombs (1961). In hydrothermal metamorphism, hot aqueous solutions or gases flow through fractured rocks causing mineralogical and chemical changes in the rock matrix. The study of hydrothermal processes has experienced a dramatic expansion during the last two decades. It has been generalized and today research on water–rock interaction is the centerpiece of modern petrology and geochemistry (e.g., Giggenbach 1981, 1984; Gillis 1995; Buick and Cartwright 1996; Droop and Al-Filali 1996; Alt 1999; Gianelli and Grassi 2001). Water–rock interaction and hydrothermal processes are particularly relevant for the making of ore deposits, rock leaching and alteration, formation of vein systems and fissure deposits. The understanding of hydrothermal processes is also important in the context of geothermal energy production, use of thermal water in spas, design of radioactive waste repositories, and many other areas.

Hydrothermal metamorphism in active geothermal areas is known from many areas including California, Iceland, Japan, and New Zealand. However, water–rock interaction takes place whenever circulating water in pores and fractures has a composition that is not in equilibrium with the rock matrix. Consequently, water–rock reactions occur at all temperatures from the surface to very hot conditions. At present, several research boreholes in the continental crust reach several km depth and temperatures in excess of $200\,^\circ$C. The fracture pore space was found to be fluid-filled in all boreholes and progressing fluid-rock reactions are evident from the surface to bottom hole (Kozlovsky 1984; Emmermann and Lauterjung 1997; Möller et al. 1997).

To study present-day metamorphism in active hydrothermal areas or in research wells is very attractive because temperature, pressure, and fluid compositions can be measured directly in boreholes. Normally, metamorphic rocks at surface outcrops represent cold and fluid-free final products of rock transformation. The conditions under which these rocks once formed, such as temperature, pressure and fluid composition, must be derived indirectly through the study of mineral assemblages (these techniques will be detailed in later chapters).

References

Alt JC (1999) Very low-grade hydrothermal metamorphism of basic igneous rocks. In: Frey M, Robinson D (eds) Low-grade metamorphism. Blackwell, Oxford, pp 169–201

Bevins RE, Robinson D (1988) Low grade metamorphism of the Welsh Basin Lower Paleozoic succession: an example of diastathermal metamorphism? J Geol Soc Lond 145:363–366

Bischoff A, Stöffler D (1992) Shock metamorphism as a fundamental process in the evolution of planetary bodies: information from meteorites. Eur J Mineral 4:707–755

Boles JR, Coombs DS (1975) Mineral reactions in zeolitic Triassic tuff, Hokonui Hills, New Zealand. Geol Soc Am Bull 86:163–173

Buick IS, Cartwright I (1996) Fluid-rock interaction during low-pressure polymetamorphism of the Reynolds Range Group, central Australia. J Petrol 37:1097–1124

Chopin C (1984) Coesite and pure pyrope in high-grade blueschists of the western Alps: a first record and some consequences. Contrib Mineral Petrol 86:107–118

Coleman RG, Wang X (eds) (1995) Ultrahigh pressure metamorphism. Cambridge University Press, Cambridge, 528 pp

Cooke RA, O'Brien PJ, Carswell DA (2000) Garnet zoning and the identification of equilibrium mineral compositions in high-pressure-temperature granulites from the Moldanubian Zone, Austria. J Metamorph Geol 18:551–569

Coombs DS (1961) Some recent work on the lower grades of metamorphism. Aust J Sci 24:203–215

Dempster TJ, Bluck BJ (1995) Regional metamorphism in transform zones during supercontinent breakup: late Proterozoic events of the Scottish Highlands. Geology 23:991–994

Dietz RS (1961) Astroblemes. Sci Am 205:50–58

Droop GTR, Al-Filali IY (1996) Interaction of aqueous fluids with calcareous metasediments during high-T, low-P regional metamorphism in the Qadda area, southern Arabian Shield. J Metamorph Geol 14:613–634

Ellis DJ (1980) Osumilite–sapphirine–quartz–granulites from Enderby Land, Antarctica: P–T conditions of metamorphism, implications for garnet-cordierite equilibria and the evolution of the deep crust. Contrib Mineral Petrol 74:201–210

Emmermann R, Lauterjung J (1997) The German Continental Deep Drilling Program KTB. J Geophys Res 102:18179–18201

Ernst WG (1999) Metamorphism, partial preservation, and exhumation of ultrahigh-pressure belts. Island Arc 8:125–153

Frey M, Kisch HJ (1987) Scope of subject. In: Frey M (ed) Low temperature metamorphism. Blackie, Glasgow, pp 1–8

Gianelli G, Grassi S (2001) Water–rock interaction in the active geothermal system of Pantelleria, Italy. Chem Geol 181:113–130

Giggenbach WF (1981) Geothermal mineral equilibria. Geochim Cosmochim Acta 45:393–410

Giggenbach WF (1984) Mass transfer in hydrothermal alteration systems – a conceptual approach. Geochim Cosmochim Acta 48:2693–2711

Gillis KM (1995) Controls on hydrothermal alteration in a section of fast-spreading oceanic crust. Earth Planet Sci Lett 134:473–489

Grieve RAF (1987) Terrestrial impact structures. Annu Rev Earth Planet Sci 15:245–270

Hacker BR, Ratschbacher L, Webb L, Shuwen D (1995) What brought them up? Exhumation of the Dabie Shan ultrahigh-pressure rocks. Geology 23:743–746

Harley SL, Motoyoshi Y (2000) Al zoning in orthopyroxene in a sapphirine quartzite: evidence for >1120 °C UHT metamorphism in the Napier Complex, Antarctica, and implications for the entropy of sapphirine. Contrib Mineral Petrol 138:293–307

Hokada T (2001) Feldspar thermometry in ultrahigh-temperature metamorphic rocks: evidence of crustal metamorphism attaining ≈ 1100 °C in the archean Napier Complex, East Antarctica. Am Mineral 86:932–938

Holmes A (1920) The nomenclature of petrology, 1st edn. Murby, London, 284 pp

Kawachi DE (1975) Pumpellyite-actinolite and contiguous facies metamorphism in the Upper Wakatipu district, southern New Zealand. N Z J Geol Geophys 17:169–208

Kerrick DM (ed) (1991) Contact metamorphism. Reviews in mineralogy, vol 26. Mineralogical Society of America, Washington, DC, 847 pp

Kienast JR, Lombardo B, Biino G, Pinardon JL (1991) Petrology of very-high-pressure eclogitic rocks from the Brossacasco-Isasca Complex, Dora-Maira Massif, Italian Western Alps. J Metamorph Geol 9:19–34

Kisch HJ (1987) Correlation between indicators of very-low-grade metamorphism. In: Frey M (ed) Low temperature metamorphism. Blackie, Glasgow, pp 227–300

Kozlovsky YA (1984) The world's deepest well. Sci Am 251:106–112

Lamb RC, Smalley PC, Field D (1986) P–T conditions for the Arendal granulites, southern Norway: implications for the roles of P, T and CO_2 in deep crustal LILE-depletion. J Metamorph Geol 4:143–160

Levi B, Aguirre L, Nystroem JO (1982) Metamorphic gradients in burial metamorphosed vesicular lavas; comparison of basalt and spilite in Cretaceous basic flows from central Chile. Contrib Mineral Petrol 80:49–58

Merriman RJ, Frey M (1999) Patterns of very low-grade metamorphism in metapelitic rocks. In: Frey M, Robinson D (eds) Low-grade metamorphism. Blackwell, Oxford, pp 61–107

Miyashiro A (1972) Pressure and temperature conditions and tectonic significance of regional and ocean floor metamorphism. Tectonophysics 13:141–159

Miyashiro A (1973) Metamorphism and metamorphic belts. Allen and Unwin, London, 492 pp

Miyashiro A, Shido F, Ewing M (1971) Metamorphism in the mid-Atlantic ridge near 24° and 30°N. Philos Trans R Soc Lond A268:589–603

Möller P, Weise SM, Althaus E, Bach W, Behr HJ, Borchardt R, Bräuer K, Drescher F, Erzinger J, Faber E, Hansen BT, Horn EE, Huenges E, Kämpf H, Kessels N, Kirsten T, Landwehr D, Lodemann M, Machon L, Pekdeger A, Pielow H-U, Reutel C, Simon K, Walther J, Weinlich FH, Zimmer M (1997) Paleo- and recent fluids in the upper continental crust – results from the German Continental deep drilling Program (KTB). J Geophys Res 102:18245–18256

Newton RC (1987) Petrologic aspects of Precambrian granulite facies terrains bearing on their origins. In: Kröner A (ed) Proterozoic Lithospheric evolution. Am Geophys Union Geodyn Ser 17:11–26

Okay AI (1993) Petrology of a diamond- and coesite-bearing metamorphic terrain: Dabie Shan, China. Eur J Mineral 5:659–675

Reinecke T (1991) Very-high-pressure metamorphism and uplift of coesite-bearing metasediments from the Zermatt-Saas zone, Western Alps. Eur J Mineral 3:7–17

Robinson D (1987) Transition from diagenesis to metamorphism in extensional and collision settings. Geology 15:866–869

Robinson D, Merriman RJ (1999) Low-temperature metamorphism: an overview. In: Frey M, Robinson D (eds) Low-grade metamorphism. Blackwell, Oxford, pp 1–9

Robinson D, Bevins RE (1999) Patterns of regional low-grade metamorphism in metabasites. In: Frey M, Robinson D (eds) Low-grade metamorphism. Blackwell, Oxford, pp 143–168

Schertl HP, Schreyer W, Chopin C (1991) The pyrope-coesite rocks and their country rocks at Parigi, Dora Maira Massif, Western Alps: detailed petrography, mineral chemistry and PT-path. Contrib Mineral Petrol 108:1–21

Schreyer W (1995) Ultradeep metamorphic rocks: the retrospective viewpoint. J Geophys Res 100:8353–8366

Seki Y (1973) Temperature and pressure scale of low-grade metamorphism. J Geol Soc Jpn 79:735–743

Shatsky VS, Sobolev NV, Vavilow MA (1995) Diamond-bearing metamorphic rocks of the Kokchetav massif (northern Kazakhstan). In: Coleman RG, Wang X (eds) Ultrahigh pressure metamorphism. Cambridge University Press, Cambridge, pp 427–455

Smith DC, Lappin MA (1989) Coesite in the Straumen kyanite-eclogite pod, Norway. Terra Nova 1:47–56

Smith RE (1969) Zones of progressive regional burial metamorphism in part of the Tasman geosyncline, eastern Australia. J Petrol 10:144–163

Stern CR, Wyllie PJ (1981) Phase relationships of I-type granite with H_2O to 35 kilobars: the Dinkey Lakes biotite-granite from the Sierra Nevada batholith. J Geophys Res 86(B11):10412–10422

Teichmüller M (1987) Organic material and very low-grade metamorphism. In: Frey M (ed) Low temperature metamorphism. Blackie, Glasgow, pp 114–161

Vernon RH (1976) Metamorphic processes. Allen and Unwin, London, 247 pp

Wain A, Waters D, Jephcoat A, Olijynk H (2000) The high-pressure to ultrahigh-pressure eclogite transition in the Western Gneiss Region, Norway. Eur J Mineral 12:667–688

Winkler HGF (1974, 1976, 1979) Petrogenesis of metamorphic rocks, 3rd, 4th, and 5th edn. Springer, Berlin Heidelberg New York

Wise DU, Dunn DE, Engelder JT, Geiser PA, Hatcher RD, Kish SA, Odon AL, Schamel S
  (1984) Fault-related rocks: suggestions for terminology. Geology 12:391–394
Yang J, Godard G, Kienast JR, Lu Y, Sun J (1993) Ultrahigh-pressure (60 kbar) magnesite-
  bearing garnet peridotites from northeastern Jiangsu, China. J Geol 101:541–554
Zhang R-Y, Liou JG (1994) Coesite-bearing eclogite in Henan Province, central China:
  detailed petrography, glaucophane stability and PT-path. Eur J Mineral 6:217–233

CHAPTER 2

Metamorphic Rocks

This chapter deals with the descriptive characterization of metamorphic rocks. Metamorphic rocks are derived from other rocks of igneous, sedimentary or metamorphic origin. The chemical composition of this primary material (= protolith) determines the chemical and mineralogical composition of metamorphic rocks to a large degree. The compositional variation found in the primary material of metamorphic rocks is reviewed in Section 2.1.

The structure of metamorphic rocks is often inherited from the precursor material. In low-grade metasedimentary rocks, for example, the sedimentary bedding and typical structures of sedimentary rocks such as cross-bedding and graded bedding may be preserved. Ophitic structure, characteristic of basaltic lava, may be found in mafic metamorphic rocks. Very coarse-grained structures of igneous origin can occasionally be found even in high-grade metamorphic rocks. Most metamorphic rocks, however, exhibit structures that are of distinct metamorphic origin. The structures typically result from a combination of deformation and recrystallization. Deformation normally accompanies large-scale tectonic processes causing metamorphism. Some descriptive terms of metamorphic structures are defined in Section 2.2. Classification principles and nomenclature of metamorphic rocks are explained in Section 2.3.

Large-scale tectono-thermal processes move rocks along unique paths in pressure-temperature-time (P–T–t) space. Rocks may undergo continuous recrystallization and new minerals may replace old ones in a complex succession. Earlier-formed minerals and groups of minerals often experience metastable survival because of unfavorable reaction kinetics. This happens particularly if an aqueous fluid phase is absent. Metamorphism may proceed episodically. The study of metamorphic rocks aims at the correct identification of the group of minerals that may have coexisted in chemical equilibrium at one stage during the evolutionary history of the metamorphic rock. This group of minerals is the **mineral assemblage**. The total succession of mineral assemblages preserved in a metastable state in the structure of a metamorphic rock is designated **mineral paragenesis**. Some aspects regarding the mineral assemblage and the mineral paragenesis are discussed in Section 2.4.

Discussion and analysis of phase relationships in metamorphic rocks is greatly facilitated by the use of composition phase diagrams. Their construction is explained in the last section of this chapter (Sect. 2.5) in a graphic representation of mineral assemblages. The quantitative computation of equilibrium composition phase diagrams is not discussed in this volume, however.

2.1
Primary Material of Metamorphic Rocks

All metamorphic rock-forming processes make rocks from other rocks. The precursor rock or **protolith** determines many attributes of the new metamorphic rock.

Metamorphism results from the addition (or removal) of heat and material to discrete volumes of the crust or mantle by tectonic or igneous processes. Metamorphism, therefore, may affect all possible types of rock present in the earth's crust or mantle. Protoliths of metamorphic rocks comprise rocks of all possible chemical compositions and include the entire range of sedimentary, igneous and metamorphic rocks.

Metamorphic processes tend to change the original composition of the protolith. Addition of heat to rocks typically results in the release of volatiles (H_2O, CO_2, etc.) that are stored in hydrates (e.g., clay, micas, amphiboles), carbonates and other minerals containing volatile components. Therefore, many metamorphic rocks are typically depleted in volatiles relative to their protoliths. Metamorphism that releases only volatiles from the protolith is, somewhat illogically, termed isochemical. On a volatile-free basis, the chemical composition of protolith and product rock is identical in isochemical metamorphism. In truly isochemical metamorphism, protolith and product rocks are of identical composition including the volatile content. Isochemical metamorphism in such a strict sense is extremely rare.

Many, if not most, metamorphic processes also change the cationic composition of the protolith. This type of metamorphism is termed allochemical metamorphism or metasomatism. The aqueous fluid released by dehydration reactions during metamorphism may contain dissolved cations. These are then carried away with the fluid and lost by the rock system. It has been found, for example, that many granulite facies gneisses are systematically depleted in alkalis (Na and K) relative to their amphibolite facies precursor rocks. This can be explained by loss of alkalis during dehydration. Silica saturation is a general feature of almost all metamorphic fluids. Pervasive or channeled regional-scale flow of silica-saturated dehydration fluids may strongly alter silica-deficient rocks (ultramafic rocks, dolomite marbles) that come into contact with this fluid. Unique metamorphic rock compositions may result from metasomatism on a local or regional scale. Efficient diffusion and infiltration metasomatism requires the presence of a fluid phase. Metasomatism is fluid–rock interaction at elevated temperature and pressure. Fluid–rock interaction is also important in sedimentary and other near-surface environments.

Interaction of rocks with externally derived fluids is referred to as allochemical metamorphism. The volatile composition of the fluid may not be in equilibrium with the rock's mineralogy and, consequently, the rock may be altered. Some examples: flushing of rocks with pure H_2O under high P–T conditions may initiate partial melting; it may form mica and amphibole in pyroxene-bearing rocks; it may induce wollastonite or periclase formation in marbles. CO_2 metasomatism is particularly common in very high-grade rocks. Metasomatism can create rocks of extreme composition that, in turn, may serve as protoliths in subsequent metamorphic cycles. Metasomatic rocks of unusual composition occur very widespread in regional metamorphic terrains and contact aureoles. However, the total volume of such types of rocks is negligible. Although interesting petrologically, these exotic rocks will not be discussed in Part II of this book where we present a systematic treatment of prograde metamorphism of the most important types of bulk rock compositions.

2.1.1
Chemical Composition of Protoliths of Metamorphic Rocks

The average composition of crust and mantle is listed in Table 2.1. The mantle constitutes the largest volume of rocks on planet Earth. From geophysical and petrophysical evidence and from mantle fragments exposed at the surface we know that the mantle is dominated by ultramafic rocks of the peridotite family. The bulk of the mantle is in a solid state and experiences continuous recrystallization as a result of large-scale convection and tectonic processes. Therefore, nearly all mantle rocks also represent metamorphic rocks. The composition of the mantle (Table 2.1) is representative for the most prominent type of metamorphic rock on this planet. However, mantle rocks can only be transported through the lid of the crust to the surface of the earth by active tectonic or igneous processes. Outcrops of ultramafic rocks are common and widespread, particularly in orogenic belts. However, the total volume of ultramafic rocks exposed on continents is small.

Crustal rocks may be divided into rocks from oceanic and continental environments. Characteristic compositions of continental and oceanic crust are listed in Table 2.1. Typical compositions of basalt (the dominant crustal rock of oceanic environments) and tonalite (quartz-diorite to granodiorite; the dominant igneous rocks of continents) are also given in Table 2.1. It is evident that the average composition of oceanic crust is well represented by an average basalt composition, and the average composition of continental crust can be described by an average tonalite composition.

Table 2.1. Composition of the earth's crust and mantle. (After Carmichael 1989)

	Peridotite mantle	Continental crust	Oceanic crust	Basalt	Tonalite
SiO_2	45.3	60.2	48.6	47.1	61.52
TiO_2	0.2	0.7	1.4	2.3	0.73
Al_2O_3	3.6	15.2	16.5	14.2	16.48
FeO	7.3	6.3	8.5	11.0	5.6
MgO	41.3	3.1	6.8	12.7	2.8
CaO	1.9	5.5	12.3	9.9	5.42
Na_2O	0.2	3.0	2.6	2.2	3.63
K_2O	0.1	2.8	0.4	0.4	2.1
H_2O	<0.1	1.4	1.1	<1.0	1.2
CO_2	<0.1	1.4	1.4	<1.0	0.1

Table 2.2. Abundance of rocks (vol%) in the earth's crust. (After Carmichael 1989)

Igneous rocks	64.7
Sedimentary rocks	7.9
Metamorphic rocks	27.4

Igneous rocks (64.7)		Sedimentary rocks (7.9)	
Granites	16	Shales	82
Granodiorites/diorites	17	Sandstones, arkoses	12
Syenites	0.6	Limestones	6
Basalts/gabbros	66		
Peridotites/dunites	0.3		

Table 2.3. Chemical composition of sedimentary and igneous rocks. (After Carmichael 1989)

	Sandstones, graywackes	Shales (platforms)	Pelites pelagic clays	Carbonates (platforms)	Tonalite	Granite	Basalt MORB
SiO_2	70.0	50.7	54.9	8.2	61.52	70.11	49.2
TiO_2	0.58	0.78	0.78	–	0.73	0.42	2.03
Al_2O_3	8.2	15.1	16.6	2.2	16.48	14.11	16.09
Fe_2O_3	0.5	4.4	7.7	1.0	–	1.14	2.72
FeO	1.5	2.1	2.0	0.68	5.6	2.62	7.77
MgO	0.9	3.3	3.4	7.7	2.8	0.24	6.44
CaO	4.3	7.2	0.72	40.5	5.42	1.66	10.46
Na_2O	0.58	0.8	1.3	–	3.63	3.03	3.01
K_2O	2.1	3.5	2.7	–	2.1	6.02	0.14
H_2O	3.0	5.0	9.2	–	1.2	0.23	0.70
CO_2	3.9	6.1	–	35.5	0.1		
C	0.26	0.67	–	0.23			

Table 2.2 lists the types of rocks that make up typical crust. These most abundant rock types will also be the predominant protoliths of metamorphic rocks. Igneous rocks of mafic composition (basalts, gabbros) are most important in oceanic crust (oceanic crust covers much larger areas than continental crust). Mafic rocks constitute, therefore, an important chemical group of metamorphic rocks (mafic schists and mafic gneisses, amphibolites, mafic granulites, eclogites). Typical compositions of basaltic protoliths of metamorphic rocks are given in Tables 2.1 and 2.3.

Granite and related rocks such as granodiorite and quartz-diorite (typical composition given in Table 2.3) dominate the continental crust. These rock types make up the family of metamorphic rocks termed metagranitoids (=quartzo-feldspathic rocks). Metagranitoids are granofelses, gneisses and schists derived from granite, granodiorite and quartz-diorite that represent 33% of all igneous rocks (Table 2.2).

On a global basis, sedimentary rocks are dominated by shales and clays of pelagic and platform (shelf) environments (82% of all sediments, Table 2.2). Compositions of typical shales from continental and oceanic settings are listed in Table 2.3. Pelagic clays (Table 2.3) represent the typical sediments of the deep oceans. The extremely fine-grained clay-rich sediments are designated as pelites and form the most important type of metamorphic rock of sedimentary origin (metapelites).

Shales deposited on continental shelfs typically contain carbonate minerals. This is reflected in the chemical composition of the average shale analysis given in Table 2.3. Carbonate-rich shales are usually referred to as marls. Metamorphic equivalents, e. g., calcareous mica-schists, are important types of metasediments in orogenic belts and will be discussed separately in this volume.

With reference to Table 2.2, sandstones (greywackes, arkoses) and limestones (carbonate rocks) are the remaining important groups of rocks. Characteristic compositions are listed in Table 2.3.

If metamorphic rocks become protoliths in a new cycle of metamorphism, their composition will comprise those of the most common sedimentary and igneous rocks, as given in Table 2.3. Thus, all metamorphic rocks can be grouped into seven classes of characteristic bulk rock composition.

2.1.2
Chemical Composition Classes of Metamorphic Rocks and Their Protoliths

The seven classes are arranged according to increasing chemical complexity.

1. *Ultramafic rocks.* Usually mantle-derived, very Mg-rich family of rocks (typical composition; Table 2.1, peridotite). Metamorphism of ultramafic rocks will be discussed in Chapter 5.

2. *Carbonate rocks.* Sedimentary rocks, modally dominated by carbonate minerals (calcite, dolomite). Examples: limestone, marble (typical composition; Table 2.3, carbonates). See Chapter 6.

3. *Pelites (shales).* Pelitic rocks and shales are the most common type of sedimentary rock. Pelagic clays (pelites) are poor in calcium compared with shales from platforms (Table 2.3). Pelites constitute a separate composition group. Their metamorphic equivalents are termed metapelites (metapelitic schists and gneisses). Metapelites are usually rich in distinctive metamorphic mineral assemblages. See Chapter 7.

4. *Marls.* Marls are shales containing a significant proportion of carbonate minerals (usually calcite). Carbonate-rich shales are characteristic sediments of platform environments (composition; Table 2.3). Metamorphism of marly rocks will be discussed in Chapter 8.

5. *Mafic rocks.* Metamorphic mafic rocks (e.g., mafic schists and gneisses, amphibolites) are derived from mafic igneous rocks, mainly basalts and, of lesser importance, gabbros. Compositions are given in Tables 2.1 and 2.3. Based on total volume, basalts are the most important group of volcanic rocks and metabasalts occur very widespread in metamorphic terrains. Metamorphic assemblages found in mafic rocks are used to define the intensity of metamorphism in the metamorphic facies concept (Chap. 4). The prograde metamorphism of mafic rocks will be presented in Chapter 9.

6. *Quartzo-feldspathic rocks.* Rocks of sedimentary (sandstone, greywacke) or igneous origin (granite, granodiorite, tonalite, monzonite, syenite, etc.) that are modally dominated by quartz and feldspar (typical compositions;

- **Joint.** A single fracture in a rock with or without a small amount (<1 cm) of either dilatational or shear displacement (joints may be sealed by mineral deposits during or after their formation).
- **Cataclasis.** Rock deformation accomplished by some combination of fracturing, rotation, and frictional sliding producing mineral grain and/or rock fragments of various sizes and often of angular shape.
- **Metamorphic differentiation.** Redistribution of mineral grains and/or chemical components in a rock as a result of metamorphic processes. Metamorphic process by which mineral grains or chemical components are redistributed in such a way to increase the modal or chemical anisotropy of a rock (or portion of a rock) without changing the overall chemical composition.

2.3
Classification and Names of Metamorphic Rocks

The names of metamorphic rocks are usually straightforward and self-explanatory. The number of special terms and cryptic expressions is relatively small. Nevertheless, in order to be able to communicate with other geologists working with metamorphic rocks, it is necessary to define commonly used names and expressions and to briefly review currently used classification principles for metamorphic rocks.

There is not one sole classification principle used for the description of metamorphic rocks, which consequently means that all metamorphic rocks may have a series of perfectly correct and accepted names. However, **modal mineralogical composition and mesoscopic structure are the main criteria for naming metamorphic rocks**. In addition, the **composition and the nature of the protolith** (original material) is an important classification criterion. Finally, some well-established **special names** are used also in metamorphic geology.

The names of metamorphic rocks consist of a **root** name and a series of **prefixes**. The root of the name may be a special name (e. g., amphibolite) or a name describing the structure of the rock (e. g., gneiss). The root name always embraces some modally dominant metamorphic minerals (amphibolite is mainly composed of amphibole+plagioclase; gneiss is mainly composed of feldspar±quartz). The rock may be further characterized by adding prefixes to the root name. The prefixes may specify some typical structural features of the rock or may give some additional mineralogical information (e. g., banded epidote-bearing garnet-amphibolite, folded leucocratic garnet-hornblende gneiss). The prefixes are optional and the name may consist of the root only.

2.3.1
Rock Names Referring to the Structure

The structure of a rock concerns the characteristic distribution of its constituents (minerals, aggregates, layers, etc.). It results from the geometrical arrangement of minerals, mineral aggregates with inequant crystal shapes and other structural features, as discussed above. This structure is largely controlled by mechanical deformation and chemical segregation processes which are almost always associated with metamorphism. Expressions that mainly characterize the structure of metamorphic rocks are often used as root names. Metamorphic rocks are named primarily by using **descriptive structural terms**. The most important of these terms [1] are:

- **Gneiss.** A metamorphic rock displaying a gneissose structure. The term gneiss may also be applied to rocks displaying a dominant linear fabric rather than a gneissose structure, in which case the term lineated gneiss may be used. This term is almost exclusively used for rocks containing abundant feldspar (±quartz), but may also be used in exceptional cases for other compositions (e.g., feldspar-free cordierite-anthophyllite gneiss). Examples: garnet-biotite gneiss, granitic gneiss, ortho-gneiss, migmatitic gneiss, banded gneiss, garnet-hornblende gneiss, mafic gneiss.
- **Schist.** A metamorphic rock displaying on the hand-specimen scale a pervasive, well-developed schistosity defined by the preferred orientation of abundant inequant mineral grains. For phyllosilicate-rich rocks the term schist is usually reserved for medium- to coarse-grained varieties, whilst finer-grained rocks are termed slates or phyllites. The term schist may also be used for rocks displaying a strong linear fabric rather than a schistose structure. Examples: epidote-bearing actinolite-chlorite schist (=greenschist), garnet-biotite schist, micaschist, calcareous micaschist, antigorite schist (=serpentinite), talc-kyanite schist (=whiteschist).
- **Slate.** A very fine-grained rock of very low metamorphic grade displaying slaty cleavage.
- **Phyllite.** A fine-grained rock of low metamorphic grade displaying a perfect penetrative schistosity resulting from parallel arrangement of phyllosilicates. Foliation surfaces commonly show a lustrous sheen.
- **Granofels.** A metamorphic rock lacking schistosity, gneissose structure, and mineral lineations.

2.3.2
Names for High-Strain Rocks

Metamorphism may locally be associated with an extremely high degree of rock deformation. Localized high strain in metamorphic terrains may produce rocks with distinctive structures. Some widely used special names for high-strain rocks are defined below:

- **Mylonite.** A rock produced by mechanical reduction of grain size as a result of ductile, non-cataclastic deformation in localized zones (shear zones, fault zones), resulting in the development of a penetrative fine-scale foliation, and often with an associated mineral and stretching lineation.

---

[1] The definition of terms given in this chapter partly follows preliminary unpublished recommendations of the "International Commission on the Nomenclature of Metamorphic Rocks".

- **Ultramylonite.** A mylonite in which most of the megacrysts or lithic fragments have been eliminated (>90% fine-grained matrix).
- **Augen mylonite (blastomylonite).** A mylonite containing distinctive large crystals or lithic fragments around which the fine-grained banding is wrapped.
- **Cataclasite.** A rock which underwent cataclasis.
- **Fault breccia.** Cataclasite with breccia-like structure formed in a fault zone.
- **Pseudotachylite.** Ultra-fine-grained vitreous-looking material, flinty in appearance, occurring as thin veins, injection veins, or as a matrix to pseudo-conglomerates or -breccias, which seals dilatancy in host rocks displaying various degrees of fracturing.

2.3.3
Special Terms

Some commonly used and approved special terms (names, suffixes, prefixes) include:
- **Mafic** minerals
  Collective expression for ferro-magnesian minerals.
- **Felsic** minerals
  Collective term for quartz, feldspar, feldspathoids and scapolite.
- **Mafic rock.** Rock mainly consisting of mafic minerals (mainly consisting of $\equiv$ modally >50%).
- **Felsic rock.** Rock mainly consisting of felsic minerals.
- **Meta-.** If a sedimentary or igneous origin of a metamorphic rock can be identified, the original igneous or sedimentary rock term preceded by "meta" may be used (e.g., metagabbro, metapelite, metasediment, metasupracrustal). Also used to generally indicate that the rock in question is metamorphic (e.g., metabasite).
- **Ortho-** and **para-.** A prefix indicating, when placed in front of a metamorphic rock name, that the rock is derived from an igneous or sedimentary rock, respectively (e.g., paragneiss).
- **Acid, Intermediate, Basic, Ultrabasic.** Terms defining the SiO_2 content of igneous and metamorphic rocks (>63, 63–52, 52–45, <45 wt% SiO_2).
- **Greenschist** and **greenstone.** Schistose (greenschist) or non-schistose (greenstone) metamorphic rocks whose green color is due to the presence of minerals such as chlorite, actinolite, epidote and pumpellyite (greenschist for, e.g., epidote-bearing actinolite-chlorite schist; greenstone for, e.g., chlorite-epidote granofels).
- **Blueschist.** Schistose rock whose bluish color is due to the presence of sodic amphibole (e.g., glaucophane schist). However, the "blue" color of a blueschist will not be recognized by a non-geologist (i.e., it is not really blue, very rare outcrops of really blue glaucophanites do exist though). Blueschists are schistose rocks containing amphibole with significant amounts of the M4 cation position in the amphibole structure occupied by Na (riebekite, crossite, glaucophane).

- **Amphibolite.** Mafic rock predominantly composed of hornblende (>40%) and plagioclase.
- **Granulite.** Metamorphic rock in or from a granulite facies terrain exhibiting characteristic granulite facies assemblages. Anhydrous mafic minerals are modally more abundant than hydrous mafic minerals. Muscovite is absent in such rocks. Characteristic is the occurrence of metamorphic orthopyroxene in mafic and felsic rocks. The term is not used for marbles and ultramafic rocks in granulite facies.
- **Charnockite, Mangerite, Jotunite, Enderbyite.** Terms applied to orthopyroxene-bearing rocks with igneous structure and granitic (charnokite), monzonitic (mangerite, jotunite), and tonalitic (enderbyite) composition, respectively, irrespective of whether the rock is igneous or metamorphic.
- **Eclogite.** A plagioclase-free mafic rock mainly composed of omphacite and garnet, both of which are present in large proportions.
- **Eclogitic rock.** Rock of any composition containing diagnostic mineral assemblages of the eclogite facies (e.g., jadeite-kyanite-talc granofels).
- **Marble.** A metamorphic rock mainly composed of calcite and/or dolomite (e.g., dolomite marble).
- **Calc-silicate rock.** Metamorphic rock which, besides 0–50% carbonates, is mainly composed of Ca-silicates such as epidote, zoisite, vesuvianite, diopside-hedenbergite, Ca-garnet (grossular-andradite), wollastonite, anorthite, scapolite, Ca-amphibole.
- **Skarn.** A metasomatic Ca-Fe-Mg-(Mn)-silicate rock often with sequences of compositional zones and bands, formed by the interaction of a carbonate and a silicate system in mutual contact. Typical skarn mineralogy includes: wollastonite, diopside-salite, grossular, zoisite, anorthite, scapolite, margarite (Ca skarns); hedenbergite, andradite, ilvaite (Ca-Fe skarns); forsterite, humites, spinel, phlogopite, clintonite, fassaite (Mg skarns); rhodonite, tephroite, piemontite (Mn skarns).
- **Rodingite.** Calc-silicate rock, poor in alkalis and generally poor in carbonates, generated by metasomatic alteration of mafic igneous rocks enclosed in serpentinized ultramafic rocks. The process of rodingitization is associated with oceanic metamorphism (serpentinization of peridotite, rodingitization of enclosed basic igneous rocks such as gabbroic/basaltic dykes). Metarodingite is a prograde metamorphic equivalent of rodingite produced by oceanic metamorphism.
- **Quartzite.** A metamorphic rock containing more than about 80% quartz.
- **Serpentinite.** An ultramafic rock composed mainly of minerals of the serpentine group (antigorite, chrysotile, lizardite), e.g., diopside-forsterite-antigorite schist.
- **Hornfels.** A non-schistose very fine-grained rock mainly composed of silicate±oxide minerals that shows substantial recrystallization due to contact metamorphism. Hornfelses often retain some features inherited from the original rock such as graded bedding and cross-bedding in hornfelses of metasedimentary origin.
- **Migmatite.** Composite silicate rock, pervasively heterogeneous on a meso- to megascopic scale, found in medium- to high-grade metamorphic ter-

rains (characteristic rocks for the middle and lower continental crust). Migmatites are composed of dark (mafic) parts (**paleosome**) and light (felsic) parts (**neosome**) in complex structural association. The felsic parts formed by crystallization of locally derived partial melts or by metamorphic segregation; the mafic parts represent residues of the inferred partial melting process or formed by metamorphic segregation. Parts of the felsic phases may represent intruded magmas from a more distant source.

- **Restite.** Remnant of a rock, chemically depleted in some elements relative to its protolith. The depletion is the result of partial melting of that rock.

2.3.4
Modal Composition of Rocks

The mineral constituents of metamorphic rocks are classified as follows: (1) Major constituent: present in amounts >5 vol%. To account for a major constituent not included in the definition of a rock name, the mineral name is set in front of the rock name (e. g., muscovite-gneiss, epidote-amphibolite). It follows from the above that an epidote-amphibolite is a rock that is mainly composed of epidote, hornblende and plagioclase. A garnet-staurolite gneiss is a metamorphic rock which is mainly composed of feldspar and quartz (included in the name root "gneiss"). In addition, it contains modally more staurolite than garnet as major constituents. (2) Minor constituent: present in amounts <5 vol%. If one wishes to include a minor constituent mineral in the rock name it is connected with the term "-bearing" (e. g., rutile-ilmenite-bearing garnet-staurolite gneiss; contains less rutile than ilmenite). (3) Critical mineral (or mineral assemblage): indicating by its presence or absence distinctive conditions for the formation of a rock (it also may indicate a distinct chemical composition of the rock). The critical mineral(s) may be present as major and/or minor constituent(s).

It is up to the geologist to decide how many and which of the minerals that are not yet included in the definition of the rock name he/she wants to prefix. It is also possible to use abbreviations of mineral names in the rock names (e. g., Bt-Ms gneiss for biotite-muscovite gneiss). The recommended standard abbreviations for mineral names are listed in the Appendix.

2.3.5
Names Related to the Origin of the Protolith

The chemical variation found in metamorphic rocks has been grouped into seven composition classes according to the most frequently occurring rock types in the crust and mantle.

Metamorphic rocks may be characterized by referring their names to the nature of the original material. Examples of such names include: metapelite, metabasite, metabasalt, metapsamite, metagranite, metagabbro, semi-pelitic gneiss, metamarl (e. g., calcareous micaschist), metaeclogite.

2.4
Mineral Assemblages and Mineral Parageneses

The petrogenesis of metamorphic rocks is often a long-lasting complicated process in time and space which perhaps may endure for 30 Ma or so (see Chap. 3). A geologist collects the end product of such a long drawn-out process at the earth's surface (1 bar and 30 °C). The geologist is, of course, interested in finding out something about the evolution, the petrogenesis of the collected rock at this sample locality, to compare it with rocks from other localities and finally draw some sound conclusions on the large-scale processes that caused the observed rock metamorphism.

One of the important methods to decipher the petrogenesis and evolution of metamorphic rocks is the application of chemical thermodynamics to the heterogeneous chemical systems exemplified by rocks. Rocks are usually composed of a number of different kinds of minerals. The mineral species represent chemical subspaces of the rock, so-called phases. Most metamorphic rocks are also composed of a number of different chemical constituents that are designated system components. The chemical constituents making up the rocks' total composition are distributed among a group of mineral species or phases, each of them with a distinct composition (heterogeneous system). The number, composition and identity of the minerals are uniquely defined by the laws of thermodynamics and depend exclusively on the prevailing intensive variables such as pressure and temperature. The group of minerals that make up a rock at equilibrium is designated **equilibrium mineral assemblage** or equilibrium phase assemblage. The succession of mineral assemblages that follow and replace one another during the metamorphic evolution of a given terrain are designated **mineral parageneses**.

In any practical work with metamorphic rocks it is impossible to demonstrate that a given mineral assemblage once coexisted in chemical equilibrium. Therefore, one uses a less rigorous definition: a **mineral assemblage** is an association of mineral species in mutual grain contact. The assemblage occurs in a chemically homogeneous portion of a rock. For practical work with thin sections it is convenient to use a matrix table for marking all observed two-phase grain contacts. An example: in a thin section of a metamorphic rock showing a coarse-grained mosaic structure the following mafic minerals have been identified: staurolite, garnet, biotite and kyanite. All observed grain contacts between two of these minerals are marked with an 'X' in Table 2.4. The thin-section observations summarized in Table 2.4 imply that the four minerals indeed represent a mineral assemblage.

In rocks displaying clear disequilibrium structures (reaction rims, symplectitic structures, replacement structures), it is often difficult to determine mineral assemblages that represent a particular stage in the rocks' evolutionary history.

One has to keep in mind that mineral assemblages are usually identified in thin sections which represent a two-dimensional section through a volume of rock. In such sections, only two and a maximum of three minerals may be in mutual grain contact. In a three-dimensional rock, a maximum of four

Table 2.4. Practical determination of a mineral assemblage

Mineral	Staurolite	Garnet	Biotite	Kyanite
Staurolite	X	X	X	X
Garnet		X	X	X
Biotite			X	X
Kyanite				X

A cross in the Grt-St cell means garnet and staurolite have mutual grain contacts in thin section.

minerals can be in contact at a point in space; three minerals form contacts along a line, and two minerals contact along surfaces (that show as lines in thin sections). It is therefore strongly recommended to study **more than one thin section** from the most significant and interesting samples. As an example, in a series of 20 thin sections of a single hand specimen of a coarse-grained sapphirine-granulite from the Alps we found unique assemblages and unique structures in nearly all of the sections. The problem is especially acute in coarse-grained samples, where the scale of chemical homogeneity may considerably exceed the size of a thin section.

In extremely fine-grained samples, the mineral assemblage must be determined by X-ray techniques. In this case it is not possible to maintain the requirement of mutual grain contact in the definition of an assemblage. However, this shortcoming may be overcome by the use of a scanning electron microscope working in the backscatter mode.

Rocks should always be examined by X-ray techniques in order to identify minerals which are difficult to distinguish under the microscope (e. g., muscovite, paragonite, talc, pyrophyllite, also quartz and untwinned albite).

Staining techniques help the distinction of some important rock-forming minerals with similar optical properties (e. g., calcite, dolomite).

Minerals occurring as inclusions in refractory minerals such as garnet but not in the matrix of the rock do not belong to the main matrix assemblage. In our example of the staurolite-garnet-biotite-kyanite assemblage, garnet may show small composite two-phase inclusions of chlorite and chloritoid. Chlorite-chloritoid-garnet constitutes in this case another, earlier assemblage of the rock.

During metamorphism, some earlier-formed minerals may become unstable and react chemically to form a new, more stable assemblage. However, metastable relics of the early assemblage may partly survive. Great care must be taken in the study of metamorphic microstructures in order to avoid mixing up mineral assemblages. The correct identification of successive series of mineral assemblages, that is the parageneses of a metamorphic rock, represents the "great art" of metamorphic petrology. It can be learned only by experience.

Figure 2.1 shows some of the aspects related to the recognition of mineral assemblages. Three fictive rocks all contain the minerals quartz, calcite and wollastonite on the scale of a thin section. The general microstructure of the three sections (Fig. 2.1) shows the distribution of the three minerals in the

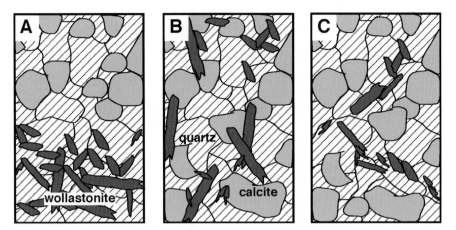

Fig. 2.1 A–C. Mineral assemblages in three different rocks all containing wollastonite, calcite and quartz

respective rocks. Rock A is clearly heterogeneous on the scale of a thin section; the upper part contains the assemblage quartz + calcite, the lower half of the section is free of carbonate and contains the assemblage quartz + wollastonite. The rock does not contain the assemblage Qtz + Cal + Wo. The two parts of the section are different in overall composition. Note, however, on a volatile-free basis the two domains of the rock may have very similar or identical compositions (same Ca/Si ratio). In rock B obviously all three minerals can be found in mutual grain contact, Qtz + Cal + Wo constitutes the mineral assemblage of this rock. In rock C, which appears to be compositionally homogeneous on the thin-section scale, quartz + calcite and quartz + wollastonite form common grain boundaries. However, no wollastonite + calcite grain boundaries can be found. The three phases Qtz + Cal + Wo do not represent a mineral assemblage in the strict sense. Yet many geologists would probably approve the assemblage, despite this. (Comment: after the publication of the sixth edition of this book this Fig. 2.1, intended to illustrate the problem of correctly identifying a mineral assemblage in thin sections, inspired a lively discussion about various subjects among petrologists on the geo-metamorphism mail base. A number of comments were made and it also was suggested that this figure should be modified. We preferred to keep it essentially in its original form. Some aspects of the debate will be discussed instead in Chapter 3.7. However, we strongly recommend readers and anyone interested in rock metamorphism to subscribe to the mail base geo-metamorphism: sometimes one captures a stimulating dispute on questions related to geo-metamorphism. The present address is: geo-metamorphism@jiscmail.ac.uk: you need to send an e-mail with nothing but the following in the body of the message: subscribe geo-metamorphism.)

2.5
Graphic Representation of Metamorphic Mineral Assemblages

Once the mineral assemblage of a rock has been identified, it is convenient or even compulsory to represent the chemical composition of the minerals that constitute the assemblage on a graphical figure. Such a figure is called a **chemograph** and represents a **composition phase diagram**. The geometric arrangement of the phase relationships on such a phase diagram is called the **topology**. Composition phase diagrams can be used just to document the assemblages found in rocks of a given metamorphic terrain or outcrop. However, such diagrams are an indispensable tool for the analysis of metamorphic characteristics and evolution of a terrain. They can be used to deduce sequences of metamorphic mineral reactions. Finally, composition phase diagrams can also be calculated theoretically from thermodynamic data of minerals that permit the quantitative calibration of field-derived chemographies.

Composition phase diagrams display the chemical composition of minerals and the topologic relationships of mineral assemblages. The variables on the diagrams are concentrations or amounts of chemical entities. All other variables that control the nature of the stable mineral assemblage such as pressure and temperature must be constant. Composition phase diagrams are isothermal isobaric diagrams. Also, not more than two composition variables can conveniently be displayed on a two-dimensional xy diagram (sheet of paper, computer screen).

2.5.1
Mole Numbers, Mole Fractions and the Mole Fraction Line

It is useful to change the scale for the compositional variables from weight percent (wt%) to mole percent (mol%), mole fractions or mole numbers. Most chemographies use mol% or mole fraction as units for the composition variables. The mineral forsterite (Fo), for example, is composed of 42.7 wt% SiO_2 and 57.3 wt% MgO. Mole numbers and mole fractions for this mineral are calculated as follows:

- SiO_2: 42.7/60.1 (molecular weight SiO_2) = **0.71** (number of moles of SiO_2 per 100 g Fo)
- MgO: 57.3/40.3 (molecular weight MgO) = **1.42** (number of moles of MgO per 100 g Fo)

This mineral has a MgO/SiO_2 ratio of 1.42/0.71 = 2. It has 2 mol MgO per 1 mol SiO_2. The composition of the mineral, forsterite, is reported as Mg_2SiO_4 or (2 MgO SiO_2) or 66.66% MgO + 33.33% SiO_2 or 2/3 MgO + 1/3 SiO_2. This is all equivalent. However, the last version has many advantages ⇒ mole fraction basis.

The mole fraction is defined as follows:

$$X_{MgO} = \frac{(\text{number of moles of MgO})}{(\text{number of moles of MgO}) + (\text{number of moles of SiO2})}$$

Fig. 2.2. Composition space and the two-component $MgO–SiO_2$ system

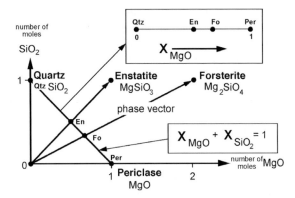

For the example above: $X_{MgO}=2/(2+1)=1.42/(1.42+0.71)=0.66$ (dimensionless quantity).

Figure 2.2 shows a graphic representation of the two-component system MgO and SiO_2 in a rectangular coordinate system. The two components MgO and SiO_2 define the two-dimensional **composition space**. The graphical representation of the compositions of forsterite, enstatite, quartz and periclase is a point in the Cartesian coordinate system with the axes: numbers of moles MgO and numbers of moles SiO_2. It is apparent from Fig. 2.2 that the composition of forsterite, e.g., can also be viewed as a vector in the composition space with the elements (2,1). Mineral compositions are unequivocally characterized by a vector, the **phase vector**, in the composition space.

Mineral compositions can also be represented in terms of mole fractions rather than mole numbers. The mole fractions of the two **system components** MgO and SiO_2 sum to unity in all four minerals represented in Fig. 2.2. This can be expressed by the general relationship:

$$X_{MgO} + X_{SiO_2} = 1$$

This equation represents a straight line in Fig. 2.2 connecting $X_{MgO}=1$ with $X_{SiO_2} = 1$.

The phase vectors of forsterite and enstatite intersect the mole fraction line at unique positions (En and Fo, respectively; Fig. 2.2). It is, therefore, not necessary to represent phase composition in a two-component system on a two-dimensional diagram. The topologic information is contained on the mole fraction line. The dimension of the graph can be reduced from 2 to 1. If the concentration of one component is known in a phase that is composed of two components, the second concentration is known as well. In an n-component system there are n–1 independent compositional variables; the remaining concentration is given by the equation ($\Rightarrow$ n composition variables and one equation relating them):

$$\sum_{i=1}^{n} X_i = 1$$

Note also, for example, that enstatite compositions expressed as $MgSiO_3$, $Mg_2Si_2O_6$, $Mg_4Si_4O_{12}$ all have the same intersection point with the mole fraction line (multiplying the phase vector for enstatite with a scalar preserves its position on the mole fraction line). Phase compositions in a two-component system can be represented on a mole fraction line. The appropriate diagram is shown in Fig. 2.2.

2.5.2
The Mole Fraction Triangle

The mineral talc is composed of three simple oxide components and its composition can be written as: $H_2Mg_3Si_4O_{12}$. Based on a total of 12 oxygen atoms per formula unit, talc consists of 1 mole H_2O, 3 moles MgO and 4 moles SiO_2. A graphic representation of the talc composition in the three-component system is shown in Fig. 2.3. The three components, MgO, SiO_2 and H_2O, define a Cartesian coordinate system with numbers of moles of the components displayed along the coordinate axes. The unit vectors of the system components span the composition space. The composition of talc is represented by a phase vector with the elements; 3× the unit vector of MgO, 4× the unit vector of SiO_2 and 1× the unit vector of H_2O.

The total number of moles of system components is 8. The composition of talc can be normalized to a total number of 8 moles of system components. In this case, the talc composition will be expressed by $\Rightarrow$ mole fractions: $X_{H_2O}=1/8$, $X_{MgO}=3/8$, $X_{SiO_2}=4/8$, or: $X_{H_2O}=0.13$, $X_{MgO}=0.38$, $X_{SiO_2}=0.50$; in Mol%: $H_2O=12\%$, $MgO=38\%$, $SiO_2=50\%$. The graphic representation of the talc composition on a mole fraction basis is given by the intersection of the talc phase vector of Fig. 2.3 and the plane $\Sigma X_i=1$. The mole fraction plane is a regular triangle with the corners $X_i=1$. This triangle is called the **mole fraction triangle**.

Fig. 2.3. Composition space and the three-component $MgO–H_2O–SiO_2$ system

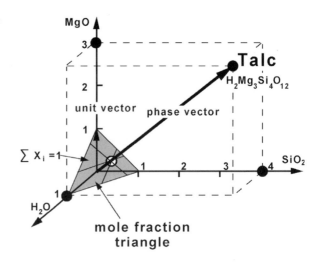

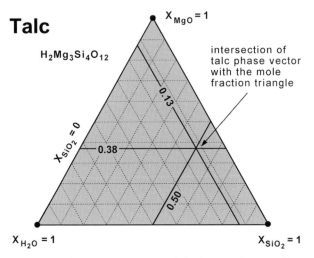

Fig. 2.4. Mole fraction triangle $MgO–H_2O–SiO_2$ and the intersection coordinates of the talc phase vector with the mole fraction triangle

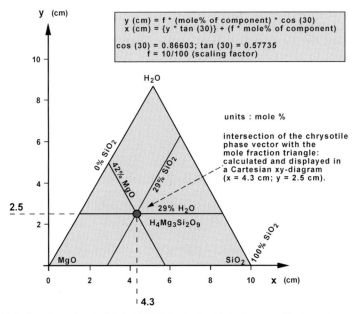

Fig. 2.5. Mole fraction triangle $MgO–H_2O–SiO_2$ in the Cartesian coordinate system

A representation of the mole fraction triangle is shown in Fig. 2.4. The lines of constant X_i are parallel with the base lines of the triangle. This follows from Fig. 2.3, where it can be seen that the planes $X_i = \text{constant}$ and $X_{j \neq i} = 0$ intersect the mole fraction plane along lines parallel to the base line

$X_i = 0$. Three rulers with $X_i = 0.1$ increments are shown in Fig. 2.4 for the three components. Triangular coordinate paper is available commercially.

One more example of a mineral composition in a three-component system $MgO\text{-}SiO_2\text{-}H_2O$ follows: the chemical formula of chrysotile (one of many serpentine minerals) can be written as: $H_4Mg_3Si_2O_9$. The sum of moles of system components is 7; $\Rightarrow$ mole fractions: $X_{H_2O} = 2/7$, $X_{MgO} = 3/7$, $X_{SiO_2} = 2/7$, or equivalent: $X_{H_2O} = 0.29$, $X_{MgO} = 0.42$, $X_{SiO_2} = 0.29$; expressed in mol%: $H_2O = 29\%$, $MgO = 42\%$, $SiO_2 = 29\%$.

Because triangular coordinate drawing paper is not always at hand for plotting composition data, the formulas for recalculation into Cartesian coordinates are given in Fig. 2.5 together with the position of the serpentine composition. The value of the scaling factor "f" depends on the desired size of the figure. The scaling factor must be multiplied by 100 when using mole fractions rather than mol%.

2.5.3
Projections

2.5.3.1
Simple Projections

On a mole fraction triangle two compositional variables of a three-component system can be depicted. Most rocks, however, require more than three components to describe and understand the phase relationships. Graphical representation of an eight-component system requires a seven-dimensional figure. Projection phase diagrams are graphical representations of complex n-component systems that show two composition variables at a time while keeping the other n–3 composition variables constant. The remaining variable is given by the mole fraction equation as outlined above.

In a suite of collected samples of similar bulk composition (e.g., 20 samples of metapelite), one often finds certain mineral species that are present in many of the samples. This circumstance permits projection of the phase compositions from the composition of such a mineral that is present in excess. For instance, in many metapelitic rocks, quartz is a modally abundant mineral whereas calcite is present in most marbles. Composition phase diagrams for metapelites can therefore be constructed by projecting through SiO_2 onto an appropriate mole fraction triangle, for marbles by projection the phase compositions from $CaCO_3$.

The $MgO\text{-}SiO_2\text{-}H_2O$ (MSH) system will be used to explain the basic principle of making projections. Some phase compositions in the MSH system are found in Table 2.5. This table represents a **composition matrix** with oxide components as defining unit vectors of the composition space and the mineral compositions as column vectors.

Figure 2.6 shows the chemographic relationships in the ternary system $H_2O\text{-}MgO\text{-}SiO_2$. The corners of the triangle represent $X_i = 1$ or 100 mol% of the system components. The lines connecting them are the binary subsystems of the three-component system. The compositions of the black dots oc-

Table 2.5. Phase compositions in the MSH system. (a) Composition matrix (moles); columns are phase vectors, composition space defined by the system components SiO_2, MgO, H_2O. (b) Composition matrix (mole fractions); columns are normalized phase vectors, values are coordinates on the mole fraction triangle. (c) Composition matrix (mole fractions); projected through H_2O, columns are normalized phase vectors, values are coordinates on the SiO_2–MgO binary (mole fraction line)

	Fo	Brc	Tlc	En	Ath	Qtz	Per	Atg	Fl
(a)									
SiO_2	1	0	4	2	8	1	0	2	0
MgO	2	1	3	2	7	0	1	3	0
H_2O	0	1	1	0	1	0	0	2	1
Sum	3	2	8	4	16	1	1	7	1
(b)									
SiO_2	0.33	0.00	0.50	0.50	0.50	1.00	0.00	0.29	0.00
MgO	0.67	0.50	0.38	0.50	0.44	0.00	1.00	0.43	0.00
H_2O	0.00	0.50	0.13	0.00	0.06	0.00	0.00	0.29	1.00
Sum	1	1	1	1	1	1	1	1	1
(c)									
SiO_2	0.33	0.00	0.56	0.50	0.53	1.00	0.00	0.40	0.00
MgO	0.67	1.00	0.43	0.50	0.47	0.00	1.00	0.60	0.00

cur as stable phases in nature (phase components). The ternary system has three binary subsystems (MgO–SiO_2, MgO–H_2O, and SiO_2–H_2O). Some phase compositions in the ternary system fall on straight lines such as Tlc–Ath–En and Brc–Atg–Tlc. This means that the phase compositions along the straight line (e. g., Tlc–Ath–En) are linearly dependent; one of these compositions can be expressed by the other two (4 En+Tlc=Ath). Therefore, there are only two components required to describe the compositions of the other phases on the straight line ⇒ **pseudobinary join**. The colinearity is also said to be a **compositional degeneracy** in the system.

Now, one might wish to analyze and discuss phase relationships among the minerals shown in Fig. 2.6 for geological situations in which an aqueous fluid phase (H_2O) is present in all rocks and in equilibrium with the solid-phase assemblage. The presence of excess water in all considered rocks permits projection of the other phase compositions through water onto the MgO–SiO_2 binary (in principle, on any pseudobinary as well). Imagine that you are standing in the H_2O corner of Fig. 2.6. What you will see from there is shown at the bottom of Fig. 2.6. The chemography on the MgO–SiO_2 binary is a projected chemography. The positions of the mineral compositions on this binary are expressed in terms of mole fractions X_{Mg} in Fig. 2.6. They can be calculated from the composition matrix (Table 2.5b) by first deleting the row containing the component one wishes to project from (in our case H_2O) and second renormalizing the column vectors to unity (Table 2.5c). It follows that projections from quartz, periclase and H_2O can be prepared using the composition matrix of Table 2.5b. It must be stressed again, however, that the projections are meaningful only if the projection component is present as a phase of fixed composition in all rocks of interest. The projected chemography in Fig. 2.6 cannot be used in H_2O-absent situations. Likewise,

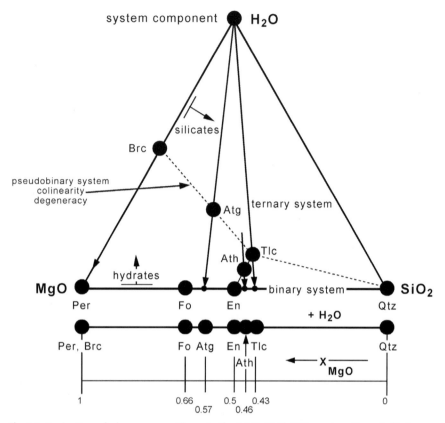

Fig. 2.6. Projection of phase compositions in the MgO–H$_2$O–SiO$_2$ system through H$_2$O onto the MgO–SiO$_2$ binary

an analogous projection through MgO onto the SiO$_2$–H$_2$O binary requires that periclase is present in all assemblages, and projection through SiO$_2$ onto the MgO–H$_2$O binary requires that quartz is present in all assemblages. It is therefore indispensable to write the projection compositions on any projected composition phase diagram. In simple projections, the projecting phase component is identical to one of the simple oxide system components.

Projection of solid-solution phases. The magnesian minerals shown in Fig. 2.6 show in nature a variable substitution of Mg by Fe. If the restriction of water saturation is maintained, the compositions in the FeO-MgO–SiO$_2$–H$_2$O system can be projected from H$_2$O onto the mole fraction triangle FeO–MgO–SiO$_2$, as shown in Fig. 2.7. In certain applications one does not wish to consider the complexity arising from the Fe-Mg substitution. In such cases, one projects the compositions onto the MgO–SiO$_2$ binary parallel to the direction of the substitution vector MgFe$_{-1}$. The resulting projection is identical to the binary chemography shown in Fig. 2.6. All solid-solution phases must be projected along exchange vectors. More examples will be given below.

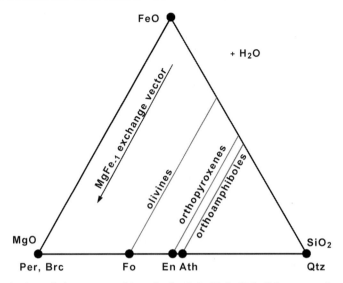

Fig. 2.7. Projection of phase compositions in the $FeO-MgO-H_2O-SiO_2$ system through H_2O and along the $MgFe_{-1}$ exchange vector onto the $MgO-SiO_2$ binary

2.5.3.2
Complex Projections

The technique of projecting the phase compositions in the m-dimensional space from one system component onto a subspace with m–1 dimensions is one of the two key steps in producing geological composition phase diagrams. In the MSH system used as an example here, the projection was made from the H_2O component in the three-component space onto the $MgO-SiO_2$ binary. However, often one wishes to project from a composition other than the components of the original system. Most commonly, it is necessary to find a projection from a composition of a phase that is present in excess. In order to make this possible, the composition matrix has to be rewritten in terms of a set of new system components. One of these new components must be the wanted new projection composition.

For example, we would like to prepare a figure representing the phase compositions in the MSH system on a mole fraction triangle with the corners Mg_2SiO_4 (Fo), $Mg(OH)_2$ (Brc), and $Mg_3Si_4O_{10}(OH)_2$ (Tlc). Secondly, we would like to study phase relationships in rocks that contain excess forsterite and need a projection from Mg_2SiO_4 (forsterite) onto the brucite-talc binary.

The solution to the problem is shown in Table 2.6. In Table 2.6a the phase compositions are expressed in terms of the new system components: Mg_2SiO_4 (Fo), $Mg(OH)_2$ (Brc), and $Mg_3Si_4O_{10}(OH)_2$ (Tlc). As an example, the composition of enstatite can be expressed by 2 Fo –1 Brc + 1 Tlc, which is equivalent to 1 MgO + 1 SiO_2. Table 2.6b gives the coordinates of the mineral compositions in the mole fraction triangle Fo–Brc–Tlc. The algebraic operation which

Table 2.6. Phase compositions in the Fo–Brc–Tlc-system. (a) Composition matrix in terms of moles; columns are phase vectors, composition space defined by the system components. (b) Composition matrix in terms of mole fractions; columns are normalized phase vectors, values are coordinates on the mole fraction triangle. (c) Composition matrix in terms of mole fractions; projection through Mg_2SiO_4 (Fo), columns are normalized phase vectors, values are coordinates on the mole fraction line

	Fo	Brc	Tlc	En	Ath	Qtz	Per	Atg	Fl
(a)									
Fo	1.00	0.00	0.00	2.00	4.00	−1.00	4.00	0.00	−4.00
Brc	0.00	1.00	0.00	−1.00	−2.00	−1.00	1.00	3.00	5.00
Tlc	0.00	0.00	1.00	1.00	5.00	1.00	−1.00	1.00	1.00
(b)									
Fo	1.00	0.00	0.00	1.00	0.57	1.00	1.00	0.00	−2.00
Brc	0.00	1.00	0.00	−0.50	−0.29	1.00	0.25	0.75	2.50
Tlc	0.00	0.00	1.00	0.50	0.71	−1.00	−0.25	0.25	0.50

Initial composition matrix (old system components) $[A][B_{OC}]$ (Table 2.5 a)
New matrix (new system components) $[I][B_{NC}]$ (Table 2.6 a)
Matrix operation: $[A^{-1}][A][B_{OC}] \Rightarrow [I][B_{NC}]$

Old basis [A]			Inverse of old basis $[A^{-1}]$			Identity matrix [I]		
1	0	4	−0.33	0.67	−0.67	1	0	0
2	1	3	−0.33	0.17	0.83	0	1	0
0	1	1	0.33	−0.17	0.17	0	0	1

	Fo	Brc	Tlc	En	Ath	Qtz	Per	Atg	Fl
(c)									
Brc	0.00	1.00	0.00	−1.00	−0.69	1.00	0.00	0.75	0.83
Tlc	0.00	0.00	1.00	1.00	1.69	−1.00	0.00	0.25	0.17

transforms the composition space expressed in terms of simple oxide components (Table 2.5 a) to the composition space expressed in terms of Mg_2SiO_4 (Fo), $Mg(OH)_2$ (Brc), and $Mg_3Si_4O_{10}(OH)_2$ (Tlc) components (Table 2.6 a) is also illustrated in Table 2.6. It can be seen that the operation is a pre-multiplication of Table 2.5 a by the inverse of the leading 3×3 square matrix in Table 2.5 a. The result of the operation is the composition matrix (Table 2.6 a) with the mineral compositions expressed by the new set of system components. Today, any standard commercial spreadsheet program running on any PC or MAC will perform these algebraic operations for you (e.g., Excel, Wingz). We are now set for the construction of the desired forsterite projection. Just as in simple projections, delete the Fo-row in Table 2.6 b and renormalize to constant sum. Table 2.6 c shows the coordinates of the phase compositions along the Brc–Tlc binary. Some of the compositions project to the negative side of talc; periclase cannot be projected onto the Brc–Tlc binary at all. This is apparent also from Fig. 2.6, where it can be recognized that, seen from forsterite, periclase projects away from the Brc–Tlc binary. It is also clear from the procedure outlined above that the compositions which one chooses to project from or which one wants to see in the apexes of the mole fraction triangle must be written as column vectors in the original A-matrix.

With the two basic operations, projection and redefinition of system components, one can construct any thermodynamically valid composition phase diagram for any geological problem.

2.5.3.3
AFM Projections

A classical example of a composite projection is the AFM projection for metapelitic rocks (see also Chap. 7). Many of the phase relationships in metapelitic rocks can be described in the six-component system K_2O–FeO–MgO–Al_2O_3–SiO_2–H_2O (KFMASH system). A graphic representation of the system requires projection from at least three fixed compositions. Many metapelitic rocks contain excess quartz and some aspects of metamorphism can be discussed for water-present conditions. Therefore, a projection from SiO_2 and H_2O can be easily prepared. However, none of the remaining four components is present as a phase in such rocks. Now, in low- and medium-grade metapelites muscovite is normally present as an excess phase; in high-grade rocks muscovite is replaced by K-feldspar. Therefore, a useful diagram could be prepared by projecting through muscovite or K-feldspar onto the plane Al_2O_3–FeO–MgO. Furthermore, under the condition of excess quartz, an Al_2SiO_5 polymorph is always more stable than corundum. This, in turn, requires that the composition matrix for minerals in metapelitic rocks is rewritten in terms of the new system components $KAl_3Si_3O_{10}(OH)_2$–Al_2SiO_5–

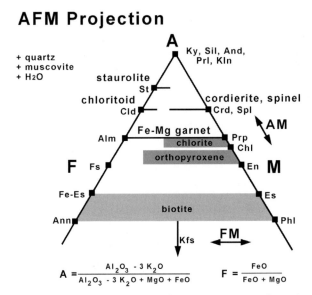

Fig. 2.8. Coordinates of phase compositions in the K_2O–FeO–MgO–Al_2O_3–SiO_2–H_2O system projected through muscovite, quartz and H_2O onto the plane Al_2O_3, FeO and MgO (AFM projection)

Table 2.7. Composition matrix for minerals in the KFMASH system. (a) Composition matrix in terms of oxide components. (b) Inverse of leading 6×6 square matrix (A-matrix⇒A^{-1}). (c) Composition matrix in terms of Ky, Qtz, Ms, H$_2$O, FeO and MgO. (d) Renormalized composition matrix, column vectors are coordinates of mineral compositions in the mole fraction triangle Ky (A), FeO (F) and MgO (M) (⇒AFM-diagram), projection through Qtz, Ms and H$_2$O

	Ky	Qtz	Ms	Fl	FeO	MgO	Alm	Prp	Ann	Phl	FEs	Es	St	Cld	Crd	OPX	Spl	Chl
(a)																		
SiO$_2$	1	1	6	0	0	0	3	3	6	6	5	5	8	2	5	1.8	0	3
Al$_2$O$_3$	1	0	3	0	0	0	1	1	1	1	2	2	9	2	2	0.2	1	1
FeO	0	0	0	0	1	0	3	0	6	0	5	0	4	2	0	0.5	0	0
MgO	0	0	0	0	0	1	0	3	0	6	0	5	0	0	2	1.3	1	5
K$_2$O	0	0	1	1	0	0	0	0	1	1	1	1	0	0	0	0.0	0	0
H$_2$O	0	0	2	1	0	0	0	0	2	2	2	2	2	2	0	0.0	0	4
(b)																		
	0.00	1.00	0.00	0.00	-3.00	0.00												
	1.00	-1.00	0.00	0.00	-3.00	0.00												
	0.00	0.00	0.00	0.00	1.00	0.00												
	0.00	0.00	1.00	0.00	-2.00	1.00												
	0.00	0.00	1.00	0.00	0.00	0.00												
	0.00	0.00	2.00	1.00	0.00	0.00												
(c)																		
Ky	1.00	0.00	0.00	0.00	0.00	0.00	1.00	1.00	-2.00	-2.00	-1.00	-1.00	9.00	2.00	2.00	0.20	1.00	1.00
Qtz	0.00	1.00	0.00	0.00	0.00	0.00	2.00	2.00	2.00	2.00	0.00	0.00	-1.00	0.00	3.00	1.60	-1.00	2.00
Ms	0.00	0.00	1.00	0.00	0.00	0.00	0.00	0.00	1.00	1.00	1.00	1.00	0.00	0.00	0.00	0.00	0.00	0.00
Fl	0.00	0.00	0.00	1.00	0.00	0.00	0.00	0.00	0.00	0.00	0.00	0.00	2.00	2.00	0.00	0.00	0.00	4.00
FeO	0.00	0.00	0.00	0.00	1.00	0.00	3.00	0.00	6.00	0.00	5.00	0.00	4.00	2.00	0.00	0.50	0.00	0.00
MgO	0.00	0.00	0.00	0.00	0.00	1.00	0.00	3.00	0.00	6.00	0.00	5.00	0.00	0.00	2.00	1.30	1.00	5.00
(d)																		
Ky	1.00	0.00	0.00	0.00	0.00	0.00	0.25	0.25	-0.50	-0.50	-0.25	-0.25	0.69	0.50	0.50	0.10	0.50	0.17
FeO	0.00	0.00	0.00	0.00	1.00	0.00	0.75	0.00	1.50	0.00	1.25	0.00	0.31	0.50	0.00	0.25	0.00	0.00
MgO	0.00	0.00	0.00	0.00	0.00	1.00	0.00	0.75	0.00	1.50	0.00	1.25	0.00	0.00	0.50	0.65	0.50	0.83

FeO–MgO–SiO$_2$–H$_2$O, as shown in Table 2.7. The coordinates of the phase compositions in the AFM mole fraction triangle can then be represented in an AFM diagram (Fig. 2.8). All Mg-Fe-free aluminum silicates project to the A-apex, pure enstatite, talc and anthophyllite are found in the M-apex, ferro-silite and Fe-anthophyllite project to the F-apex. Biotites have negative A-co-ordinates. The iron-magnesium substitution (FeMg$_{-1}$ exchange) in the ferro-magnesian minerals is parallel to the AM binary, the Mg-Tschermak substitu-tion (2 Al Si$_{-1}$ Mg$_{-1}$ exchange) is parallel to the AM binary. Minerals such as staurolite, chloritoid, garnet, cordierite and spinel do not show any Tscher-mak variation, their compositional variation is exclusively parallel to the FeMg$_{-1}$ exchange. Other minerals such as chlorites, biotites, orthopyroxenes and orthoamphiboles show both FeMg$_{-1}$ exchange and 2 Al Si$_{-1}$ Mg$_{-1}$ ex-change. The composition of these minerals is represented by fields in an AFM diagram. The AFM projection (Fig. 2.8) can be used exclusively for rocks with excess quartz and muscovite and for water-present conditions. Similar projections can readily be prepared from the composition matrix in Table 2.7 c if one wants to analyze phase relationships for rocks containing excess muscovite, quartz and alumosilicate but treating H$_2$O as compositional variable on the diagram. Such a WFM diagram can be constructed by delet-ing the rows Ky, Qtz and Ms and renormalizing the remaining three rows. A QFM diagram (Qtz–FeO–MgO) projecting through Ms, Ky and H$_2$O can be constructed from Table 2.7c for discussing rocks with excess alumosilicate and muscovite but not excess quartz. To project from K-feldspar rather than muscovite, the Ms column in Table 2.7 a has to be replaced by the column vector of K-feldspar; the subsequent procedure is as described above. If, for example, in low-grade rocks muscovite contains much Tschermak component (phengite), one may replace the Ms column vector (end-member muscovite) in Table 2.7 a by the analyzed mica composition.

Figure 2.8 shows the general projection coordinates of AFM phases. How-ever, composition phase diagrams are used to represent phase relationships at a given pressure (P) and temperature (T). A specific example of an AFM diagram at a distinct P and T is shown in Fig. 2.9. Present in excess is quartz, muscovite and H$_2$O. The AFM surface is divided into a number of subregions. Three types can be distinguished:

1. One-phase fields: if the total rock composition projects inside, for exam-ple, the one-phase field for biotite, it will be composed of quartz, musco-vite and biotite. The composition of biotite is given by the composition of the rock. One-phase fields are divariant fields because the composition of the mineral freely changes with the two composition variables of the bulk rock.

2. Two-phase fields: if the total rock composition projects inside the two-phase field biotite+cordierite it will contain the assemblage Ms + Qtz + Crd + Bt. The composition of coexisting biotite and cordierite can be con-nected with a tie line (isopotential line) passing through the projection point of the bulk rock composition (e.g., Bt1–Crd1 in Fig. 2.9). Through any bulk rock inside the two-phase field a tie line can be drawn which connects the two coexisting minerals (tie line bundle). It follows that all

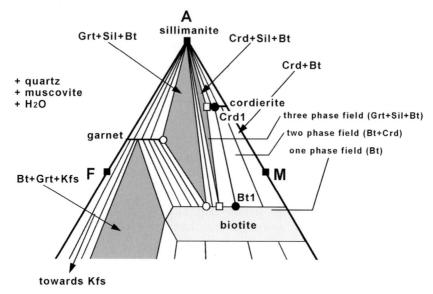

Fig. 2.9. AFM diagram at a specified pressure and temperature showing phase relationships among some typical AFM phases

bulk rock compositions on a specific Crd-Bt tie line (at a given P and T) contain minerals of identical composition. If the total rock projects close to cordierite it will contain much cordierite and little biotite (and vice versa). The mineral compositions are controlled by the Fe–Mg exchange equilibrium between cordierite and biotite (see Chap. 3). Also the Tschermak exchange in biotite is controlled by the assemblage. Consider, for example, a bulk rock which projects in the garnet + biotite field. For a given biotite composition, the composition of the coexisting garnet is fixed by the tie line passing through the given biotite composition and the projection point of the bulk rock.

3. Three-phase fields: if the total rock composition projects inside the three-phase field biotite + cordierite + sillimanite it will contain the assemblage Ms + Qtz + Crd + Bt + Sil. Here the compositions of all minerals are entirely controlled by the assemblage (open squares in Fig. 2.9). All bulk rock compositions that project in the three-phase field Crd + Bt + Sil will be composed of these three minerals of identical composition in modal proportions depending on the rock composition. Three-phase fields are invariant fields at constant P and T, because the mineral compositions do not vary with the rock composition. The cordierite coexisting with biotite and sillimanite has the most Fe-rich composition of all cordierite at the P–T conditions of the AFM diagram. Note that, in general, all three-phase fields on an AFM diagram must be separated from each other by two-phase fields. In the three-phase field Grt + Sil + Bt the compositions of the minerals are controlled by the assemblage as well (open circles in

Fig. 2.9). It follows that biotite in this assemblage must be more Fe-rich than biotite in the assemblage Crd + Sil + Bt. The garnet coexisting with Bt + Sil has the most Mg-rich composition of all garnets at the given pressure and temperature.

AFM-type diagrams will be extensively used for discussing metamorphism in metapelitic rocks in Chapter 7.

2.5.3.4
ACF Projections

Phase relationships in marbles, calc-silicate rocks, calcareous metapelites and other rocks with calcic phases can be analyzed, for example, in a simple CFMAS system. Calcic phases may include amphiboles, plagioclase, epidote, diopside and carbonate minerals. A graphic representation of the phase relationships can, for instance, be made by projecting from quartz (if present in excess) and from a CO_2–H_2O fluid phase of constant composition onto the mole fraction triangle Al_2O_2–CaO–FeO (ACF diagram). The ACF diagram is also a projection parallel to the $FeMg_{-1}$ exchange vector. All minerals of the AFM system can also be represented on ACF diagrams provided that one also projects through muscovite or K-feldspar. The coordinates of mineral compositions can be calculated as explained above for the MSH and KFMASH systems, respectively. Figure 2.10 is a typical ACF diagram, showing phase relationships at a certain pressure and temperature. The Tschermak variation is parallel to the AF binary and affects mainly chlorite and amphibole (and pyroxene which is not present at the P–T conditions of the figure). The three-phase fields Pl + Am + Grt and Ky + Grt + Pl are separated by a two-phase field Grt + Pl because of $CaMg_{-1}$ substitution in garnet (grossular component). Both garnet and plagioclase show no compositional variation along the TS vector. Any mineral that can be described by the components K_2O–CaO–FeO–MgO–Al_2O_3–SiO_2–H_2O–CO_2 can be represented on an ACF diagram such as that shown in Fig. 2.10. However, the consequences of $FeMg_{-1}$ substitution in minerals cannot be discussed by means of ACF diagrams. Therefore, any discontinuous reaction relationship deduced from an ACF diagram is in reality continuous and dependent on the Fe–Mg variation (if it involves Fe–Mg minerals, of course). For example, the replacement of the garnet-plagioclase tie line by a more stable tie line between kyanite and amphibole can be related to the reaction: $Pl + Grt \Rightarrow Am + Ky \pm Qtz \pm H_2O$. Equilibrium of the reaction, however, depends not only on pressure and temperature, but also on the Fe-Mg variation in garnet and amphibole. The projection coordinates of a selection of mineral compositions are shown on the chemograph at the top of Fig. 2.10.

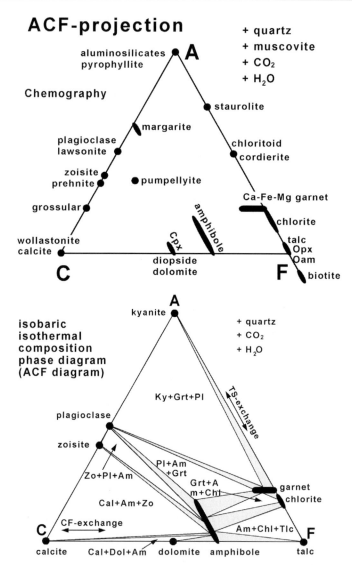

Fig. 2.10. Coordinates of phase compositions in the $(K_2O)–CaO–FeO–MgO–Al_2O_3–SiO_2–H_2O–CO_2$ system projected through (muscovite), quartz, CO_2 and H_2O and parallel to the $MgFe_{-1}$ exchange vector onto the plane Al_2O_3, FeO and MgO (ACF projection). *Top* Coordinates of some ACF phase compositions. *Bottom* Phase relationships among some typical ACF phases at a specified pressure and temperature

2.5.3.5
Other Projections

Any other graphic representation of phase relationships on composition phase diagrams can be prepared by the procedure outlined above. The type of graphic representation of assemblages is entirely dictated by the material one is working with and by the problem one wishes to solve. The following steps may be a guide to the production of adequate phase composition diagrams:

1. Group the collected rocks into populations with similar bulk compositions ("normal" metapelites, metabasalts, and so on).
2. Identify minerals that are present in the majority of a given group of rocks (e.g., muscovite and quartz in metapelites). Special assemblages require a special treatment (e.g., in quartz-absent corundum-bearing metapelites one may project through corundum onto a QFM plane).
3. If the excess minerals are not simply composed of oxide components, rewrite the composition matrix in terms of the compositions of the excess phases and the desired compositions at the corners of the mole fraction triangle selected as a new projection plane.
4. Delete rows in the composition matrix containing the excess phases and renormalize the column vectors. Draw the diagram and keep in mind the proper distribution of one-, two- and three-phase fields. Do not forget to write the compositions of the projection phases on the diagram (without this information your figure is worthless).

In Part II of this volume we will make extensive use of various kinds of composition phase diagrams.

References

Ashworth JR (1985) Migmatites. Blackie, Glasgow, 301 pp
Bell TH, Johnson SE (1992) Shear sense: a new approach that resolves conflicts between criteria in metamorphic rocks. J Metamorph Geol 10:99–124
Brodie KH, Rutter EH (1985) On the relationship between deformation and metamorphism with special reference to the behaviour of basic rocks. Advances in physical geochemistry. In: Thompson AB, Rubie DC (eds) Advances in physical geochemistry. Springer, Berlin Heidelberg New York, pp 138–179
Carmichael RS (1989) Practical handbook of physical properties of rocks and minerals. CRC Press, Boca Raton
Cooke RA, O'Brien PJ, Carswell DA (2000) Garnet zoning and the identification of equilibrium mineral compositions in high-pressure-temperature granulites from the Moldanubian Zone, Austria. J Metamorph Geol 18:551–569
Greenwood HJ (1975) Thermodynamically valid projections of extensive phase relationships. Am Mineral 60:1–8
Kisch HJ (1992) Development of slaty cleavage and degree of very low-grade metamorphism: a review. J Metamorph Geol 9:735–750
Orville PM (1969) A model for metamorphic differentiation origin of thin-layered amphibolites. Am J Sci 267:64–68
Selverstone J (1993) Micro- to macroscale interactions between deformational and metamorphic processes, Tauern Window, Eastern Alps. Schweiz Mineral Petrogr Mitt 73:229–239
Spear FS (1988) Thermodynamic projection and extrapolation of high-variance assemblages. Contrib Mineral Petrol 98:346–351

Spear FS, Rumble D III, Ferry JM (1982) Linear algebraic manipulation of n-dimensional
    composition space. In: Ferry JM (ed) Characterization of metamorphism through miner-
    al equilibria, vol 10. Reviews in mineralogy. Mineralogical Society of America, Washing-
    ton, DC, pp 53–104
Spry A (1969) Metamorphic textures. Pergamon Press, New York
Thompson JB (1957) The graphical analysis of mineral assemblages in pelitic schists. Am
    Mineral 42:842–858
Thompson JB (1982) Composition space; an algebraic and geometric approach. In: Ferry
    JM (ed) Characterization of metamorphism through mineral equilibria, vol 10. Reviews
    in mineralogy. Mineralogical Society of America, Washington, DC, pp 1–31
Tracy RJ, Robinson P (1983) Acadian migmatite types in pelitic rocks of Central Massachu-
    setts. Migmatites, melting and metamorphism. Shiva, Nantwich, pp 163–173
Williams ML (1994) Sigmoidal inclusion trails, punctuated fabric development and interac-
    tions between metamorphism and deformation. J Metamorph Geol 12:1–21
Zeck HP (1974) Cataclastites, hemiclastites, holoclastites, blasto-ditto and myloblastites –
    cataclastic rocks. Am J Sci 274:1064–1073

CHAPTER 3

Metamorphic Processes

Rock metamorphism is always associated with processes and changes. Meta-morphism reworks rocks in the earth's crust and mantle. Typical effects of rock metamorphism include:

- Minerals and mineral assemblages originally not present in the rock may form; the new mineral assemblages form at the expense of old ones, consequently older minerals may disappear (e.g., a metamorphic rock may originally contain Grt + Qtz + Sil; a metamorphic event transforms this rock into one that contains Crd (cordierite) in addition to the minerals previously present in the rock).
- The relative abundance of minerals in a rock may systematically change and the new rock may have a different modal composition (metamorphism may increase the amount of Crd present in the rock and decrease the volume proportion of Grt + Qtz + Sil).
- Metamorphic minerals may systematically change their composition (e.g., the X_{Fe} of Grt and Crd may simultaneously increase during metamorphism).
- The structure of rocks in crust and mantle may be modified (e.g., randomly oriented sillimanite needles may be aligned parallel after the process).
- The composition of the bulk rock may be altered during metamorphism by adding or removing components to or from the rock from a source/sink outside the volume of the rock considered (e.g., adding K_2O dissolved in an aqueous solution to a Grt + Crd + Sill + Qtz rock may result in the formation of biotite).

The typical changes in the modal composition of rocks and in the composition of minerals that constitute the rocks are caused by chemical reactions. The principles of metamorphism are, therefore, strongly related to the principles of chemical reactions. Mineral- and rock-forming metamorphic processes are mainly controlled by the same parameters that control chemical reactions. Metamorphic petrology studies reaction and transport processes in rocks. Metamorphic processes are caused by transient chemical, thermal and mechanical disequilibrium in confined volumes of the earth's crust and mantle. These disequilibrium states ultimately result from large-scale geological processes and the dynamics of the Earth's planetary system as a whole. Metamorphic processes always result from disequilibrium and gradients in parameters that control reaction and transport in rocks. Metamorphic processes cease when the rocks reach an equilibrium state. Chemical reaction is always inherent in the term metamorphism. The term metamorphosis actually means transformation, modification, alteration, and conversion and thus is clearly a process-related expression.

Metamorphism is a very complex occurrence that involves a large number of chemical and physical processes at various scales. Metamorphic processes can be viewed as a combination of (1) chemical reactions between minerals and between minerals and gasses, liquids and fluids (mainly H_2O), and (2) transport and exchange of substances and heat between domains where such reactions take place. The presence of an aqueous fluid phase in rocks undergoing metamorphism is critical to the rates of both chemical reactions and chemical transport. Consequently, an advanced understanding of metamorphism requires a great deal of insight into the quantitative description of chemical reactions and chemical transport processes, especially reversible and irreversible chemical thermodynamics.

The term metamorphism as it is related to processes, changes and reactions clearly also includes the aspect of time. Metamorphism occurs episodically and is particularly related to mountain-building episodes at convergent plate margins (collision zones) and during subsequent uplift and extension of continental crust, as well as during seafloor spreading and continental rifting.

3.1
Principles of Metamorphic Reactions

In the following we briefly explain some basic aspects of metamorphic reactions and introduce some elementary principles that are essential for a basic understanding of metamorphism. The treatment is not sufficient for a thorough understanding of metamorphic processes and chemical reactions in rocks. For the reader who needs to know more we recommend the study of textbooks about chemical thermodynamics (e.g., Prigogine 1955; Lewis and Randal 1961; Moore 1972; Guggenheim 1986) or textbooks that deal particularly with application of thermodynamics to mineralogy and petrology (Wood and Fraser 1976; Fraser 1977; Greenwood 1977; Powell 1978; Lasaga and Kirkpatrick 1981; Ferry 1982; Saxena and Ganguly 1987). We particularly recommend Chatterjee (1991) and the superb book by Fletcher (1993).

Now let us consider, for example, a rock that contains the minerals albite and quartz. The Ab-Qtz rock is located at a certain depth in the crust (e.g., at point $T_h = 12.5$ kbar, 550 °C in Fig. 3.1). At that given pressure and temperature, the two minerals (phases) are associated with unique values of molar Gibbs free energy. The Gibbs free energy, usually abbreviated with the symbol G, is a thermodynamic potential with the dimension Joules/mole (energy/mole) and it is a function of pressure and temperature. The free energy of minerals and their mixtures are negative quantities (because they refer to the free energy of formation from the elements or oxides rather than absolute energies, e.g., G of albite at 900 K and 1 bar: -3257.489 kJ mol^{-1}). The Ab + Qtz rock under consideration can be formed by mechanically mixing,

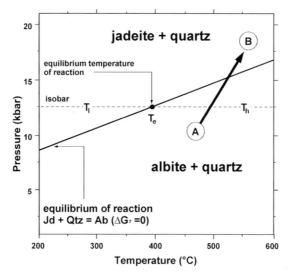

Fig. 3.1. Pressure–temperature diagram showing equilibrium conditions of the reaction jadeite+quartz = albite

e.g., 1 mol albite and 1 mol quartz. Thermodynamically speaking, all rocks represent mechanical mixtures of phases. Rocks are, therefore, heterogeneous thermodynamic systems[1]. The phases, in turn, can be viewed as chemically homogeneous subspaces of the considered system, e.g., a volume of rock. Minerals, aqueous fluids, gasses and melts are the thermodynamic phases in rocks. The phases that make up rocks are usually chemical mixtures of a number of phase components (most minerals are solid chemical solutions and show a wide range in composition). In our example, albite and quartz shall be pure $NaAlSi_3O_8$ and SiO_2, respectively. The total free energy of the rock is the sum of the free energies of its parts that is in our case $n_{Ab}G_{Ab} + n_{Qtz}G_{Qtz}$ (where n_i = number of moles of substance i). The rock is characterized by a unique value of G and, by taking 1 mol of each substance, its total composition is $NaAlSi_4O_{10}$. However, the composition of such a mechanical mixture ($\equiv$ rock) can also be obtained by mixing (powders of) jadeite ($NaAlSi_2O_6$) and quartz in the appropriate proportions. It is also clear that this rock is associated with a unique Gibbs free energy at the given pressure and temperature and that it corresponds to the sum of $G_{Jd} + 2\ G_{Qtz}$. The free energy of the Ab-Qtz rock may be designated G_{AQ} and that of the Jd-Qtz rock G_{JQ}. The Gibbs free energies of the two rocks at P and T can be calculated from Eqs. (1) and (2) provided that the free energy values of the three minerals can be calculated for those values of P and T:

$$G_{AQ} = G_{Ab} + G_{Qtz} \qquad\qquad (1)$$

---

[1] The thermodynamic description of heterogeneous systems was developed and formulated mainly by W. Gibbs (1878, 1906). Gibbs' scientific contributions were fundamental for the development of modern quantitative petrology.

$$G_{JQ} = G_{Jd} + 2\,G_{Qtz} \tag{2}$$

The free energies of the three minerals at P and T can be calculated from thermodynamic data and equations of state given by the laws of chemical thermodynamics. The question one may ask now is which one of the two possible rocks, the albite + quartz rock or the jadeite + quartz rock, will be present at the conditions T_h (Fig. 3.1). According to the thermodynamic laws, it is always the mixture with the lowest total free energy at the prevailing conditions that is the **stable** mixture (assemblage), while the other mixture is **metastable**. These laws can be summarized and expressed by the following statement:

The chemical components (constituents) making up a rock's total composition are distributed into a group of homogeneous phases, minerals and fluid(s), that constitute the assemblage with the lowest free energy for the system at that given pressure and temperature. This assemblage is called the equilibrium phase assemblage.

At T_h G_{AQ} is more negative than G_{JQ} and thus the albite + quartz rock is stable, while the jadeite + quartz assemblage is metastable or less stable than Ab + Qtz. The free energy difference of the two rocks can be calculated by subtracting Eq. (2) from Eq. (1):

$$\Delta G = G_{AQ} - G_{JQ} = G_{Ab} - G_{Jd} - G_{Qtz} \tag{3}$$

ΔG is the free energy difference of the two rocks with identical composition but different mineralogy. The mineralogy of the two rocks is related by a reaction that can be expressed by the stoichiometric equation:

$$
\begin{array}{ccc}
1\,NaAlSi_2O_6 & + \quad 1\,SiO_2 = & 1\,NaAlSi_3O_8 \\
\text{jadeite} & \text{quartz} & \text{albite}
\end{array} \tag{4}
$$

The stoichiometric coefficients in this particular reaction equation are all equal to 1. The equation suggests that albite in the feldspar + quartz rock that is stable at T_h may decompose to jadeite + quartz at some other conditions than T_h. Similarly, the reaction shown in Eq. (4) describes the formation of 1 mol Na-feldspar from 1 mol jadeite + 1 mol quartz. The free energy change of the reaction ΔG_r is calculated from:

$$\Delta G_r = G_{Ab} - G_{Jd} - G_{Qtz} \tag{5}$$

By convention, the stoichiometric coefficients on the right-hand side of a reaction equation are always taken as positive, while those on the left-hand side are taken as negative. In the example made here the energy difference of the two rocks [Eq. (3)] is identified as the free energy change of the albite-forming reaction. ΔG_r is, like G of the individual phases, a function of P and T. Three basic cases may be distinguished:

$$\Delta G_r < 0 \tag{6}$$

At all pressure and temperature conditions that satisfy Eq. (6) the products of the reaction are stable and the reactants are metastable. At T_h, ΔG_r is negative and albite + quartz constitutes the stable assemblage. A rock with jadeite + quartz is metastable at T_h.

$$\Delta G_r > 0 \tag{7}$$

At all conditions that satisfy Eq. (7) the reactants of the reaction are stable and the products are metastable. At T_l, ΔG_r is positive and jadeite + quartz constitutes the stable assemblage (Fig. 3.1). A rock with albite + quartz is metastable at T_l.

$$\Delta G_r = 0 \tag{8}$$

At all conditions that satisfy Eq. (8) the reactants and the products of the reaction are simultaneously stable. These conditions are referred to as the **equilibrium conditions of the reaction**. In Fig. 3.1, the equilibrium condition of the reaction expressed by Eq. (8) is represented by a straight line. Along that line all three minerals are simultaneously stable. At a pressure of 12.5 kbar, the equilibrium temperature of the reaction is 400 °C and corresponds to the point T_e along the isobar of Fig. 3.1. At this unique temperature the rock may contain all three minerals in stable equilibrium. At any temperature other than T_e the reaction is not at equilibrium and it will proceed in such a way as to produce the assemblage with the most negative free energy for that given temperature. In Fig. 3.1, the equilibrium line of the reaction divides the P–T space into two half spaces with two distinct stable assemblages. The P–T diagram represents in this way a free energy map of the considered system. Like a topographic map where the surface of the earth is projected along a vertical axis, a P–T phase diagram represents a map where the lowest free energy surface of the system is projected along the G-axis onto the P–T plane.

At equilibrium, the mineralogical composition of a rock is entirely dictated by its composition, the pressure and the temperature. If temperature and pressure change, new assemblages may have a lower free energy, and chemical reaction will replace the old assemblage by a new, more stable assemblage. A rock always tries to reach a state of equilibrium by minimizing its free energy content. This is accomplished by readjusting the mineralogy or the composition of minerals if necessary.

Stable assemblages cannot be distinguished from metastable ones by any petrographic technique, and metastable equilibria cannot be separated from stable equilibria. This becomes obvious if we look at a rock sample collected on a rainy day that contains the three aluminosilicate minerals, kyanite, andalusite, and sillimanite in addition to quartz (a rock type that is relatively common in the central Alps). All three Al-silicates are metastable in the presence of quartz and water under surface conditions relative to the hydrous Al-silicate kaolinite. In general, at some arbitrary metamorphic P–T condition, only one Al-silicate + quartz can be stable, the other two possible two-phase assemblages must be metastable. Metastable persistence of metamorphic minerals and assemblages is common in metamorphic rocks. This is, of course, an extremely fortunate circumstance for metamorphic petrologists. Rocks that formed in the deep crust or mantle with characteristic high-pressure and high-temperature mineral assemblages may be collected at the earth's surface. If metastable assemblages were not common in metamorphic rocks we would find only low P-T rocks at the surface. Metastable assemblages may even survive over geological time scales (hundreds of millions of years). It is obvious from Fig. 3.1 that a rock that contains jadeite + quartz is metastable under conditions at the surface of the earth. However, rocks containing Jd + Qtz can be found at the surface and witness that the rocks were exposed to very high pressures earlier in their geological history. Other examples of metastable persistence of minerals that are stable at high pressure or temperature are the presence of minerals such as sillimanite, coesite, aragonite, and diamond in rocks collected at the earth's surface.

Some criteria and methods to detect **disequilibrium** in rocks do exist, however. One may distinguish two main kinds of disequilibrium: structural disequilibrium and chemical disequilibrium. Structural disequilibrium is indicated by distinct shapes and forms of crystals and spatial distribution and arrangement of groups of minerals in heterogeneous systems. Structural disequilibrium may be found in rocks that underwent chemical reactions or successive series of reactions that all ceased before equilibrium structures developed and overall chemical equilibrium was reached. Many different kinds of compositional features of metamorphic rocks and minerals may indicate chemical disequilibrium. As an illustration we may use our example again. We may have collected a rock that contains omphacite + quartz (sodium-rich clinopyroxene where jadeite is present as a phase component) and independent information suggests that this rock formed at about 600 °C. The same rock may also contain pure albite in domains, local patches or veins. The feldspar composition in this rock represents a clear chemical disequilibrium feature. The minerals omphacite + albite + quartz never coexisted in stable or metastable equilibrium because the pyroxene contains a calcic component (diopside) and the feldspar does not (the phase component anorthite is not present in the feldspar). However, at the inferred temperature of 600 °C, plagioclase in a mafic omphacite-bearing rock should, in an equilibrium situation (stable or metastable), contain some anorthite component. The presence of albite in such a rock must be related to some metamorphic process that progressed under conditions other than the equilibration of the omphacite + quartz assemblage. It is, on the other hand, important to note that the assemblage omphacite + quartz + albite represents an overall disequilibrium but the assemblages omphacite + quartz and albite + quartz may well be equilibrium assemblages that equilibrated at different times at different P-T conditions. The question of equilibrium is always related to the scale of equilibrium domains. Disequilibrium may exist between large rock bodies in the crust, layers of rocks of different composition in an outcrop, between local domains of a thin section or between two minerals in mutual grain contact. Any chemically zoned mineral represents disequilibrium. At some scale there is disequilibrium always and at any time.

Fig. 3.2. Schematic figure showing the Fe–Mg distribution between pairs of coexisting garnet and biotite from one specific rock sample

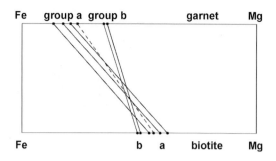

Chemical properties of coexisting minerals in a rock are often used to reason about the equilibrium question. As an example, a series of garnet-biotite pairs have been analyzed from a homogeneous Grt-Bt-schist and the data are shown schematically in Fig. 3.2. The data pairs can be arranged into two groups. *Group a* represents matrix biotite in contact with the rim of garnet, whilst *group b* represents analyses of biotite grains included in the core of the garnet. Overall chemical equilibrium (stable or metastable) between garnet and biotite requires that both minerals have a uniform composition in the rock. Crossing tie lines are inconsistent with overall equilibrium. However, it is evident that the Grt-Bt pairs from each group are not in conflict with the requirements of chemical equilibrium. They may constitute two different **local equilibrium** systems. The Grt-Bt pair connected with a dashed line in Fig. 3.2 shows clear crossing tie-line relationships within *group a* pairs and clearly represents a disequilibrium pair. Note that the absence of crossing tie-line relationships does not necessarily prove that the rock was in a state of overall equilibrium. The lack of obvious disequilibrium phenomena can be taken as evidence for, but not proof of, equilibrium.

Metastable persistence of minerals and mineral assemblages as well as the obvious disequilibrium in many rocks follows from the controlling factors and circumstances of reaction kinetics. The rate of a mineral reaction may be slower than the rate of, e.g., cooling of a volume of rock. The absence of activation energy for reaction in cooling rocks and nucleation problems of more stable minerals typically affect reaction kinetics. The kinetics of reactions in rocks is extremely sensitive to the presence or absence of H_2O. Aqueous fluids serve as solvent and reaction medium for mineral reactions. For example, if kyanite-bearing rocks are brought to pressure and temperature conditions where andalusite is more stable than kyanite, kyanite may be replaced by andalusite in rocks containing a free aqueous fluid in the pore space or along grain boundaries, whilst andalusite may fail to form in fluid-absent dry rocks.

However, it is commonly reasonable to assume that during **prograde** metamorphism rocks pass through successive sequences of equilibrium mineral assemblages. These sequences can be viewed as a series of stages, each of them characterized by an equilibrium assemblage and the different stages are connected by mineral reactions. This assumption is founded on

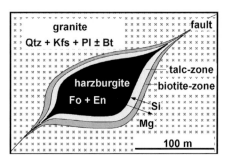

Fig. 3.3. Large-scale disequilibrium between a lens of harzburgite (ultramafic rock fragment from the upper mantle) that has been tectonically embedded in granitic (gneissic) crust. The harzburgite lens has been enveloped by protective shells of talc and biotite. These shells were produced by chemical reaction between incompatible rocks and their formation required transfer of, e.g., Si from granite to harzburgite and Mg from harzburgite to granite

much convincing evidence that prograde metamorphism takes place under episodic or continuous water-present conditions. One would therefore expect to find disequilibrium and metastable assemblages particularly in rocks that were metamorphosed under fluid-absent or fluid-deficient conditions. Aqueous fluids are typically not present in cooling rocks after they have reached maximum metamorphic pressure and temperature conditions. Structural and chemical disequilibrium is also widespread in very high-grade rocks of the granulite facies that lost the hydrous minerals and aqueous fluids during earlier stages of prograde metamorphism. Microstructures such as reaction rims, symplectites, partial replacement, corrosion and dissolution of earlier minerals are characteristic features of granulite facies rocks. They indicate that, despite relatively high temperatures (700–900 °C), equilibrium domains were small and chemical communication and transport were hampered as a result of dry or H_2O-poor conditions.

To further illustrate some aspects of disequilibrium, we may consider large bodies of incompatible rock types in the crust. In many orogenic fold belts (e.g., Alps, Caledonides) mantle-derived ultramafic rock fragments were emplaced in the continental crust during the collision phase and stacking of nappes. Mantle fragments (harzburgites, lherzolites, serpentinites) of various dimensions can be found as lenses in granitic crust (Fig. 3.3). Forsterite (in the mantle fragment) + quartz (in the crustal rocks) is metastable relative to enstatite at any geologically accessible pressure and temperature conditions. The presence of Fo-bearing mantle rocks in Qtz-rich crustal rocks represents a large-scale disequilibrium feature. Chemical mass transfer across the contact of the incompatible rock types results in the formation of shells of reaction zones that encapsulate the mantle fragment. The nature of the minerals found in the reaction zones depends on the P–T conditions at the reaction site. In our example, the reaction shells are made up of talc-rich and biotite-rich zones that are often nearly monomineralic (Fig. 3.3). Such biotite- or amphibole-rich shells that grew as a result of disequilibrium on a large scale and subsequent chemical transport and reaction are also known as black-wall formations. The process is one of contact metasomatism. Chemical communication, transport and reaction are most efficient if aqueous fluids are present in excess and wet the grain boundaries in the incompatible rock bodies. Metasomatism proceeds until the incompatible mineralogy has been used up by the process or the rocks lose their aqueous fluid phase. In rocks with dry grain boundaries, diffusional mass transfer is extremely slow and inefficient even over geological time spans.

Let us return to the jadeite-quartz-albite example (Fig. 3.1). At the conditions of point A, a rock containing albite + quartz is more stable than a rock made of jadeite + quartz. Consider an albite + quartz rock that has reached a state of chemical and structural equilibrium at conditions of point A. The assemblage of this rock may be replaced by jadeite + quartz by reaction (4) if the rock is brought to the conditions at point B. It is obvious that a transfer from A to B would bring the Ab + Qtz rock into the field where Jd + Qtz represents the stable assemblage.

3.2
Pressure and Temperature Changes in Crust and Mantle

3.2.1
General Aspects

Changes in pressure and temperature are the prime causes of metamorphism. This leads to the question: what kind of geologic processes lead to changes in pressure and temperature in the crust or the mantle.

In a general way, such changes are caused by some force that acts on rocks (driving force: driving the process or change). Any kind of force applied to a rock will cause some flow or transfer of a property in such a way as to reduce the size of the applied force.

Temperature differences between volumes of rocks result in the transfer of heat (heat flow) from the hot rock to the cold rock until both rocks reach the same temperature. A simple linear equation describes this process:

$$J_Q = -L_Q \nabla T \tag{9}$$

Equation (9) states that a non-uniform temperature distribution in a system will transfer heat (Q) in the direction normal to the isotherms of the temperature field and from high temperature to low temperature in order to decrease the temperature difference. The linear equation (known as Fourier's law) describes heat flow in systems with small temperature differences (such as most geologic systems). L_Q is a material-dependent constant related to the thermal conductivity of the material κ. J_Q is the rate of heat transfer per unit area (heat flow vector) parallel to the highest gradient in the temperature field ($J_Q = \partial Q / \partial t$). ∇T represents the gradients in temperature in the three-dimensional space. If there are only linear temperature gradients in one direction, the ∇T in Eq. (9) reduces to ($T_{x_1} - T_{x_2}$), the temperature difference between two points. The general expression presented above shows that gradients in intensive variables in a system ultimately cause "processes" and "changes" geological or other. As metamorphism is mainly related to chemical reactions in rocks that are largely controlled by changes in temperature, pressure and rock composition, it is useful to discuss geological aspects of heat transfer, pressure changes and chemical mass transfer in the crust and mantle in some detail.

3.2.2
Heat Flow and Geotherms

Fourier's law states that heat will be transported from a place at high temperature to an area at low temperature. In the case of the earth, and looking at it on a large scale, there is a hot interior and a cold surface. This necessarily results in transport of heat from the center to the surface. The surface of the earth is at a nearly constant temperature (about 10 °C) and the core mantle boundary is probably also at a constant temperature as a result of the liquid state of the outer core that permits very rapid convection and heat transport. The general situation is depicted in Fig. 3.4. A consequence of the temperature difference between the two surfaces is a steady and continuous heat flow from the interior to the surface.

Heat flow is measured in $mW\,m^{-2}$ (milli-Watt per square meter); however, in the literature, heat flow data are often given in "HFU" (heat flow units). A HFU is defined as $\mu cal\,cm^{-2}\,s^{-1}$, that is equivalent to $4.2\,\mu J\,s^{-1}\,cm^{-2}$ or $0.042\,J\,s^{-1}\,m^{-2}$. Because $J\,s^{-1}$ is equivalent to W (Watt), it follows that 1 HFU is equal to $42\,mW\,m^{-2}$.

The heat flow from the interior of the earth is in the order of $30\,mW\,m^{-2}$. However, heat flow measurements at the surface of the earth vary between about 30 and $120\,mW\,m^{-2}$. The total heat flow at the surface is composed of a number of contributions: (1) heat flow from the interior resulting from conductive heat transport as described by Fourier's law, (2) transport of heat by convection in the mantle (Fig. 3.5), and (3) transport of heat generated by decay of radioactive elements.

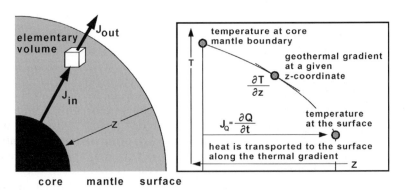

Fig. 3.4. Conductive heat transfer between hot interior and cold surface of the earth

The continental crust consists mostly of granitoid rocks that produce about 30 mJ heat per kilogram and year. Oceanic crust that is generally composed of basaltic rocks produces about 5 mJ heat per kilogram and year, whereas mantle rocks produce only a small amount of radioactive heat (ca. 0.1 mJ heat per kilogram and year). The extra heat produced in the crust contributes significantly to the observed heat flow at the surface.

Heat flow through a volume element of the crust may occur under the following conditions:

1. The heat flow into the crustal volume is equal to the heat flow out from that volume. In this case the temperature in the volume element remains constant. The temperature profile along z is independent of time (**steady-state geotherm**).

2. The heat flow into the crustal volume is greater than the heat flow out from that volume. The excess heat put into the crustal volume will be used in two different ways: (1) to increase the temperature of the rock volume, and (2) to drive endothermic chemical reactions in rocks.

3. The heat flow into the crustal volume is lower than the heat flow out from the volume. In this case the heat loss of the volume of rock results in a temperature decrease (the rock cools). In this situation exothermic chemical reactions may produce extra heat to some extent. These reactions have the effect of preventing the rocks from cooling.

Different heat flow at the surface also has the consequence that rocks at the same depth in the crust and upper mantle may be at different temperatures

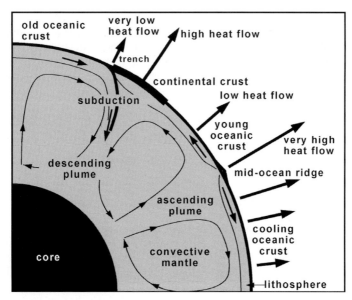

Fig. 3.5. Steady-state heat flow resulting from conductive heat transfer is modified by active tectonic processes. Heat flow at the surface varies by a factor of 4

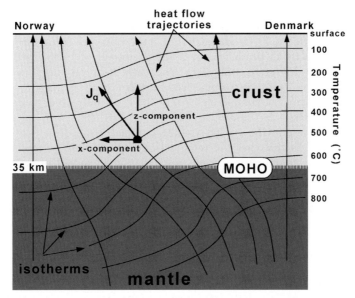

Fig. 3.6. Modeled temperature field along a cross section from Denmark to Norway. (After Balling 1985)

leading to lateral heat transport parallel to the earth's surface (parallel to xy surfaces). This is shown in Fig. 3.6. Along a profile from Denmark to southern Norway, the observed surface heat flows have been used by Balling (1985) to model the temperature field in the crust and mantle. The temperature field is shown in Fig. 3.6. Because flow vectors are always normal to the force field from which the flow results, the heat flow trajectories will roughly look like the flow vectors shown in Fig. 3.8. It is obvious from this two-dimensional section that at a given point in the crust the heat flow has a vertical *and* a horizontal component. Also note that the MOHO under the mid-Paleozoic continental crust of central Europe (Denmark) is at about 700 °C, whereas the MOHO under the Precambrian crust of the Baltic shield in the north is only at about 350 °C (!). It follows that the base of a continental crust may be at largely different temperatures depending on the state and evolutionary history of a lithosphere segment and on the thermal regime deeper in the mantle. Observed surface heat flows can be translated into geotherms that may be time-dependent geotherms or steady-state geotherms. Such model geotherms are shown in Fig. 3.7 together with typical associated surface heat flows and temperatures at an average MOHO depth beneath continents of about 35 km (10 kbar). The geotherms represent curves that relate the temperature with depth (or pressure). The geothermal gradient (expressed as dT/dp or dT/dz) is the slope of the geotherm at a given depth in the crust or mantle and characterizes the temperature increase per pressure (or depth) increment. In this volume, you will see many P–T diagrams with T on the *x*-axis and p (depth) on the *y*-axis. Steep geotherms (or paths of

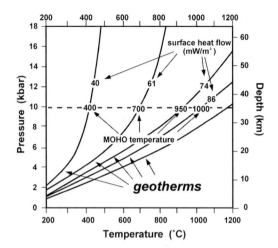

Fig. 3.7. Model geotherms with associated surface heat flow values and MOHO temperatures

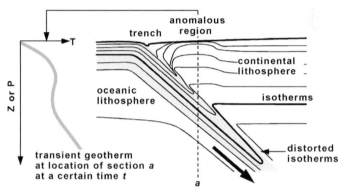

Fig. 3.8. Temperature field along a cross section through an Andean-type subduction zone and a transient geotherm at location *a*. (Peacock 1990)

metamorphism) on these figures correspond to low geothermal gradients; flat geotherms represent high geothermal gradients. Geotherms have high gradients near the surface and the gradients become progressively lower with increasing depth.

3.2.2.1
Transient Geotherms

Temperature changes in the crust and mantle are caused by adding or removing extra heat to or from the rocks. Heat flow changes may have a number of geological causes. Deep-seated causes include changes in the relative positions of lithosphere plates and their continental rafts relative to mantle convection systems (Fig. 3.5). Collision of lithosphere plates may lead to subduction of oceanic crust and cause abnormal thermal regimes. Collision of continental rafts leads to extreme crustal thickening (from 35 km of normal continental thickness to 70 km in active collision zones, active example: Himalayas). Smaller volumes of crust can be moved relative to each other, for instance during a continent-continent collision. The formation of nappes and the stacking of slices of crust are typical examples of redistribution of rocks in crust and mantle. A very efficient way to transfer large amounts of heat in the crust is the transport by melts (magmatic bodies). Large volumes of mantle-derived basaltic melt may underplate continental areas and release enormous amounts of latent heat of fusion during solidification. Heat transported to the crust by rising intrusive magmatic bodies locally disturbs the thermal regime and the temperature field in the crust.

Also, all tectonic transport disturbs the pre-tectonic steady-state geotherm. At any one instant during active tectonics, the crust is characterized by a specific geotherm that relates temperature to depth. The geotherm is now changing with time. These **transient geotherms** describe the temperature-depth relationship at a given time and for a given geographic location (xy coordinates). Some aspects of tectonic transport and its effect on the instantaneous geotherm are depicted in Fig. 3.8. The figure shows schematically the temperature field along a cross section through an Andean-type destructive margin. The down-going slab of oceanic lithosphere is relatively cold. Thermal relaxation is a slow process compared with plate movement. The continental lithosphere above the subducted oceanic plate is characterized by a complex, thermally anomalous region that results from magmatic heat transfer and from smaller-scale deformation in the crust (nappes). The geotherm at location **a** at a given instant (instantaneous geotherm) in the active history of the subduction event may look like the geotherm shown on the left-hand side of Fig. 3.8. The tectonic transport may even be fast enough to create temperature inversion in a transient geotherm. Such temperature inversions will, however, be very short-lived phenomena and they will be eliminated rapidly if subduction ceases.

3.2.3
Temperature Changes and Metamorphic Reactions

If a crustal volume receives more heat than required to maintain a steady-state geotherm, the volume of rock will experience an increase in temperature and it will undergo **prograde metamorphism**. The heat capacity of the rock determines how much the temperature rises in relation to the heat added. Typical heat capacities for silicate rocks are in the order of 1 kJ/kg rock and degree C. Suppose a quartzite layer in the crust receives 100 kJ/kg rock extra heat, this heat will increase the temperature of the quartzite by 100 °C. When heated, some rocks must adjust their mineralogy by chemical reactions in order to maintain a state of minimized total free energy (as explained above). These endothermic reactions will consume a part of the heat added to the rock. Dehydration reactions are typical during prograde metamorphism. The reactions replace hydrous minerals (such as zeolites, micas, chlorite, amphibole) with less hydrous or anhydrous minerals and release

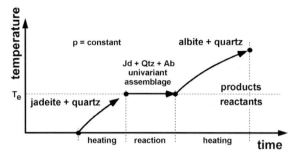

Fig. 3.9. Diagram showing the phase relationships among jadeite, quartz and albite as a function of time during a period of increased heat flow

H_2O to a fluid phase. This fluid may or may not escape from the site of production. Such reactions are strongly endothermic and consume a large amount of heat (about 90 kJ/mol H_2O). Consider now a layer of mica schist in the crust containing about 1 mol H_2O/kg rock. If a temperature increase of 100 °C (as in the quartzite layer) results in complete destruction of the hydrous minerals in this rock, a total of 190 kJ/kg schist is required to obtain the same temperature increase (100 kJ for the temperature increase and 90 kJ for the reaction).

The heat effect of mineral reactions in rocks also influences the temperature-time relationship in a volume of reactive rock. The case is illustrated in Fig. 3.9 using the Qtz + Jd ⇒ Ab reaction as an example. Supply of excess heat (in addition to the steady-state heat flow) will cause an increase in temperature until the equilibrium conditions (T_e) of the chemical reaction is reached. The heat input will then be used for the reaction[2] over a certain period of time and the temperature remains constant until one of the two reactant minerals (in this case Jd) is used up. After completion of the reaction, the temperature of the rock will continue to increase. The effect is analogous to heating water to the boiling point. The temperature of the water increases to the boiling temperature as heat is added to the water. During the boiling of the water the temperature remains constant and the heat is used to drive the reaction (phase transition) liquid water ⇒ steam. After all the water is boiled off, further addition of heat will increase the temperature of the steam.

If a crustal volume receives less heat than it requires to maintain a steady-state geotherm, it will cool and be potentially affected by **retrograde metamorphism**. During cooling, it may become necessary for the rock to change its mineralogical composition in order to maintain a state of minimum free energy. In our example, for a rock consisting of albite and quartz, the reaction Ab ⇒ Qtz + Jd converts all albite into jadeite and quartz as the rock cools to T_e (Fig. 3.9) under equilibrium conditions. The heat released by the reaction buffers the temperature to a constant value (T_e) until all the albite has been used up and further cooling can take place. The situation is analogous

---

[2] Note: the heat required for this type of H_2O-absent reaction is much less then the heat consumption for a dehydration reaction discussed above.

to removing heat from water. The temperature will decrease until water begins to freeze. As long as both reactants (liquid water) and products (ice) are
present in the system, the temperature is confined to the freezing temperature. After conversion of all water into ice the temperature will continue to
fall if further heat is removed from the ice.

3.2.4
Pressure Changes in Rocks

Normally, the pressure in the crust and mantle is characterized by isobaric
surfaces that are about parallel to the surface of the earth. As outlined above,
the free energy of minerals and associations of minerals (rocks) is also a
function of pressure. Pressure changes in rocks are related to the change of
position along the vertical space coordinate (z-axis in Fig. 3.4). The prevailing pressure at a given depth in a crustal profile on a steady-state geotherm
is given by the average density of the material above the volume of interest.
It can be calculated from ($dp/dz = -g\rho$):

$$P_{(z)} = -g \int_0^z \rho_{(z)} dz + P_{(z=0)} \tag{10}$$

where g is the acceleration due to gravity (9.81 m s^{-2}), ρ is the density of the
rock at any value of z (e.g., $2.7 \text{ g cm}^{-3} = 2700 \text{ kg m}^{-3}$ and taken as independent of depth) and $P_{(z=0)}$ is the pressure at the surface (e.g., $10^5 \text{ Pa} =$
$1 \text{ bar} = 10^5 \text{ N m}^{-2} = 10^5 \text{ kg m}^{-1} \text{ s}^{-2}$). The z-axis direction is always taken as
negative. The pressure, e.g., at 10 km (10,000 m) depth is then calculated as
264.97 MPa or 2.6497 kbar. The pressure at the base of a continental crust of
normal thickness (35–40 km) is about 10 kbar. The pressure at z results from
the weight of the rock column above the volume of interest and is designated
lithostatic pressure. This pressure is usually nearly isotropic. Non-isotropic
pressure (stress) may occur as a result of a number of geologic feasible processes and situations as discussed below.

   A pressure difference of about 100 bar at a depth of, e.g., 10 km
($P \sim 2700$ bar) will occur if a column of nearby rocks has a different average
density (e.g., 2700 instead of 2800 kg m^{-3}). If the density contrast of two columns of rock is 300 kg m^{-3}, a ΔP of ~ 900 bar at a depth of 30 km is possible.
This is nearly one kbar at a total pressure of ~ 8 kbar. However, because crustal material has similar densities, possible pressure differences between columns of rock at the same depth are limited to relatively small fractions of
the lithostatic pressure (typically <10%). However, such pressure gradients
may cause flow of fluids stored in the interconnected pore space of rocks according to Darcy's law. The transport of fluids may cause chemical reactions
in the rocks receiving these fluids if the fluids are not in equilibrium with
the solid assemblage of the rock.

   Non-isotropic pressure (stress) also results from tectonic forces and from
volume changes associated with temperature changes of rocks. Stress is one

of the major controlling factors of the structure of metamorphic rocks. It is also a very important force for small-scale migration and redistribution of chemical components in metamorphic rocks. The processes of pressure solution, formation of fabric (such as foliation) and formation of metamorphic banding from homogeneous protolith rocks are caused by non-isotropic pressure distribution in rocks and associated transport phenomena. The redistribution of material in stressed rocks attempts to reduce the non-isotropic pressure and to return to isotropic lithostatic pressure conditions.

In general, the lithostatic pressure acting on a volume of rock changes as a result of a change of the depth of that volume of rock (change of the position along the z-direction). In almost all geological occurrences, depth changes are caused by tectonic processes (that ultimately are also the result of thermal processes, e.g., mantle convection, density changes resulting from temperature changes). A lithosphere plate can be subducted in a collision zone and be transported to great depth in the mantle (or even to the core-mantle boundary) before it is resorbed by the mantle. Some of the subducted material may return to the surface before a normal steady-state geotherm is established. Metamorphic rocks of sedimentary or volcanic origin with mineral assemblages that formed at pressures in excess of 30 kbar $(3 \times 10^9 \, Pa = 3 \, GPa = 30 \, kbar)$ have been reported from various orogenic belts; 30 kbar pressure is equivalent to about 100 km subduction depth $[z = -P/(g\rho)]$. In continental collision zones the continental crust often is thickened to double that of its normal pre-collision thickness. At the base of the thickened crust the pressure increases from 9 to 18 kbar. The changes in mineralogical composition of the rocks undergoing such dramatic pressure changes depend on the nature and composition of these rocks and will be discussed in Part II. Variations in depth and associated pressure changes also can be related to subsiding sedimentary basins where a given layer **a** of sediment is successively overlain and buried by new layers **b, c** ... of sediments. Layer **a** experiences a progressive increase in pressure and temperature. If sedimentation and subsidence are slow, layer **a** may in fact follow a steady-state geotherm. Metamorphism experienced by layer **a** is commonly described under the collective term **burial metamorphism**. Crustal extension or lithosphere thinning is often the cause for the required subsidence creating deep sedimentary basins. Burial metamorphism is therefore also caused by tectonic processes.

3.3
Gases and Fluids

Sedimentary rocks such as shales often contain large modal proportions of hydrous minerals. In fact, sediments deposited in marine environments can be expected to contain, under equilibrium conditions, a mineralogy that corresponds to a **maximum hydrated state**. Adding heat to the hydrous mineralogy (clays) of a sediment during a metamorphic event will drive reactions of the general form:

Fig. 3.10. Isochores of H_2O as a function of pressure and temperature. (Helgeson and Kirkham 1974)

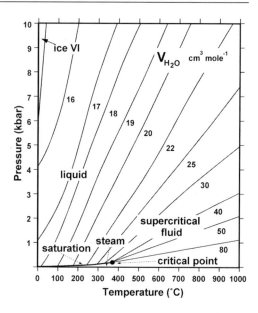

hydrous assemblage $\Rightarrow$ less hydrous or anhydrous assemblage $+ H_2O$ (11)

The general reaction (11) is a description of dehydration processes taking place during prograde metamorphism. The important feature of dehydration reactions is the release of H_2O. Steam is, in contrast to solid minerals, a very compressible phase and its volume is strongly dependent on pressure and temperature. Figure 3.10 shows the molar volume of H_2O as a function of pressure and temperature. At low temperatures, H_2O occurs either as a high-density liquid phase or as low-density steam phase depending on the prevailing pressure. The phase boundary between liquid and steam H_2O is the saturation or boiling curve. With increasing temperature the density contrast between steam and liquid H_2O continuously decreases along the boiling curve. At the critical point steam and liquid have the same density and the distinction between steam and liquid becomes obsolete. At temperatures above the critical point the H_2O phase is therefore a supercritical fluid phase or simply a fluid. Water in metamorphic systems at temperatures above 374 °C is referred to as the (aqueous) **fluid phase** or **the fluid**. The critical point of H_2O is at 374 °C and 217 bar (21.77 MPa).

Metamorphic fluids are usually dominated by H_2O. The H_2O of aqueous fluids in metamorphic rocks has its origin from various sources: (1) relic formation waters of sedimentary rocks, (2) dehydration water from H_2O originally stored in hydrous minerals, (3) meteoric water, and (4) magmatic water released from solidifying magmas. In the lower crust and in crustal segments with abundant carbonate rocks, CO_2 may become an additional important component in the fluid phase. Aqueous fluids also may contain significant amounts of dissolved salts (NaCl) and other solutes (e.g., aqueous silica complexes). At temperatures below $265\,^{\circ}C$ (at 2 kbar), H_2O and CO_2 mixtures form two separate phases. At higher temperatures, H_2O and CO_2 form continuous solutions ranging from pure H_2O to pure CO_2. However, extremely NaCl-rich brines may unmix into a dense aqueous brine and a low-density CO_2-rich vapor phase even under fairly high metamorphic P–T conditions. The composition of the most common binary metamorphic CO_2-H_2O fluids is normally reported as mole fraction X_{H_2O} (or X_{CO_2}). The mole fraction is defined as:

$$X_{H_2O} = \frac{n_{H_2O}}{n_{H_2O} + n_{CO_2}} \tag{12}$$

The dimensionless quantity X_{H_2O} is the ratio of the number of moles of H_2O and the total number of moles of H_2O and CO_2 in the fluid phase. An X_{H_2O} of 0.5 indicates, for example, a fluid with equal amounts of H_2O and CO_2 on a mole basis. Other molecular gas species found in metamorphic fluids include: CH_4, N_2, HCl, HF, and many others.

The low-density fluid produced during prograde dehydration is transported away from the site of production through interconnected pore space and lost by the system. If the rate of H_2O production exceeds the rate of transport, then the local pore pressure increases. However, the fluid pressure that may build up this way does not greatly exceed the lithostatic pressure imposed on the solid rock. The mechanical strength of the rocks is exceeded and failure occurs. This mechanism of **hydraulic fracturing** produces the necessary transport system for dehydration water. The fluid released is quickly transported and channeled away from the area undergoing dehydration. Suppose a rock undergoes dehydration at 5 kbar and $600\,^{\circ}C$. The produced fluid (Fig. 3.10) has a molar volume of about $22\ cm^3$ that corresponds to a density of about $0.82\ g\ cm^{-3}$. This is a remarkably high density for H_2O at $600\,^{\circ}C$ but it is still much lower than the typical density of rock-forming minerals of about $3 \pm 0.5\ g\ cm^{-3}$. The large density contrast between fluid and solids results in strong buoyancy forces and a rapid escape of the fluid from the rocks. Most metamorphic rocks are probably free of fluid during periods without reaction with the exception of fluids trapped in isolated pore space and as inclusions in minerals (**fluid inclusions**).

3.4
Time Scale of Metamorphism

Some simple considerations allow for an estimate of the time scale of typical orogenic metamorphism (see also Walther and Orville 1982). The range of reasonable heat flow differences in the crust that can drive metamorphic processes is constrained by maximum heat flow differences observed at the surface. The highest heat flows are measured along mid-ocean ridges ($120\ mW\ m^{-2}$), whilst the lowest are measured on old continental shields ($30\ mW\ m^{-2}$). Consider now a layer of crust with a heat flow difference between bottom ($75\ mW\ m^{-2}$) and top ($35\ mW\ m^{-2}$) of $40\ mW\ m^{-2}$. This means that the rock column receives every second 0.04 Joules heat (per m^2). The layer shall consist of shales (heat capacity $= 1$ kJ kg^{-1} °C^{-1}) with a volatile content (H$_2$O and CO$_2$) of about 2 mol kg^{-1}. The heat received by the layer of rock will be used to increase the temperature of the rock and to drive endothermic devolatilization reactions. About 90 kJ heat are required to release 1 mol H$_2$O or CO$_2$. If complete devolatilization occurs in the temperature interval 400–600 °C, a total of 380 kJ will be consumed by the model shale (180 kJ for the reactions and 200 kJ for the temperature increase from 400 to 600 °C). It takes 9.5×10^6 s (0.3 years) to supply 1 kg of rock with the necessary energy. Using a density of 2.63 g cm^{-3}, 1 kg of rock occupies a volume of 380 cm^3 and it represents a column of 0.38 mm height and 1 m^2 ground surface. From this, it follows that metamorphism requires about 8 years to advance by 1 cm. Eight million years (Ma) are required to metamorphose a shale layer of 10 km thickness. Similarly, if the heat flow difference is only 20 mW m^{-2}, the layer thickness 20 km and the initial temperature 200 °C, it will then take about 48 Ma to metamorphose the rock from a hydrous 200 °C shale to an anhydrous 600 °C metapelitic gneiss.

These crude calculations show that typical time spans for regional scale metamorphic processes are on the order of 10–50 million years. Similar time scales have been derived from radiometric age determinations.

3.5
Pressure – Temperature – Time Paths and Reaction History

During tectonic transport, any given volume of rock follows its individual and unique path in space and time. Each volume of rock may experience loss or gain of heat, and changes in its position along the z-coordinate result in changes in the lithostatic pressure loaded on the rock. Figure 3.11 shows again a simple model of a destructive plate margin. The situation here depicts a continent-continent collision with the formation of a continental crust twice its normal thickness. At the depth level **c** there is a volume element of rock (indicated by a filled square) at time t_1 (0 Ma). In Fig. 3.11 B–E, a series of schematic P–T diagrams shows the position of the element of rock in P–T space during tectonic transport. The time slices are arbitrary and have been chosen in accordance with time scales of the formation of Alpine-type mountain chains. The volume element at t_1 is on a stable steady-state geotherm. At time t_2 (10 Ma), active tectonic transport moves the crust together with the volume element underneath another continental crust of normal 35 km thickness. The increasing depth position of the volume element is accompanied by an increasing pressure on the volume element [Eq. (10)]. At the same time, the volume element begins to receive more heat than at its former position at t_1. However, because heat transport is a slow process compared with tectonic transport, dp/dT tends to be much steeper than the corresponding dp/dT slope of the initial steady-state geotherm. The consequence is shown in Fig. 3.11 C. The volume element has traveled in P–T space along a path that is at the high-pressure side of the initial steady-state geotherm. The rocks of the volume element are now on a geotherm that changes its shape

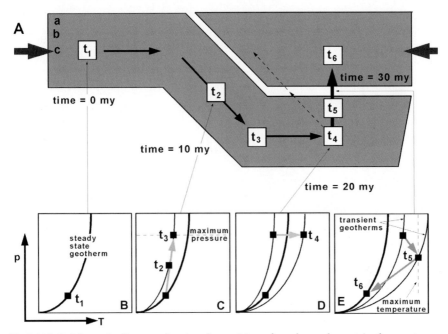

Fig. 3.11 A–E. Schematic diagram showing the position of a volume element in the crust as a function of time during a continent-continent collision and the corresponding paths followed by the element in P–T space

as time progresses, a transient geotherm. At t_3 the crust has reached twice its normal thickness (about 70 km), which is about the maximum thickness in continent-continent collision zones. The rocks have reached their maximum depth and consequently their maximum pressure point. The base of the double crust (70 km) is at a maximum pressure of about 2 GPa (20 kbar). Continued plate motion does not increase the thickness of the crust and pressure remains constant for all elements as long as thrusting is going on. On the other hand, heat transfer to the rocks of the volume element in question increases the temperature, as shown in Fig. 3.11 D. From this point, a number of feasible mechanisms may control the path of the volume element. Continued tectonic transport may return slices and fragments of rock to shallower levels in a material counter current (indicated with a dashed arrow in Fig. 3.11 A), or simply, after some period of time, the plate convergence stops for a number of possible reasons (e.g., because frictional forces balance the force moving the plate). The thick crust starts uplift, and erosion removes the double crust and restores the original thickness. By this mechanism the volume element may return to its original depth position and, given enough time, the stable steady-state geotherm will be re-established. The path between t_4 and t_5 is characterized by decompression (transport along the z-axis). If initial uplift rates are slow compared with heat transport rates, the rocks will experience a continued temperature increase during uplift (as

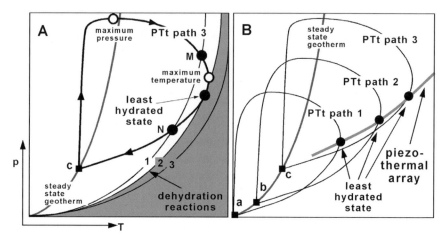

Fig. 3.12 A, B. A P–T–t path of a volume element at depth *c* of Fig. 3.11. The mineral assemblage stable at the tangent point of the path with dehydration reaction *2* will, in general, be preserved in metamorphic rocks. The tangent point corresponds to the least hydrated state of the rock. **B** Clockwise P–T–t loops for volumes of crust from different depth levels (Fig. 3.11). The mineral assemblage of the individual volumes corresponds to the least hydrated state of that volume. The P–T points of the least hydrated state from all samples of a metamorphic terrain define a curve that is known as piezo-thermal array (~metamorphic field gradient)

shown in Fig. 3.11 E). However, at some stage along the path, the rocks must start to lose more heat to the surface than they receive from below, and consequently cooling begins. The point t_5 in Fig. 3.11E represents the maximum temperature position of the path traveled by the volume element. At t_6 the volume element has returned to its former position on the steady-state geotherm.

The consequences for pressure and temperature of the geologic process illustrated in Fig. 3.11 are summarized in Fig. 3.12 . The rock at level **c** (P–T–t-path 3) follows a clockwise pressure-temperature loop. Such clockwise P–T–t paths are a characteristic feature of orogenic metamorphism. Such P–T–t paths have been documented from such diverse mountain belts as: Scandinavian Caledonides, Alps, Appalachian, and the Himalayas. Clockwise loops may in detail show a number of additional complications and local features. Counter-clockwise P–T–t paths have been reported from granulite facies terrains where an event of heating from igneous intrusions precedes crustal thickening. They may also occur in terrains that experience an initial phase of crustal extension. Very often "normal" orogenesis is characterized by the following sequence of P–T–t path sections: isothermal thickening, isobaric heating, isothermal decompression and isobaric cooling. Returning to Fig. 3.12A, it is evident that the maximum temperature point along the P–T–t path followed by a metamorphic rock does not necessarily coincide with the maximum pressure point of the path. This means that maximum pressure and maximum temperature will be generally diachronous.

What do we see of the path traveled by a rock in P–T–t space when we collect and study a metamorphic rock? If the rock always maintains an equilibrium state and metamorphism is strictly isochemical, we will not find any relics of its metamorphic history. However, rocks that formed at, e.g., 800 °C and 10 kbar can be collected at the surface and they *do* show characteristic high P–T mineral assemblages! This immediately tells us that at some stage during the reaction history a certain high-grade mineral assemblage is not converted to low-grade assemblages on the way back to the surface. The question is: which point of the P–T path is preserved and recorded by the mineralogy of the rock? The answer may be difficult to give in a specific geologic situation and for specific samples. However, low-grade rocks (or sediments, e.g., a shale) often contain modally large amounts of hydrates and are, therefore, affected by a series of discontinuous and continuous dehydration reactions during prograde metamorphism. In the pressure-temperature range typical for crustal metamorphism, dehydration reactions have equilibrium conditions as shown in Fig. 3.12 A (reactions 1, 2 and 3). A rock traveling along path 3 will be shaped by dehydration reactions. When crossing dehydration reaction 1 at point **M** it will partially lose water and continue as a rock containing a smaller amount of hydrates (mica, chlorite, amphibole). Dehydration of the rock will continue until the tangent point of the P–T path and last active dehydration reaction is reached. This point corresponds to the **least hydrated state** in the metamorphic history of the rock. It does not necessarily correspond to the maximum temperature point reached by the rock. Particularly if the P–T–t path shows a pronounced period of rapid uplift and decompression, the P–T coordinates of the least hydrated state may be dramatically different from maximum P and T. At this point the rock contains a prograde assemblage in equilibrium with an aqueous fluid phase saturating the mineral grain boundaries and filling the pore space. The total amount of free fluid is extremely small in most metamorphic rocks, however. Typical flow porosities of metamorphic rocks at P and T are in the order of 0.2 vol% (e.g., 2 cm^3 H_2O/1000 cm^3 rock). After having passed the point of the least hydrated state, the rock crosses the dehydration reactions in the reverse direction. The reactions consume water and form hydrates from anhydrous (or less hydrous) minerals. The minute amount of free fluid will be used up readily by the first rehydration reaction. For example, 2‰ free H_2O in a rock of appropriate composition and mineralogy at 9 kbar and 700 °C (corresponding to 0.1 mol H_2O per 1000 cm^3 of rock) can be used to form 15 cm^3 biotite per 1000 cm^3 rock (1.5 vol% biotite). The rock will then be devoid of a free fluid phase that effectively prevents the mineralogy of the rock adjusting to changing P–T conditions. Water is not only essential in dehydration reactions, but also plays a central role as a transport medium and catalyst for chemical reactions in rocks. Its final loss marks the closure of the reaction history of a rock. Even reconstructive phase transitions (e.g., kyanite = sillimanite) require the presence of water in order to proceed at finite rates even in geological times. Consequently, by studying the phase relationships of metamorphic rocks, one will in general determine the pressure-temperature conditions corresponding to the least hydrated state. This is not far off from the conditions when a free aqueous fluid was last present in the rock.

On its way to the surface (or point c) the rock crosses reaction 1 at point N. However, the reaction will affect the rock only if water is available for the back reaction. Rehydration (retrograde metamorphism) is widespread and very common in metamorphic rocks. However, the rehydration reactions very often do not run to completion, and assemblages from the least hydrated state may survive as relics. The necessary water for retrograde metamorphism is usually introduced to the rocks by late deformation events. Deformation may occur along discrete shear zones that also act as fluid conduits. The shear zone rocks may contain low-grade mineral assemblages whereas the rocks unaffected by late deformation still show the high-grade assemblage. With some luck, it is possible to identify in a single rock sample a series of mineral assemblages that formed at consecutive stages of the reaction history. The reaction history of rocks can often be well documented for all stages that modified the assemblage of the least hydrated state.

Relics from earlier portions of the P–T path (before it reached its least hydrated state) are seldom preserved in metasedimentary rocks. Occasionally, early minerals survive as isolated inclusions in refractory minerals such as garnet. Garnet in such cases effectively shields and protects the early mineral from reacting with minerals in the matrix of the rock with which it is not stable at some later stage of the rock's history. Rocks with water-deficient protoliths such as basalts, gabbros and other igneous rocks may better preserve early stages of the reaction history. In some reported case studies, meta-igneous rocks recorded the entire P–T–t path from the igneous stage to subduction metamorphism and subsequent modifications at shallower crustal levels (see Chap. 9.8.2.1).

Rocks collected in a metamorphic terrain may originate from different depth levels in the crust (Fig. 3.11, levels a, b, and c). All rocks follow their individual P–T path as shown in Fig. 3.12 b. Analysis of phase relationships and geologic thermobarometry may therefore yield a series of different P–T coordinates corresponding to the least hydrated states of the individual samples. The collection of all P–T-points from a given metamorphic terrain define a so-called **piezo-thermic array**. Its slope and shape characterize the specific terrain. The details of a piezo-thermic array of a given terrain depend on the geologic history and dynamic evolution of the area. The name "**metamorphic field gradient**" is preferred by some petrologists over the expression piezo-thermic array. Both expressions mean essentially the same thing. It is an arrangement of P–T points of least-hydrated state conditions reached during a now **fossil** metamorphism by rocks that are exposed **today** at the erosion surface. This two-dimensional array of P–T points should *not* be confused with a geotherm.

3.6
Chemical Reactions in Metamorphic Rocks

Chemical reactions in rocks may be classified according to a number of different criteria. Below follows a brief presentation of various kinds of reactions that modify the mineralogy or mineral chemistry of metamorphic rocks.

3.6.1
Reactions Among Solid-Phase Components

These are often termed "solid–solid" reactions because only phase components of solid phases occur in the reaction equation. Examples of typical "solid–solid" reactions follow.

3.6.1.1
Phase Transitions, Polymorphic Reactions

Al_2SiO_5 Ky = And; Ky = Sil; Sil = And
$CaCO_3$ calcite = aragonite
C   graphite = diamond
SiO_2 α-Qtz = β-Qtz; α-Qtz = coesite, ...
$KAlSi_3O_8$ microcline = sanidine

3.6.1.2
Net-Transfer Reactions

Such reactions transfer the components of reactant minerals to minerals of the product assemblage.
Reactions involving anhydrous phase components only:
   Jd + Qtz = Ab
   Gros + Qtz = An + 2 Wo
   2 Alm + 4 Sil + 5 Qtz = 3 Fe − Crd
Volatile-conserving "solid–solid" reactions:
   Tlc + 4 En = Ath
   Lws + 2 Qtz = Wa
   2 Phl + 8 Sil + 7 Qtz = 2 Ms + Crd

3.6.1.3
Exchange Reactions

These reactions exchange components between a set of minerals.
Reactions involving anhydrous phase components only
   Fe–Mg exchange between olivine and orthopyroxene: Fo + Fs = Fa + En
   Fe–Mg exchange between clinopyroxene and garnet: Di + Alm = Hed + Prp
Volatile-conserving "solid–solid" reactions
   Fe–Mg exchange between garnet and biotite: Prp + Ann = Alm + Phl

Cl–OH exchange between amphibole and biotite: Cl-Fpa + OH-Ann = OH-Fpa + Cl-Ann

3.6.1.4
Exsolution Reactions/Solvus Reactions

High-T alkali feldspar = K-feldspar + Na-feldspar
Ternary high-T feldspar = meso-perthite + plagioclase
Mg-rich calcite = calcite + dolomite
High-T Cpx = diopside + enstatite
Aluminous Opx = enstatite + garnet

The common feature of all "solid–solid" reactions is that the equilibrium conditions of the reaction are independent of the composition of the fluid phase, or, more generally speaking, of the chemical potentials of volatile phase components during metamorphism. It is for this reason that all "solid–solid" reactions are potentially useful geologic thermometers and barometers. The absence of volatile components in the reaction equation of "solid–solid" reactions should not be confused with general fluid-absent conditions during reaction progress. Metamorphic reactions generally require the presence of an aqueous fluid phase in order to achieve significant reaction progress even in geological time spans, attainment of chemical equilibrium in larger scale domains and chemical communication over several grain-size dimensions. Although P–T coordinates of the simple reconstructive phase transition kyanite = sillimanite are independent of H_2O, the detailed reaction mechanism may involve dissolution of kyanite in a saline aqueous fluid and precipitation of sillimanite from the fluid at nucleation sites that can be structurally unrelated to the former kyanite (e.g., Carmichael 1968). It is clear that the chlorine-hydroxyl exchange between amphibole and mica represents a process that requires the presence of a saline aqueous fluid. How-

Fig. 3.13. Schematic equilibrium conditions of solid–solid and dehydration reactions in p–T space and the Clausius-Clapeyron equation (ΔS_r = entropy change and ΔV_r = volume change of the reaction)

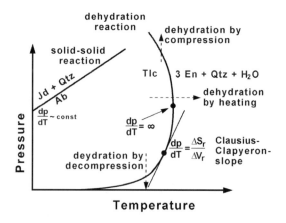

ever, the equilibrium of the exchange reaction is independent of the composition of that fluid.

Phase transitions, "solid–solid" net transfer and exchange reactions often show near linear equilibrium relationships in P–T space. The slope of the equilibrium curves of such reactions can be readily calculated (estimated) from the Clausius-Clapeyron equation (Fig. 3.13).

3.6.2
Reactions Involving Volatiles as Reacting Species

3.6.2.1
Dehydration Reactions

Reactions involving H_2O are the most important metamorphic reactions. Low-grade metasediments contain modally abundant hydrates. Pelitic and mafic rocks from the subgreenschist facies consist mainly of clay minerals (and zeolites). Serpentine minerals make up lower-grade ultramafic rocks. These sheet silicates contain up to about 12 wt% H_2O. The hydrates are successively removed from the rocks by continuous and discontinuous dehydration reactions as heat is added to the rocks. The release of H_2O during prograde metamorphism of hydrate-rich protoliths ensures that a free H_2O fluid is present in the rocks either permanently or periodically during the progress of dehydration reactions. This in turn often permits discussion of metamorphism for conditions where the lithostatic pressure acting on the solids also applies to a free fluid phase. If the fluid is pure H_2O (and choosing an appropriate standard state), the condition is equivalent to $a_{H_2O} = 1$. The general shape of dehydration equilibria is shown schematically in Fig. 3.13. Using the talc breakdown reaction as an example, the general curve shape of dehydration reactions on P–T diagrams can be deduced as follows:

$$Tlc = 3\,En + 1\,Qtz + 1\,H_2O(V) \tag{13}$$

The dP/dT slope of a tangent to the equilibrium curve is given at any point along the curve by the Clapeyron equation. All dehydration reactions have a positive ΔS_r and the curvature is largely controlled by the volume change of the reaction. The volume term can be separated into a contribution from the solids and the volume of H_2O, respectively:

$$\Delta V_r = \Delta V_{solids} + \Delta V_{H_2O} \tag{14}$$

The volume change of the solids can be calculated from:

$$\Delta V_{solids} = 3\,V_{En} + 1\,V_{Qtz} - 1\,V_{Tlc} \tag{15}$$

ΔV_{solids} is a small and negative quantity and varies very little with P and T. Consequently, the curve shape of dehydration reactions largely reflects the volume function of H_2O that is shown in Fig. 3.10. At low pressures, V_{H_2O} is large and the dP/dT slope small. In the pressure range of about 2–3 kbar, the molar volume of H_2O rapidly decreases and dehydration reactions tend to be strongly curved. At pressures greater than about 3 kbar and up to 10–15 kbar, the volume of H_2O is usually comparable to the volume change of the solids and ΔV_r tends to be very small. Perceptively, dP/dT slopes are very large in this pressure range. At even higher pressures, the volume of H_2O cannot compensate the negative values of ΔV_{solids} and ΔV_r and the dP/dT slope also become negative. From the general shape of dehydration equilibria in P–T space shown in Fig. 3.13, it is evident that talc, for instance, can be dehydrated by heating (temperature increase), decompression or compression.

The dehydration of zeolite minerals is often accompanied by large negative ΔV_{solids} and the dehydration equilibria have negative slopes already at moderate or low pressures. For example, the reaction analcim + quartz $\Rightarrow$ albite + H_2O has a ΔV_{solids} of –20.1 cm^3 mol^{-1} and the equilibrium has a negative slope above pressures of about 2 kbar.

In contrast to dehydration reactions, dP/dT slopes of "solid–solid" reactions (and phase transitions) are nearly constant (Fig. 3.13).

The equilibrium constant of the talc breakdown reaction (13) can be expressed by the mass action equation:

$$K = \frac{a_{En}^3\, a_{Qtz}\, a_{H_2O}}{a_{Tlc}} \tag{16}$$

The equilibrium constant K of a chemical reaction has a fixed value at a given P and T (a constant as the name suggests). It can be calculated from the fundamental (and also for geological applications extremely useful) equation:

$$-R\,T\ln K = \Delta G_r^0 \tag{17}$$

The free energy change of the reaction on the right-hand side of Eq. (16) can be calculated for any pressure and temperature pair of interest provided the thermodynamic data for all phase components in the reaction are known (R = universal gas constant; T = temperature in K). For our talc reaction, Eq. (17) takes the form:

$$\ln K = \left\{\ln a_{H_2O} + \left[\ln a_{En}^3 + \ln a_{Qtz} - \ln a_{Tlc}\right]\right\} = -\Delta G_r^0/(R\,T) \tag{18}$$

By considering pure end-member solids i ($a_i = 1$), the expression in square brackets becomes zero and the equilibrium is dependent on the activity of H_2O. The equilibrium is shown in Fig. 3.13 for the case $P_{H_2O} = P_{lithostatic}$ (pure H_2O fluid present, $a_{H_2O} = 1$). Equation (18) is graphically represented in Fig. 3.14 for three values of a_{H_2O}. It can be seen from Fig. 3.14 that the equilibrium conditions for dehydration reactions are displaced to lower temperatures (at P = constant) by decreasing a_{H_2O}. Solutions of Eq. (18) for different values of a_{H_2O} are shown in Fig. 3.15. It follows from Fig. 3.14 that the maximum temperature for the talc breakdown reaction is given by the condition $a_{H_2O} = 1$ that is equivalent to $P_{H_2O} = P_{lithostatic}$ and the presence of a pure H_2O fluid.

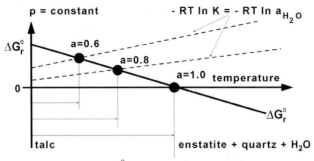

Fig. 3.14. Solutions of the equation $\Delta G_r^0 = -RT \ln K$ for dehydration reactions and three different values of a_{H_2O}

Fig. 3.15. Schematic equilibrium conditions of dehydration reactions for various a_{H_2O} conditions

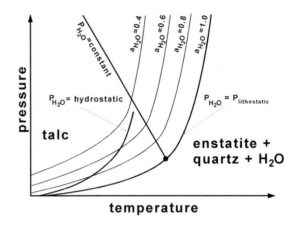

It also follows from Fig. 3.14 that talc dehydrates to enstatite and quartz at lower temperatures if $a_{H_2O} < 1.0$ at any pressure. The P–T space can, therefore, be contoured with a series of dehydration curves of constant a_{H_2O}. Geologically, the condition of $a_{H_2O} < 1.0$ can be realized basically in two different situations: (1) the fluid is not a pure H_2O fluid but is rather a mixture of H_2O and some other components not taking part in the reaction (e.g., CH_4, N_2, CO_2), or (2) there is no free fluid phase present at all. If P_{H_2O} is held constant, dehydration occurs along a curve with a negative slope intersecting the $a_{H_2O} = 1$ curve where $P_{H_2O} = $ constant $= P_{lithostatic}$. Towards higher pressure, a_{H_2O} decreases along the curve. This condition is of little geological significance.

The equilibrium curve for the condition $P_{H_2O} = P_{hydrostatic}$ has a shape similar to that for $P_{H_2O} = P_{lithostatic}$. The two curves converge at very low pressures. Note, however, that here $a_{H_2O} = 1$ along the curve (standard state: pure H_2O at T and $P_{hydrostatic}$) and it does not vary with increasing pressure. Also note that under the condition $P_{H_2O} = P_{hydrostatic}$ the pressure is different for the solids and the fluid. This consequently means that such a system is not in mechanical equilibrium. Hydrostatic pressure conditions may occur in very porous or strongly fractured rocks at shallow crustal levels. Typical geologic settings where such conditions may occur include hydrothermal vein formation, oceanic metamorphism, shallow level contact aureoles, and deep groundwater in basement rocks. Hydrostatic pressures usually cannot be maintained at typical metamorphic pressures of some kbar. Pressure solution and local chemical redistribution of material rapidly isolate the fluid phase and the system returns to mechanical equilibrium. Geologically then, the situation $P_{H_2O} = P_{lithostatic}$ is the most important one and the curves $a_{H_2O} < 1$ may be relevant in rocks with an impure aqueous fluid or in "dry" environments.

3.6.2.2
Decarbonatization Reactions

This type of reaction describes the decomposition of carbonate minerals such as calcite and dolomite. For example, adding heat to a rock containing calcite and quartz will eventually cause the following reaction: $Cal + Qtz = Wo + CO_2$. The form of the equilibrium conditions in P–T space is analogous to the one of the dehydration reactions. Also, the effects of variations in a_{CO_2} (P_{CO_2}) are analogous to the ones discussed above for dehydration reactions. In metamorphic carbonate rocks the fluid phase rarely consists of pure CO_2, it is usually a mixture of CO_2 and H_2O (and other volatile species).

3.6.2.3
Mixed Volatile Reactions

In rocks containing both hydrates (sheet silicates and amphiboles) and carbonates reactions are important that produce or consume both CO_2 and H_2O simultaneously. In such rock types, the fluid phase contains at least the two volatile species, CO_2 and H_2O. Under certain conditions, for instance if graphite is present in the rocks, other species such as H_2 and CH_4 may be present in significant amounts. However, the fluid composition in carbonate-bearing rocks can be assumed to be a binary mixture of CO_2 and H_2O in most cases. One compositional variable is therefore sufficient to describe the fluid composition. Usually, this variable is defined as the mole fraction of CO_2 in the fluid (X_{CO_2}). X_{CO_2} is zero in pure H_2O fluids and equal to one in pure CO_2 fluids. Consider the following characteristic mixed volatile reaction:

$$\text{margarite} + 2\,\text{quartz} + \text{calcite} = 2\,\text{anorthite} + 1\,CO_2 + 1\,H_2O \tag{19}$$

For pure solid phases the equilibrium constant is: $K_{P,T} = a_{CO_2} a_{H_2O}$. The equilibrium conditions depend on the activities of CO_2 and H_2O in addition to pressure and temperature. However, if we assume that the fluid is a binary CO_2-H_2O mixture (and using an ideal solution model for the fluid) the equilibrium constant reduces to: $K_{P,T} = X_{CO_2}(1 - X_{CO_2})$ or $K_{P,T} = X_{CO_2} - (X_{CO_2})^2$. Depending on the actual value of the equilibrium constant at P and T, this quadratic equation may have no, one or two solutions. The equilibrium conditions of the margarite breakdown reaction is therefore a complex surface in the $P - T - X_{CO_2}$ space. Because it is inconvenient to work graphically in

Fig. 3.16. Schematic equilibrium conditions of a
mixed volatile reaction
(isobaric T–X diagram)

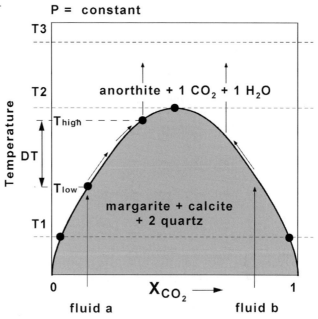

a three-dimensional space, mixed volatile reactions are most often represented
on isobaric T–X diagrams or isothermal P–X diagrams depending on the geological problem one wants to solve. It is also possible to select a certain metamorphic P–T field gradient that best describes the regional metamorphic area
under consideration and construct $(P - T) - X_{CO_2}$ sections (examples are presented in Chap. 6). For instance, let us keep the pressure constant. We can then
solve the quadratic equation above for a series of temperatures and display the
solution on an isobaric T–X section (Fig. 3.16). At T_1 the four-mineral assemblage margarite + quartz + calcite + anorthite may coexist with either very H_2O-
rich or very CO_2-rich fluids. At one unique temperature, T_2, the equation has
only one solution. The temperature T_2 represents the maximum temperature
for the reactant assemblage; above it, no margarite, calcite and quartz may coexist in rocks. At T_3 no (real) solution for the equation exists and the product
assemblage may coexist with binary CO_2–H_2O fluids of any composition. The
heavy solid curve in Fig. 3.16 connects all solutions of the equilibrium constant
equation and represents the reaction equilibrium of the margarite breakdown
reaction in quartz- and calcite-bearing rocks. The curve separates a T–X field
where the reactants are stable (shaded) from an area where the products are
stable. At any given temperature $<T_2$, the products are stable in H_2O-rich or
CO_2-rich fluids, respectively. The reactants may coexist with fluids of intermediate compositions.

Consider now a margarite-bearing rock containing excess quartz and calcite and a pore fluid of composition "*a*". Adding heat to this rock raises its
temperature until the reaction boundary is reached at T_{low}. Further increase

of the temperature causes the production of anorthite and a fluid with equal proportions of CO_2 and H_2O. Because the produced fluid is more CO_2-rich than the original pore fluid of the rock (fluid a), the total fluid composition is driven to more CO_2-rich compositions. The original fluid becomes more and more diluted with the fluid produced by the reaction; ultimately, the fluid is entirely dominated by the reaction fluid and has the composition $X_{CO_2} = 0.5$ (1 mol CO_2 and 1 mol H_2O). This will be the case at T_2, the maximum temperature of the reactant assemblage. However, the reactant assemblage may become exhausted with respect to one or more reactant minerals before reaching T_2. For example, if the original rock contains 1 mol margarite, 3 mol calcite and 5 mol quartz per 1000 cm³, the reaction will consume all margarite during reaction progress at some temperature along the isobaric univariant curve (T_{high}). The rock consists then of anorthite, calcite and quartz, and a fluid of the composition at T_{high}. Further addition of heat will increase the temperature of the rock but the fluid composition will remain unchanged. The univariant assemblage margarite + anorthite + calcite + quartz, therefore, occurs in a temperature interval ΔT and coexists with fluids with increasing CO_2 content. The temperature interval ΔT is also associated with gradual changes in modal composition in the rock. At T_{low} the first infinitesimally small amount of anorthite appears and it is continuously increasing in modal abundance as the reaction progresses towards T_{high} where the last small amount of margarite disappears. The fluid composition at the maximum temperature, T_2, depends exclusively on the reaction stoichiometry. Consider the reaction:

$$11 \text{ dolomite} + \text{tremolite} = 13 \text{ calcite} + 8 \text{ forsterite} + 9 \text{ CO}_2 + 1 \text{ H}_2\text{O}. \qquad (20)$$

Here, the produced fluid has the composition $X_{CO_2} = 0.9$ (9 mol CO_2 and 1 mol H_2O) and the temperature maximum of the equilibrium is at the same fluid composition. The general shape of reaction equilibrium for reactions of the type: $a = b + m\,H_2O + n\,CO_2$ (both fluid species are products by the reaction) is shown in Fig. 3.17 together with mixed volatile reactions with other reaction stoichiometries.

The equilibrium constant of the reaction:

$$2 \text{ zoisite} + 1 \text{ CO}_2 = 3 \text{ anorthite} + \text{calcite} + 1 \text{ H}_2\text{O} \qquad (21)$$

can be written as: $K = (1/X_{CO_2}) - 1$. It varies between zero and $+\infty$ (ln K $-\infty$, $+\infty$). Consequently, the univariant assemblage coexists with CO_2-rich fluids at low temperatures and with H_2O-rich fluids at high temperatures. The general shape of reactions $a + CO_2 = b + H_2O$ is shown in Fig. 3.17. The reaction:

$$8 \text{ quartz} + 5 \text{ dolomite} + 1 \text{ H}_2\text{O} = \text{tremolite} + 3 \text{ calcite} + 7 \text{ CO}_2 \qquad (22)$$

represents a reaction of the type: $a + H_2O = b + CO_2$ and has, compared to reaction (21), an inverse shape in T–X-diagrams (Fig. 3.17). It follows from the earlier discussion on pure dehydration and decarbonatization reactions (see

Fig. 3.17. Schematic equilibrium conditions of mixed volatile reactions with different reaction stoichiometries (isobaric T–X diagram)

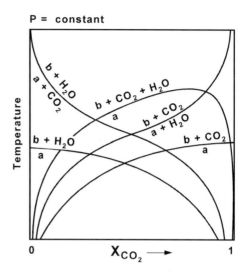

Figs. 3.14 and 3.15) that they must have general curve shapes as shown in Fig. 3.17. The figures presented so far also permit a deduction of the general shape of mixed volatile reaction equilibria in isothermal P–X sections (you may try to draw a P–X figure at constant temperature similar to Fig. 3.17).

3.6.2.4
Oxidation/Reduction Reactions

A number of cations that make up minerals occur in different oxidation states. Examples are: Fe^{2+}/Fe^{3+}, Cu^+/Cu^{2+}, Mn^{2+}/Mn^{3+}. The most important redox couple in common rock-forming silicates and oxides is Fe^{2+}/Fe^{3+}. The two iron oxides hematite and magnetite are related by the reaction:

$$6\,Fe_2O_3 = 4\,Fe_3O_4 + O_2 \tag{23}$$

The equilibrium conditions of this reaction are given (for pure hematite and magnetite) by: $K = a_{O_2}$. Again, the equilibrium depends on three variables: P, T and a_{O_2}. The pressure dependence is very small for this redox reaction. The equilibrium may be displayed on isobaric T versus $\ln a_{O_2}$ diagrams. In the literature, the activity is often replaced by f_{O_2} (fugacity of O_2). $\ln f_{O_2}$ is numerically identical to $\ln a_{O_2}$ if an appropriate standard state is chosen. The fugacity is close to the partial pressure of O_2 under low-pressure conditions. However, fugacity can be related to the partial pressure of O_2 at any pressure. The partial pressure of O_2 in crustal rocks is extremely small (in the order of 10^{-20}–10^{-40} bar). However, coexistence of hematite and magnetite may also be formulated in the presence of water as:

$$6\,Fe_2O_3 + 2\,H_2 = 4\,Fe_3O_4 + 2\,H_2O \tag{24}$$

and the equilibrium can be shown on T versus $\ln a_{H_2}$ diagrams. The assemblage biotite + K-feldspar + magnetite is common in high-grade metamorphic and igneous rocks. The redox reaction:

$$2\,KFe_3AlSi_3O_{10}(OH)_2 + O_2 = 2KAlSi_3O_8 + 2Fe_3O_4 + 2\,H_2O \tag{25}$$
$$\text{biotite} \qquad\qquad \text{K-feldspar} \quad \text{magnetite}$$

depends on O_2 (and H_2O). If O_2 is increased then biotite is oxidized to Kfs + Mag; if O_2 is lowered, then Kfs + Mag is reduced to biotite. Or the other way round, the three-mineral assemblage Bt + Kfs + Mt defines the activity of O_2 in the presence of water. Redox reactions do not necessarily involve volatiles. Redox reactions may also be formulated in ionic form (e.g., Mag–Hem reaction):

$$2\,Fe_3O_4 + H_2O = 3\,Fe_2O_3 + 2\,H^+ + 2\,e^- \tag{26}$$

Oxidation of magnetite to hematite in the presence of water produces two H^+ and two electrons. The magnetite–hematite reaction (26) depends on the redox potential and the activity of the hydrogen ion (p_H). Graphical representation of equilibria of the type (26) is often performed by means of Eh (oxidation potential) versus p_H diagrams (see Garrels and Christ 1965).

3.6.2.5
Reactions Involving Sulfur

Sulfides are widespread accessory minerals in metamorphic rocks. Most common are pyrrhotite (FeS) and pyrite (FeS_2). If both common Fe-sulfides occur in a metamorphic rock, the stable coexistence of pyrite and pyrrhotine requires equilibrium of the reaction:

$$2\,FeS_2 = 2\,FeS + S_2 \tag{27}$$

The dominant sulfur species in metamorphic fluids is either H_2S or SO_2. If one wishes to discuss sulfide-involving reactions in terms of the most abundant species, one may rewrite Eq. (27) in the presence of water ($2\,H_2O = 2\,H_2 + O_2$; $2\,H_2S = 2\,H_2 + S_2$) by:

$$FeS_2 + H_2 = FeS + H_2S \tag{28}$$

Graphic representation of sulfide-involving reactions depends on the actual problem, but partial pressure (activity, fugacity) of a sulfur species in the fluid versus temperature diagrams are popular. If sulfide reactions are combined with redox reactions, oxygen (hydrogen) versus sulfur diagrams are useful. For example, Fig. 3.18 shows the phase relationships among some iron oxides and sulfides in terms of a_{H_2} and a_{H_2S}. The figure shows that the presence of one Fe-sulfide and one Fe-oxide in a metamorphic rock (e.g, magnetite and pyrrhotite) fixes the activities of H_2 and H_2S to distinct values along

Fig. 3.18. Qualitative phase relationships among hematite, magnetite, pyrrhotite and pyrite in a log a_{H_2} versus log a_{H_2S} activity–activity diagram

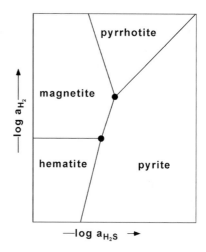

an univariant line at a given pressure and temperature in the presence of water. Rocks may be zoned with respect to oxides and sulfides depending on gradients in a_{H_2S} (e.g., under reducing conditions, rocks may contain magnetite, pyrrhotite or pyrite with increasing a_{H_2S}).

3.6.2.6
Reactions Involving Halogens

Fluorine and chlorine may replace OH groups in all common rock-forming hydrates, notably, mica, talc and amphibole (less so in serpentine minerals and chlorite). The halogen content of common rock-forming minerals may be related to exchange reactions of the type (e.g., fluorine-hydroxyl exchange between biotite and fluid):

$$KMg_3AlSi_3O_{10}(OH)_2 + 2\,HF = KMg_3AlSi_3O_{10}(F)_2 + 2\,H_2O \tag{29}$$

Also in this case, equilibria of the type (29) are commonly displayed and discussed in terms of partial pressure (fugacity, activity) of a gaseous halogen species (HCl, HF) versus temperature diagrams or in terms of μ–μ diagrams (activity–activity diagrams).

3.6.2.7
Complex Mixed Volatile Reactions and Fluids

It is a common situation in metamorphic rocks that they contain simultaneously halogen-bearing silicate-hydrates, carbonates, sulfides, and oxides. Dealing with such rocks consequently means that all types of metamorphic reactions must be considered together. For example, consider an ordinary calcareous mica schist containing the following minerals: calcite, quartz,

K-feldspar, plagioclase, muscovite, biotite, graphite and pyrite. Among the minerals and the nine potentially important fluid species the following linearly independent equilibria can be written:

1. Equilibria involving silicates, sulfides and carbonates

$$KAl_2AlSi_3O_{10}(OH)_2 + CaCO_3 + 2\,SiO_2 = KAlSi_3O_8 + CaAl_2Si_2O_8 + H_2O + CO_2$$

$$(30\,a)$$

$$KFe_3AlSi_3O_{10}(OH)_2 + 3\,S_2 = 2\,KAlSi_3O_8 + 3\,FeS_2 + H_2O \qquad (30\,b)$$

2. Equilibria involving graphite

$$C + O_2 = CO_2 \qquad\qquad\qquad\qquad\qquad\qquad\qquad\qquad (31\,a)$$

$$C + 2\,H_2 = CH_4 \qquad\qquad\qquad\qquad\qquad\qquad\qquad\qquad (31\,b)$$

$$2\,C + O_2 = 2\,CO \qquad\qquad\qquad\qquad\qquad\qquad\qquad\qquad (31\,c)$$

3. Equilibria involving fluid species only

$$2\,H_2O = 2\,H_2 + O_2 \qquad\qquad\qquad\qquad\qquad\qquad\qquad (32\,a)$$

$$S_2 + 2\,H_2 = 2\,H_2S \qquad\qquad\qquad\qquad\qquad\qquad\qquad (32\,b)$$

$$S_2 + 2\,O_2 = 2\,SO_2 \qquad\qquad\qquad\qquad\qquad\qquad\qquad (32\,c)$$

A mass action equation can be formulated for all eight equilibria [see, e.g., Eqs. (16) and (17)]. Mass action equations express the equilibrium constant in terms of activities at constant pressure and temperature. In addition, in the presence of a free fluid phase in the rock, the following mass balance equation (total pressure equation) can be written:

$$P_{total} = P_{fluid} = P_{H_2O} + P_{CO_2} + P_{O_2} + P_{H_2} + P_{CH_4} + P_{CO} + P_{S_2} + P_{H_2S} + P_{SO_2} \qquad (33)$$

Equation (33) states that the sum of the partial pressures of the fluid species is equal to the total lithostatic rock pressure. Together with activity (fugacity)–partial pressure functions for the species in the fluid, the set of nine equations can be solved iteratively for the nine unknown partial pressures and hence the fluid composition. This means that the mineral assemblage of the calcareous mica schist defines and completely controls the composition of the coexisting fluid phase at any given pressure and temperature. It is also said that the rock **buffers** the fluid composition. In most metamorphic fluids the partial pressures of O_2 and CO are extremely low compared with the total pressure. H_2 is usually low also in most metamorphic fluids. The dominant sulfur species in the fluid is H_2S at low temperatures and SO_2 at high temperatures. CH_4 may be high in some fluids at lower temperatures. CO_2 is dominant at higher temperatures.

3.6.2.8
Reactions Involving Minerals and Dissolved Components in Aqueous Solutions

The fluids that are associated with rock metamorphism in crust and mantle inevitably migrate from the source area. They come into contact with other rocks with which they may not be in equilibrium. The chemical reactions that try to establish equilibrium between the solid rock assemblage and the fluid are known under the term **"fluid–rock interaction"**. Fluid–rock interaction is important in the formation of hydrothermal ore deposits, contact metasomatism, large-scale infiltration and reaction in orogenesis, in shear zone metasomatism and in geothermal fields (to name a few examples). As an example, imagine a crustal volume containing rocks with the assemblage margarite + calcite + quartz (Fig. 3.16). A foreign externally derived fluid with the composition $X_{CO_2} = 0.01$ arrives in this rock at T_{high}. The fluid is in equilibrium with the product assemblage of reaction (19) rather than with the reactant assemblage that actually is present. The consequence of fluid infiltration is that the fluid will drive the reaction until it reaches equilibrium with the solid-phase assemblage. In this case the rock cannot buffer the fluid composition; the rock is controlled by an external fluid.

Commonly, metamorphic fluids are saline brines containing high concentrations of metal ions and complexes. Therefore, infiltrating fluids may also have a cation composition that is not in equilibrium with the solid-phase assemblage receiving this fluid. An essential aspect of fluid-rock interaction is the reaction of phase components of the solid mineral assemblage of the rock and dissolved species in the foreign fluid. Some examples:

Formation of talc schists by interaction of serpentinite with quartz-saturated fluids:

$$Mg_3Si_2O_5(OH)_4 + 2\,SiO_{2_{aq}} = Mg_3Si_4O_{10}(OH)_2 + H_2O \tag{34}$$

where $SiO_{2,aq}$ represents an uncharged hydrous silica complex in the aqueous fluid. SiO_2 saturation of the aqueous fluid infiltrating the serpentinite may result from interaction of the fluid with granite or granite-gneiss enclosing a serpentinite body.

Formation of soapstone (Tlc + Mgs rock) by interaction of serpentinite with CO_2-bearing fluids:

$$2\,Mg_3Si_2O_5(OH)_4 + 3\,CO_2 = Mg_3Si_4O_{10}(OH)_2 + 3\,MgCO_3 + 3\,H_2O \tag{35}$$

Sericitization of K-feldspar is a widespread process in retrograde metamorphism:

$$3\,KAlSi_3O_8 + 2\,H^+ = KAl_3Si_3O_{10}(OH)_2 + 2\,K^+ + 6\,SiO_{2_{aq}} \tag{36}$$
$$\text{K-feldspar} \qquad \text{muscovite (sericite)}$$

This last type of reaction is of great importance in geology. It is referred to as a **hydrolysis reaction.** Any other kind of metamorphic mineral reaction

Initial $X_{CO_2}^o = n_{CO_2}^o / (n_{CO_2}^o + 0.42)$
Final $X_{CO_2}^f = n_{CO_2}^f / (n_{CO_2}^f + 0.42)$
Solving for:
Number of moles of CO_2 in the initial fluid: $n_{CO_2}^o = 0.004209$ moles
Number of moles of CO_2 in the final fluid: $n_{CO_2}^f = 0.09655$ moles
CO_2 produced by the reaction: $\Delta n_{CO_2} = n_{CO_2}^f - n_{CO_2}^o = 0.09655 - 0.004209 = 0.09234$ moles
With reference to Eq. (40): 0.09234 mol CO_2 is equivalent to 0.09234 mol wollastonite produced by the reaction per 1000 cm^3 rock
The molar volume of wollastonite is: $V^o = 40$ cm^3 mol^{-1}
Wollastonite produced by buffered reaction: 3.7 cm^3 (0.367 vol%)

The amount of wollastonite is very small and will hardly be detected under a petrographic microscope.

Let us now consider a calcite- and quartz-bearing rock with 20 vol% wollastonite that equilibrated at 600 °C and 2 kbar. From the calculations above it is evident that isochemical metamorphism can produce only a very small modal amount of wollastonite (at 600 °C and 2 kbar). Consequently, a very wollastonite-rich rock that formed under these conditions must be the result of interaction with a foreign, externally derived H_2O-rich fluid that pushed the rock across the reaction equilibrium into the wollastonite field shown in Fig. 3.19. The question is how much H_2O must be added to the rock in order to produce the observed modal proportion of wollastonite? This can be readily calculated:

20 vol% wollastonite $(V^o = 40$ cm^3 mol$^{-1}) = 0.5$ mol/100 cm^3
The rock contains 5 mol wollastonite/1000 cm^3 rock. It follows from the reaction stoichiometry of reaction (40) that the reaction also produced 5 mol CO_2
However, the final $X_{CO_2}^f = n_{CO_2}^f / (n_{CO_2}^f + n_{H_2O})$ is equal to 0.19 (Fig. 3.19)
Solving for: $n_{H_2O} = n_{CO_2}^f (1 - X_{CO_2}^f)(1/X_{CO_2}^f)$
gives: $n_{H_2O} = 5 \times 0.81 \times (^1/_{0.19}) = 21.315$ moles H_2O
The molar volume of H_2O is: V^o (600 °C, 2 kbar) = 31 cm^3 (Fig. 3.10)
The total volume of H_2O that reacted with the rock is: $V_{H_2O} = 660$ cm^3/1000 cm^3 rock

In conclusion, a rock with 20 vol% wollastonite that formed from calcite + quartz at 600 °C and 2 kbar requires interaction with 660 cm^3 H_2O/1000 cm^3 rock. This represents a minimum value. If the composition of the interacting fluid is more CO_2-rich, more fluid is required to produce the observed amount of wollastonite in the rock. The deduced amount of interacting external fluid may be expressed in terms of a time-integrated fluid to rock ratio: fluid-rock ratio = 660 cm^3/1000 cm^3 = 0.66.

A brief comment may be added here with reference to Fig. 2.1. This fictive wollastonite + calcite + quartz rock was used in Chapter 2 to illustrate how mineral assemblage is related to the structure of the rock; i.e., how the three minerals are distributed in the rock. We now relate the three rocks to reaction (40) and Fig. 3.19.

The rock in Fig. 2.1a is chemically layered with a marble layer on top and a calc-silicate rock below. The two rocks may have similar Ca/Si ratios. However, it follows from Fig. 3.19 that, at a given temperature (e.g., 600 °C), the marble layer is consistent with a fluid with X_{CO_2} greater than the X_{CO_2} corresponding to equilibrium of reaction (40). The Wo + Qtz rock of the lower layer is consistent with X_{CO_2} smaller than the X_{CO_2} corresponding to equilibrium of reaction (40). Consequently, if a fluid is present in the rock, it is of different composition in the two chemically distinct parts of the rock. The two layers are isolated with respect to the fluid and the two minerals in the top layer do not form an assemblage with the two minerals in the bottom layer. At the interface between the two layers two different fluids may be present at a very close distance. Such "fluid discontinuities" are commonly observed in hydrothermal veins, fine-banded impermeable metasediments, and other rocks showing large compositional heterogeneity.

The rock in Fig. 2.1b contains the true assemblage Cal + Qtz + Wo. The assemblage is stable at any temperature along the equilibrium curve of reaction (40) shown in Fig. 3.19. It is also clear that at a given temperature (e.g., 600 °C), the three minerals buffer the fluid composition; they fix the composition of the fluid to a specific X_{CO_2} value at P and T.

The rock in Fig. 2.1c is difficult to interpret on the basis of Fig. 3.19. A plausible explanation you would probably hear from petrologists is that very small-scale fluid heterogeneity existed in this rock with water-rich fluid probably channeled along small-scale pathways. Actually, two very small-scale veinlets can be discerned from the structure shown in Fig. 2.1c. Along them wollastonite was produced from reaction of the Cal + Qtz rock with H_2O-rich fluid, however, you remember, it was a fictive rock with a structure drawn on the computer screen.

3.8
Phase Diagrams

3.8.1
Phase Diagrams, General Comments and Software

Phase diagrams display the equilibrium relationships among phases and phase assemblages in terms of intensive, extensive or mixed variables. A widely used type of phase diagram in metamorphic petrology is the P–T diagram. It shows the equilibrium relationships among minerals (and fluids) as a function of the intensive variables pressure and temperature. The chosen ranges of the parameter values define a P–T window that is appropriate for the problem of interest. The mineral and fluid compositions may be constant or variable on such a diagram. Composition phase diagrams show the phase relationships in terms of extensive variables at specified values of pressure and temperature. AFM and ACF diagrams are examples of composition phase diagrams. Phase diagram software is routinely used to compute and display petrologic phase diagrams from published sets of thermodynamic data of phase components. All phase diagrams in this book were computed by using two different programs: **PTXA** (or **GEØ-CALC**) is a convenient very user-friendly program that runs on DOS PC's and MAC's (Berman et al. 1987; Brown et al. 1988). It can be used for diagrams that display any combination of the variables pressure, temperature, composition of a binary CO_2–H_2O fluid, and activity of phase components. **PERPLEX** is an extremely powerful multipurpose phase diagram generator that permits the construction of remarkably complex phase diagrams if necessary (Connolly and Kerrick 1987; Connolly 1990). PERPLEX has been developed for UNIX-based workstations, but it also runs on MAC's and PC's. Other petrologic programs to be mentioned here include: THERMOCALC by Powell and Holland (2001); SUPCRT92 by Johnson et al. (1992); PTPATH by Spear et al. (1991) and THERMO by Perkins et al. (1987). Compilations of thermodynamic data for substances of geologic interest are found in: Clark (1966); Burnham et al. (1969); Stull and Prophet (1971); Helgeson et al. (1978); Robie et al. (1978); Holland and Powell (1985, 1990, 1998); Powell and Holland (1985, 1988); Berman (1988); Johnson et al. (1992). The thermodynamic data sets of Berman (1988) [RB88] and Holland and Powell (1990, 1998) [HP90; HP98] are classified as so-called internally consistent data sets because of the specific data retrieval technique used in deriving the data. These two sets are particularly well suited for use in conjunction with the two mentioned programs and are widely used by metamorphic petrologists. Thermodynamic data can be derived from calorimetric measurements and, especially, from experimental phase equilibrium data. Most phase diagrams shown in this volume have been generated by PERPLEX using the RB88 or the HP98 database (or updated versions of them).

Phase diagrams are indispensable tools for the description, analysis and interpretation of metamorphic rocks and metamorphism. The ability to read and understand phase diagrams is central and essential in metamorphic petrology. The capability to calculate and construct phase diagrams is important for all those who are in need of phase diagrams for phase assemblages and phase compositions that are unique for the geologic problem in question. It may also be necessary to modify published phase diagrams and adapt them to the actual problem (e. g., change the P–T frame or other parameter values). It is beyond the scope of this volume to provide a comprehensive treatment of the computation and construction of phase diagrams; however, it is probably useful to briefly introduce two important aspects of phase diagrams: the phase rule and Schreinemakers rules.

3.8.2
The Phase Rule

The state of a heterogeneous system, such as a rock, depends on a number of state variables (such as pressure, temperature and chemical potentials). The requirements of heterogeneous equilibrium impose some restrictions on these variables (equilibrium constraints). The phase rule relates the number of variables in a system and the number of equations that can be written

among them at equilibrium. In a system with a total number of c components and composed of p phases the number of variables is:
- for each phase: T, P, and c–1 compositional variables $= c+1$
- for the system of p phases: $p(c+1)$

Equilibrium constraints:
The temperature of all phases must be the same (thermal equilibrium) $\Rightarrow$
- p–1 equations of the type: T of phase $\alpha = $ T of phase β.
The pressure on all phases must be the same (mechanical equilibrium) $\Rightarrow$
- p–1 equations of the type P on phase $\alpha = $ P on phase β.
The chemical potential of component i must be the same in all phases
(chemical equilibrium) $\Rightarrow$
- $c(p-1)$ equations of the type $\mu_{i\alpha} = \mu_{i\beta}$; p–1 equations for each component
from 1 to c.
Hence, the number of variables n_v is: $p(c+1)$, and the number of equations n_e: $2(p-1) + c(p-1)$.

Now let us define a variable f as the difference between n_v and n_e. The term f
simply expresses how many variables can be changed independently in a system at equilibrium.

$f = n_v - n_e = [p(c+1)-2(p-1)-c(p-1)]$; or simply

$$f = c + 2 - p \tag{41}$$

This equation is known as the **phase rule**. The term f is often referred to as
variance or **degrees of freedom** of the system. A consequence of the phase
rule is that the number of phases cannot exceed the number of components
by more than 2.

In a one-component system (e.g., Al_2SiO_5) consisting of one phase (e.g.,
kyanite), there are two variables which can be varied independently (e.g., P
and T). If there are two phases present (sillimanite and kyanite), only one
variable can be specified independently. If there are three phases present (andalusite, sillimanite and kyanite), the number of variables equals the number
of equations ($f = 0$) and the system of equations has one unique solution. In
other words, coexistence of all three aluminosilicates completely defines the
state of the system; it is only possible at a unique value of pressure and temperature (the triple point, invariant point). Another well-known, one-component system is the system H_2O. Ice has $f = 2$, melting ice has $f = 1$, and ice,
water and steam ($f = 0$) may be in equilibrium at the unique P and T values
of the triple point.

3.8.2.1
Phase Rule in Reactive Systems

Consider a system with N phase components (chemical species), some inert,
some at reaction equilibrium. R is the number of independent reactions and
p the number of phases.

Number of variables: $p(N+1)$
Number of equations: $(N+2)(p-1)+R$
R = number of conditions of reaction equilibrium $\Sigma v\mu = 0$

$$f = n_v - n_e = N - R + 2 - p \tag{42}$$

An example; the mineral assemblage Grt + Crd + Sil + Qtz consists of four phases and we may consider seven phase components (garnet: Grs, Prp, Alm; cordierite: Fe-Crd, Mg-Crd; sillimanite: Sil; quartz: Qtz). The seven phase components are related by two independent reactions [net transfer: 2 Alm + 5 Qtz + 4 Sil = 3 Fe-Crd; exchange: $FeMg_{-1}$ (Grt) = $FeMg_{-1}$ (Crd)]. The variance of the assemblage calculated from Eq. (42) is 3. This means that the system of equations has a unique solution if three variables can be specified independently. If, for example, the Alm and Prp content of the garnet and the Fe-Crd content of the cordierite were measured in a rock containing the four minerals in equilibrium, then the system is uniquely defined and the equilibrium pressure and temperature can be calculated.

By defining $c = N-R$, the two forms of the phase rule become formally identical. The number of components in the sense of the phase rule is therefore to be taken as the total number of phase components (chemical species) minus the number of independent reactions between them. In the example made above, the number of system components is five (seven phase components – two independent reactions). Simple oxide components are often selected as system components. The five components $CaO-MgO-FeO-Al_2O_3-SiO_2$ (CMFAS) define the composition space of the Grt + Crd + Sil + Qtz rock used as an example here. For a comprehensive treatment of the composition space and reaction space, see Thompson (1982 a, b). The phase rule derivation has been adapted from Denbigh (1971).

3.8.3
Construction of Phase Diagrams for Multicomponent Systems
After the Method of Schreinemakers

A binary system of four phases is, according to the phase rule, invariant. The four minerals may occur at a unique pair of temperature and pressure (or any other two intensive variables). Graphically, the four-phase assemblage occurs at a point (invariant point) on a P–T diagram. From this point four univariant assemblages represented by univariant lines (curves) radiate into the P–T plane (surface). The univariant lines are characterized by the absence of one of the phases occurring at the invariant point and divide the P–T plane into four divariant sectors. Each of these sectors is characterized by one unique two-phase assemblage that occurs in one, but only one, sector. The geometric distribution of univariant and divariant assemblages follows strict rules that may be derived from chemical thermodynamics. These rules were formulated by Schreinemakers in a series of articles around 1915 and are known today as Schreinemakers' rules. Figure 3.20a shows the geometry of the general binary four-phase system and Fig. 3.20b represents a specific geo-

Fig. 3.20 A–C. Schreinemakers'
diagram for general binary
four-phase assemblages (**A**),
an example assemblage Sil,
Fe–Crd, Fe–Grt, and Fe–Bt
projected from K-feldspar,
quartz and H_2O (**B**), and the
translation of the relationships shown in **B** into P–T
space (**C**)

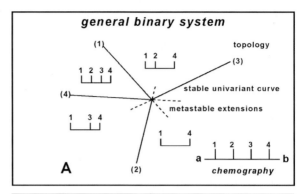

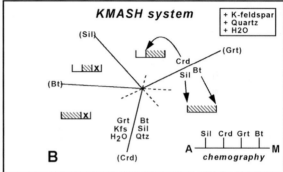

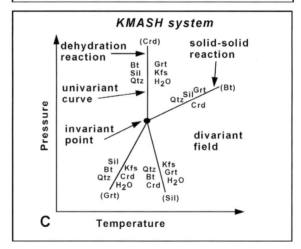

logic example. For three-component five-phase systems and complications arising from compositional degeneracies, the reader is referred to Zen (1966).

The general geometry of the binary system (Fig. 3.20a) shows an invariant
assemblage of four phases with the compositions 1, 2, 3 and 4 in the two-
component system a–b. The space around the invariant point is divided into

four sectors by four univariant curves. Each of these curves is characterized by the absence of one phase of the invariant four-phase assemblage. The name of the absent phase is put into round brackets. The sectors contain either two, one or no metastable extensions of the four univariant curves. The sector between the curves (2) and (3) contains two metastable extensions and is characterized by the presence of the divariant assemblage $1+4$ which occurs only in this sector. The assemblage contains the most a-rich and the most b-rich phase, respectively. If the sector boundary (3) is crossed, $1+4$ form phase 2, that is a new phase of intermediate composition [3 does not participate in the reaction, (3) is absent]. If the sector boundary (1) is crossed, 3 forms from $2+4$ (1 is absent). In the sector between (1) and (4), three two-phase assemblages may be present depending on the relative proportions of the components a and b in the rock. However, only one of the three assemblages is unique for the sector, namely $2+3$. Crossing the sector boundary (4), phase 2 decomposes to $1+3$ (4 does not participate in the reaction). Finally, crossing the univariant curve (2) removes 3 by the reaction $3 = 1+4$.

Figure 3.20b shows a specific geologic example. In rocks containing excess quartz and K-feldspar and for H_2O-saturated situations, the four minerals garnet, cordierite, biotite and sillimanite may be represented in a binary chemography. The chemography is a projection through quartz, K-feldspar and H_2O (and parallel to $MgFe_{-1}$). The geometry shown for the general binary case in Fig. 3.20a may be readily translated into Fig. 3.20b. The unique divariant assemblage for each sector is shaded in Fig. 3.20b. Note, however, that some assemblages (e.g., garnet+biotite; Fig. 3.20b) occur in more than one sector. Any assemblage occurring in two sectors (in the binary case) contains the phase that is absent at the sector boundary (biotite in our case). This is so because biotite+garnet is not affected by the biotite-absent reaction (Crd = Sil + Grt). Also note that, for instance, in the cordierite-absent reaction (Bt + Sil = Grt), the assemblage containing biotite must be in the sector that is not bounded by the biotite-absent reaction, and garnet must be in the sector that is not bounded by the garnet-absent reaction. This is a general rule and must be obeyed by all univariant assemblages along any of the curves.

Once the correct sequence and geometry of univariant assemblages is established around an invariant point the sequence may be translated into pressure-temperature space (Fig. 3.20c).

Three of the four reactions in our example are dehydration reactions. It is fairly safe to assume that H_2O is released from biotite as temperature increases. The reactions (Crd), (Grt) and (Sil) should therefore be arranged in such a way that H_2O appears on the high-temperature side of the univariant curve. In addition, dehydration reactions commonly have steep slopes on P–T diagrams. Two solutions are possible: (1) the (Crd) reaction extends to high pressures from the invariant point [the reactions (Grt) and (Sil) extend to the low-pressure side], or (2) (Crd) extends to low pressures. Both solutions are geometrically correct. Version (1) is shown in Fig. 3.20c because cordierite is a mineral that typically occurs in low- to medium-pressure metapelites. In solution (2), cordierite is restricted to high pressures, which is not consistent with geological experience. The slope of the solid–solid reaction (Bt) cannot be constrained in a way similar to that for the dehydration reactions. However, it must extend from the invariant point into the sector between (Sil) and the metastable extension of (Grt). This follows from Fig. 3.20b. Therefore, it may have a positive or a negative slope. The equilibrium of the reaction is in reality rather independent of temperature and it represents an example of a geologic barometer.

The geometric arrangement of univariant curves around invariant points on any kind of phase diagram must be strictly consistent with Schreinemakers' rules. Part II of this volume, on progressive metamorphism of various rock types, provides many examples of Schreinemakers' bundles in two- and three-component systems on various types of phase diagrams.

Once the correct arrangement of univariant curves around an invariant point is known the connection of this invariant point to other (if any) invariant points in the system must be examined. If, for example, spinel needs to be considered in the problem used above, five invariant assemblages exist in the system. The one already considered above was Grt + Crd + Bt + Sil and is characterized by the absence of spinel. The invariant point of Fig. 3.20 is consequently the [Spl]-absent invariant point. According to the phase rule the system has a variance of –1. Systems with negative f-values are called multisystems and more than one invariant assemblage exists. Some of the invariant assemblages may be stable, some others metastable. In general, the minerals of a multisystem cannot all simultaneously coexist at equilibrium. On phase diagrams multisystems are characterized by the presence of more than one, usually several invariant points.

A simple example can be made by reconsidering the phase relationships shown in Fig. 3.18. There, we considered the minerals pyrite, pyrrhotite, hematite and magnetite and H_2O fluid in the component system Fe–O–H–S. It follows from the phase rule that $f = 1$. Under isobaric isothermal conditions (we fix P and T), $f = -1$ and five invariant assemblages exist. In the presence of an H_2O fluid, there are four invariant assemblages involving three minerals each and lacking one mineral. Consequently, the invariant assemblages are [Po],[Py],[He] and [Mag]. As shown in Fig. 3.21, the pair [Po][He] is stable and [Mag][Py] is metastable under the selected P–T conditions. Note that the univariant curves change their stability level at invariant points, as indicated by different dashes of the metastable extensions of the stable curves. Schreinemakers' rules are also observed by the metastable invariant points, e.g., [Py]. Also note that the metastable invariant point [Mag] is inside the field of stable magnetite and that the metastable invariant point [Py] is inside the field of stable pyrite. Furthermore, there is one forbidden assemblage, namely pyrrhotite + hematite. The Po + He equilibrium has no stable sector on Fig. 3.21; it connects as a metastable curve the two metastable invariant points [Mag] and [Py].

Metastable phase relations are not normally shown on geologic phase diagrams. However, these are sometimes useful to consider and most phase diagram software permits the construction of diagrams showing the metastable phase patterns.

Fig. 3.21. Qualitative phase relationships among hematite, magnetite, pyrrhotite and pyrite in a log a_{H_2} versus log a_{H_2S} activity–activity diagram (see also Fig. 3.18). *Solid lines* Stable equilibria; *dashed lines* metastable level 1 equilibria; *dotted lines* metastable level 2 equilibria. Invariant points [Po][He] are stable, [Mag][Py] are metastable

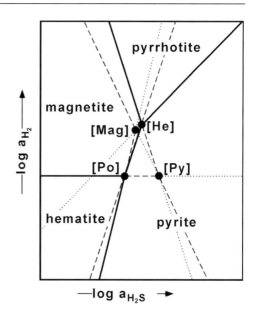

3.8.4
Use of Phase Diagrams, an Example

A phase diagram relevant for discussing metamorphism of carbonate rocks is used as an example to explain general properties and the use of phase diagrams. Marbles and metamorphism of carbonate rocks will be presented in some detail in Chapter 6.

The isobaric $T - X_{CO_2}$ diagram (Fig. 3.22 A) shows the fields of stable minerals in siliceous dolomites. Calcite and dolomite are present in excess (however, dolomite is replaced by brucite or periclase in the upper left area of the diagram). The diagram can be understood as a map showing at each point the mineral with the lowest free energy that coexists with calcite and dolomite and a fluid of the composition given along the x-axis and at the temperature given along the y-axis at a pressure of 2 kbar. The field boundaries shown in Fig. 3.22 represent reaction equilibria of mixed volatile reactions (see Sect. 3.6.2.3). The shapes of the curves correspond to those shown in Fig. 3.17 and they are given by the stoichiometry of the reaction. The equilibrium curves terminate at intersections with other equilibria (invariant points, see also Sect. 3.8.2) or they end at the boundaries of the selected parameter window. Only stable sections of reaction equilibria are shown; metastable equilibria have been omitted.

How can this figure be used for practical purposes? Before using this figure one must be certain or convinced that the pressure of 2 kbar is significant to the actual problem and that a binary CO_2–H_2O fluid was present during the petrogenesis of interest. If this is secured, then two types of simple applications are obvious:

1. given a marble with calcite + dolomite + diopside; at what possible T–X pairs was this rock stable? The answer would be: temperature above 520 °C and below 630 °C and X_{CO_2} greater than 0.7. An independent estimate of the temperature at which the marble equilibrated can be made from Cal-Dol thermometry (see Chap. 4). If the temperature estimate for our sample rock is, e.g., 550 °C, then it follows that the diopside marble indicates an extremely CO_2-rich fluid (>93 mol% CO_2). If the diopside marble is the product of prograde metamorphism, what were the possible precursor assemblages? The answer: either Tr + Cal + Dol or Qtz + Cal + Dol.
2. given that a temperature of about 500 °C has been reached during contact metamorphism at an outcrop that has a certain distance to the contact of a granite pluton. The temperature may be known, e.g., from theoretical thermal modeling, from assemblages in other types of rocks, from fluid inclusion data, from Cal-Dol thermometry or from stable isotope data. What kind of marbles can be expected at this outcrop? The answer: tremolite marble is certainly the first candidate. Quartz may still be present in rocks with an unusually CO_2-rich fluid. Forsterite marble is possible but it would require an extremely H_2O-rich fluid to form. If present at this locality, one would suspect that the Fo-marble formed as a result of infiltration with such a fluid from an external source (see again Fig. 3.19). This is even more so the case for the assemblage Fo + Brc + Cal close to the $X_{CO_2} = 0$ axis.

The phase diagram (Fig. 3.22 a) can also be used to deduce complete paths of prograde metamorphism taken by some chosen marble sample. An example can be made using a marble containing initially 88 wt% dolomite, 10 wt% quartz and 2 wt% fluid (with the composition 20 mol% CO_2 and 80 mol% H_2O). The question is, what will happen to this rock if it is progressively heated to 800 °C? The following discussion assumes that the volatile components released by the reactions are taken up by the fluid and remain in the rock. When heated, the model rock starts its reaction history at point (1) where talc becomes stable. The rock will be buffered along the talc field boundary by the talc-producing reaction exactly analogous to what has been explained in Sections 3.6.2.3 and 3.7. How far the metamorphic path follows the talc field boundary can be calculated from the rock's initial composition and the reaction stoichiometry (see Sect. 3.7). In the present example, the path reaches the invariant point Tlc + Tr + Qtz [point (2) on Fig. 3.22 a]. At this point all previously formed talc is converted into tremolite. The path continues then along the tremolite field boundary until all quartz is used up by the tremolite-producing reaction at point (3). The Tr + Dol + Cal rock is then simply heated and crosses the divariant field until diopside production sets in at point (4). All tremolite is then converted to diopside along the diopside field boundary. At point (5), the rock runs out of tremolite. At point (6), the rock reaches equilibrium with forsterite. Forsterite replaces all diopside between points (6) and (7). Finally, at point (7), diopside disappears from the marble. Fo-Dol-Cal marble is then simply heated through the divariant forsterite field to the final temperature of 800 °C.

The modal composition of the rock at any point along the reaction path in Fig. 3.22 a can be calculated in the same way as in Section 3.7 for the wollastonite reaction. Figure 3.22 b shows the modal composition and its changes with increasing temperature. The specific metamorphic mileposts along the path are labeled from (1) through (7). The markers correspond to those in Fig. 3.22 a: (1) onset of continuous talc production; (2) talc disappears and significant modal amounts of tremolite are produced by the reaction at the invariant point; (3) last quartz disappears; (4) onset of continuous diopside production; (5) the rock runs out of tremolite; (6) onset of continuous forsterite production; and (7) rock runs out of diopside. The end product of metamorphism is a high-grade forsterite marble with 25% Fo, 38% Cal, 32% Dol and fluid.

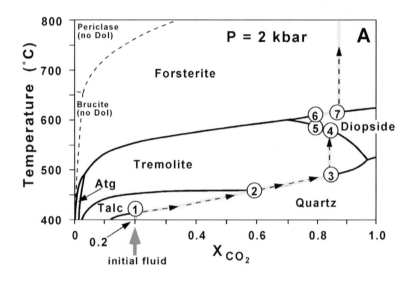

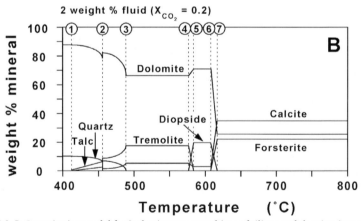

Fig. 3.22 A, B. Quantitative model for isobaric metamorphism of siliceous dolomites in terms of **A** a quantitative isobaric T–X diagram and **B** a modal composition versus temperature diagram

The two different types of diagrams are helpful tools for the study of metamorphic rocks: The equilibrium phase diagram (Fig. 3.22 a) relates the assemblages to the conditions under which they have formed, the composition phase diagram illustrates modal changes resulting from reactions in a rock of specific composition (Fig. 3.22 b). Note, however, that Fig. 3.22 b is a special variant of a composition phase diagram because it combines composition with the intensive variable temperature; it is in fact a mixed variable diagram. On the other hand, a $T - X_{CO_2}$ diagram is not a mixed variable diagram because the CO_2 and H_2O represented along the x-axis are actually related to the chemical potential of the two volatile species. Chemical potentials are intensive variables like temperature and pressure.

The model metamorphism outlined above is truly progressively prograde. This means that the rock along the path advances from low to high temperature (= prograde). In which case, the rock increases in metamorphic grade and metamorphism becomes more intense (see Chap. 4). Along the reaction path, the rock maintains equilibrium at all times and one stage follows the next in a continuous series. The highest-grade end product of metamorphism is a forsterite marble. It went through all lower grades in a succession of continuous transformations; in short, the rock experienced progressive prograde metamorphism. An inverse equilibrium path from high to low temperature would be progressively retrograde. However, retrograde metamorphism very rarely follows a strict equilibrium path in contrast to prograde metamorphism which produces its own fluid phase. Progressive metamorphism, prograde or retrograde, follows a continuous equilibrium reaction path.

A final remark, the model metamorphism illustrated by Fig. 3.22 can be understood in a temporal or spatial sense. In the first sense it describes the metamorphic reactions experienced with time by a volume of rock at a given position in space. In the second sense it models the distribution of the maximum grade reached by all rocks in an area. An example of the spatial viewpoint would be: from Fig. 3.22 it follows that a contact metamorphic aureole (2 kbar) in dolomitic country rock will show a regular systematic zonation with minor talc in the outermost (coldest) parts, followed by a zone of tremolite marble, a narrow zone with diopside marble and, closest to the contact with the intrusive rocks (the heat source), a zone with forsterite marbles. In contrast, the time aspect is contained in the following: a specific small outcrop of forsterite marble formed from diopside marble, which earlier was a tremolite marble that formed initially from talc marble which originated from the sedimentary siliceous dolomitic limestone.

References

Abers GA (2000) Hydrated subducted crust at 100–250 km depth. Earth Planet Sci Lett 176:323–330
Armstrong TR, Tracy RJ, Hames WE (1992) Contrasting styles of Taconian, eastern Acadian and western Acadian metamorphism, central and western New England. J Metamorph Geol 10:415–426
Balling NP (1985) Thermal structure of the lithosphere beneath the Norwegian-Danish basin and the southern Baltic shield: a major transition zone. Terra Cognita 5:377–378

Banno S, Sakai C, Higashino T (1986) Pressure-temperature trajectory of the Sanbagawa metamorphism deduced from garnet zoning. Lithos 19:51–63

Bell TH, Cuff C (1989) Dissolution, solution transfer, diffusion versus fluid flow and volume loss during deformation/metamorphism. J Metamorph Geol 7:425–448

Bell TH, Hayward N (1991) Episodic metamorphic reactions during orogenesis: the control of deformation partitioning on reaction sites and reaction duration. J Metamorph Geol 9:619–640

Bell TH, Mares VM (1999) Correlating deformation and metamorphism around orogenic arcs. Am Mineral 84:1727–1740

Berman RG (1988) Internally-consistent thermodynamic data for minerals in the system: $Na_2O-K_2O-CaO-MgO-FeO-Fe_2O_3-Al_2O_3.-SiO_2-TiO_2-H_2O-CO_2$. J Petrol 29:445–522

Berman RG, Brown TH, Perkins EH (1987) GEØ-CALC: software for calculation and display of pressure–temperature–composition phase diagrams. Am Mineral 72:861

Bickle MJ, McKenzie D (1987) The transport of heat and matter by fluids during metamorphism. Contrib Mineral Petrol 95:384–392

Brady JB (1988) The role of volatiles in the thermal history of metamorphic terranes. J Petrol 29:1187–1213

Brodie KH, Rutter EH (1985) On the relationship between deformation and metamorphism with special reference to the behaviour of basic rocks. In: Advances in physical geochemistry. Springer, Berlin Heidelberg New York, pp 138–179

Brown GC, Mussett AE (1981) The inaccessible earth. Allen and Unwin, London, p 152, Fig. 8.15. 235 pp

Brown TH, Berman RG, Perkins EH (1988) GEØ-CALC: software package for calculation and display of pressure-temperature-composition phase diagrams using an IBM or compatible computer. Comput Geosci 14:279–289

Burnham CW, Holloway JR, Davis NF (1969) Thermodynamic properties of water to 1000 °C and 10 000 bars. Geol Soc Am Spec Pap 132:96

Cermak V, Rybach L (1987) Terrestrial heat flow and the lithosphere structure. Terra Cognita 7:685–687

Chapman DS (1986) Thermal gradients in the continental crust. In: Dawson JB, Carswell DA, Hall J, Wedepohl KH (eds) The nature of the lower continental crust. Geological Society Special Publication. Blackwell, London, pp 63–70

Chatterjee ND (1991) Applied mineralogical thermodynamics. Springer, Berlin Heidelberg New York, 321 pp

Clark SP (1966) Handbook of physical constants. Geological Society of America Memoir, Washington, DC, 587 pp

Connolly JAD (1990) Multivariable phase diagrams: an algorithm based on generalized thermodynamics. Am J Sci 290:666–718

Connolly JAD, Kerrick DM (1987) An algorithm and computer program for calculating computer phase diagrams. CALPHAD 11:1–55

Davy P, Gillet P (1986) The stacking of thrust slices in collision zones and its thermal consequences. Tectonics 5:913–929

Day HW (1972) Geometrical analysis of phase equilibria in ternary system of six phases. Am J Sci 272:711–734

Day HW, Chamberlain CP (1989) Implications of thermal and baric structure for controls on metamorphism, northern New England, USA. In: Daly S, Cliff RA, Yardley BWD (eds) Evolution of metamorphic belts. Geological Society Special Publication. Blackwell, London, pp 215–222

de Capitani C, Brown TH (1987) The computation of chemical equilibrium in complex systems containing non-ideal solutions. Geochim Cosmochim Acta 51:2639–2652

Denbigh K (1971) The principles of chemical equilibrium. Cambridge University Press, London, 494 pp

Durney DW (1972) Solution transfer, an important geological deformation mechanism. Nature 235:315–317

Dyar MD, Guidotti CV, Holdaway MJ, Colucci M (1993) Nonstoichiometric hydrogen contents in common rock-forming hydroxyl silicates. Geochim Cosmochim Acta 57:2913–2918

England PC, Richardson SW (1977) The influence of erosion upon the mineral facies of rocks from different metamorphic environments. J Geol Soc Lond 134:201–213

England PC, Thompson AB (1984) Pressure-temperature-time paths of regional metamorphism. I. Heat transfer during the evolution of regions of thickened continental crust. J Petrol 25:894–928

England PC, Thompson AB (1986) Some thermal and tectonic models for crustal melting in continental collision zones. In: Coward MP, Ries AC (eds) Collision tectonics. Geological Society Special Publication. Blackwell, London, pp 83–94

Ernst WG (1976) Petrologic phase equilibria. Freeman, San Francisco, 333 pp

Eugster HP (1977) Compositions and thermodynamics of metamorphic solutions. In: Fraser DG (ed) Thermodynamics in geology. Reidel, Dordrecht, pp 183–202

Eugster HP (1986) Minerals in hot water. Am Mineral 71:655–673

Eugster HP, Gunter WD (1981) The compositions of supercritical metamorphic solutions. Bull Mineral 104:817–826

Ferry JM (1980) A case study of the amount and distribution of heat and fluid during metamorphism. Contrib Mineral Petrol 71:373–385

Ferry JM (1982) Characterization of metamorphism through mineral equilibria. Reviews in mineralogy, vol 10. Mineralogical Society of America, Washington, DC, 397 pp

Ferry JM (1983) Application of the reaction progress variable in metamorphic petrology. J Petrol 24:343–376

Ferry JM (2000) Patterns of mineral occurrence in metamorphic rocks. Am Mineral 85:1573–1588

Fletcher P (1993) Chemical thermodynamics for earth scientists. Longman Scientific and Technical, Essex, 464 pp

Fraser DG (1977) Thermodynamics in geology. NATO Advanced Study Institutes Series, vol 30. Reidel, Dordrecht, 410 pp

Froese E (1977) Oxidation and sulphidation reactions. In: Greenwood HJ (ed) Application of thermodynamics to petrology and ore deposits. Short course 2. Mineralogical Association of Canada, Vancouver, pp 84–98

Frost BR (1988) A review of graphite–sulfide–oxide–silicate equilibria in metamorphic rocks. Rend Soc Ital Mineral Petrol 43:25–40

Frost BR, Tracy RJ (1991) P–T paths from zoned garnets: some minimum criteria. Am J Sci 291:917–939

Fyfe WS, Turner FJ, Verhoogen J (1958)) Effect of temperature on equilibrium entropy of solids. In: Metamorphic reactions and metamorphic facies. Geol Soc Am Bull 73:25–34

Garrels RM, Christ CL (1965) Solutions, minerals and equilibria. Freeman and Cooper, San Francisco, 450 pp

Gibbs JW (1878) On the equilibrium of heterogeneous substances. Am J Sci (Trans Conn Acad) 16:343–524

Gibbs JW (1906) The scientific papers of J. Willard Gibbs. Thermodynamics. Longmans and Greed, London

Goldstein RH (2001) Fluid inclusions in sedimentary and diagenetic systems. Lithos 55:159–193

Greenwood HJ (1975) Buffering of pore fluids by metamorphic reactions. Am J Sci 275:573–594

Greenwood HJ (ed) (1977) Application of thermodynamics to petrology and ore deposits. Short course 2. Mineralogical Association of Canada, Vancouver, 230 pp

Guggenheim EA (1986) Thermodynamics. North-Holland Physics Publ, Amsterdam, 390 pp

Harker A (1932) Metamorphism. A study of the transformation of rock masses. Methuen, London

Harte B, Dempster TJ (1987) Regional metamorphic zones: tectonic controls. Philos Trans R Soc Lond A 321:105–127

Helgeson HC, Kirkham DH (1974) Theoretical prediction of the thermodynamic behaviour of aqueous electrolytes at high pressures and temperatures. I. Summary of the thermodynamic/electrostatic properties of the solvent. Am J Sci 274:1089–1098

Helgeson HC, Delany JM, Nesbitt HW, Bird DK (1978) Summary and critique of the thermodynamic properties of rock-forming minerals. Am J Sci 278-A:229 pp

Helgeson HC, Kirkham DH, Flowers GC (1981) Theoretical prediction of the thermodynamic behaviour of aqueous electrolytes at high pressures and temperatures. IV. Calculation of activity coefficients, osmotic coefficients, and apparent molal and standard and relative partial molal properties to 600 °C and 5 kb. Am J Sci 281:1249–1516

Holland TJB, Powell R (1985) An internally consistent dataset with uncertainties and correlations. 2. Data and results. J Metamorph Geol 3:343–370

Holland TJB, Powell R (1990) An enlarged and updated internally consistent thermodynamic dataset with uncertainties and correlations: the system $K_2O–Na_2O–CaO–MgO–MnO–FeO–Fe_2O_3–Al_2O_3–TiO_2–SiO_2–C–H_2–O_2$. J Metamorph Geol 8:89–124

Holland TJB, Powell R (1998) An internally-consistent thermodynamic dataset for phases of petrological interest. J Metamorph Geol 16:309–343

Jamieson RA, Beaumont C, Hamilton J, Fullsack P (1996) Tectonic assembly of inverted metamorphic sequences. Geology 24:839–842

Johnson JW, Oelkers EH, Helgeson HC (1992) SUPCRT92: a software package for calculating the standard molal thermodynamic properties of minerals, gases, aqueous species, and reactions from 1 to 5000 bars and 0 to 1000 °C. Comput Geosci 18:899–947

Jones KA, Brown M (1990) High-temperature 'clockwise' P–T paths and melting in the development of regional migmatites: an example from southern Brittany, France. J Metamorph Geol 8:551–578

Karabinos P, Ketcham R (1988) Thermal structure of active thrust belts. J Metamorph Geol 6:559–570

Kretz R (1991) A note on transfer reactions. Can Mineral 29:823–832

Krogh EJ, Oh CW, Liou JG (1994) Polyphase and anticlockwise p–T evolution for franciscan eclogites and blueschists from Jenner, California, USA. J Metamorph Geol 12:121–134

Kukkonen IT, Cermàk V, Hurtig E (1993) Vertical variation of heat flow density in the continental crust. Terra 5:389–398

Lasaga AC, Jianxin J (1995) Thermal history of rocks: P-T-t paths from geospeedometry, petrologic data, and inverse theory techniques. Am J Sci 295:697–741

Lasaga AC, Kirkpatrick RJ (1981) Kinetics of geochemical processes. Reviews in mineralogy. Mineralogical Society of America, Washington, DC, 398 pp

Lasaga AC, Rye DM (1993) Fluid flow and chemical reaction kinetics in metamorphic systems. Am J Sci 293:361–404

Lasaga AC, Lüttge A, Rye DM, Bolton EW (2000) Dynamic treatment of invariant and univariant reactions in metamorphic systems. Am J Sci 300:173–221

Lenardic A, Kaula WM (1995) Mantle dynamics and the heat flow into the Earth's continents. Nature 378:709–711

Lewis GN, Randall M (1961) Thermodynamics (revised by Pitzer KS, Brewer L). McGraw-Hill, New York, 723 pp

Lux DR, DeYoreo JJ, Guidotti CV, Deker ER (1986) Role of plutonism in low-pressure metamorphic belt formation. Nature 323:794–797

Macfarlane AM (1995) An evaluation of the inverted metamorphic gradient at Langtang National Park, central Nepal Himalaya. J Metamorph Geol 13:595–612

Manning CE, Ingebritsen SE (1999) Permeability of the continental crust: implications of geothermal data and metamorphic systems. Rev Geophys 37:127–150

Miyashiro A (1961) Evolution of metamorphic belts. J Petrol 2:277–311

Miyashiro A (1964) Oxidation and reduction in the earth's crust, with special reference to the role of graphite. Geochim Cosmochim Acta 28:717–729

Moore WL (1972) Physical chemistry. Longman, London, 977 pp

Müntener O, Hermann J, Trommsdorff V (2000) Cooling history and exhumation of lower-crustal granulite and upper mantle (Malenco, eastern central Alps). J Petrol 41:175–200

Nisbet EG, Fowler CMR (1988) Heat, metamorphism and tectonics. Short course 14. Mineralogical Association of Canada, St. John's, Newfoundland, 319 pp

Norris RJ, Henley RW (1976) Dewatering of a metamorphic pile. Geology 4:333–306

Norton D, Knight J (1977) Transport phenomena in hydrothermal systems: cooling plutons. Am J Sci 277:937–981

Olsen KH (1995) Continental rifts: evolution, structure, tectonics. Elsevier, Amsterdam, 466 pp

Oxburgh ER (1974) The plain man's guide to plate tectonics. Proc Geol Assoc 85:299–358

Oxburgh ER, England PC (1980) Heat flow and the metamorphic evolution of the Eastern Alps. Eclogae Geol Helv 73:379–398

Oxburgh ER, Turcotte DL (1971) Origin of paired metamorphic belts and crustal dilation in island arc regions. J Geophys Res 76:1315–1327

Oxburgh ER, Turcotte DL (1974) Thermal gradients and regional metamorphism in overthrust terrains with special reference to the Eastern Alps. Schweiz Mineral Petrogr Mitt 54:642–662

Peacock SM (1990) Numerical simulation of metamorphic pressure-temperature-time paths and fluid production in subducting slabs. Tectonics 9:1197–1211

Perkins D, Essene EJ, Wall VJ (1987) THERMO: a computer program for calculation of mixed-volatile equilibria. Am Mineral 72:446–447

Petrini K, Podladchikov Y (2000) Lithospheric pressure-depth relationship in compressive regions of thickened crust. J Metamorph Geol 18:67–77

Platt JP (1986) Dynamics of orogenic wedges and the uplift of high-pressure metamorphic rocks. Geol Soc Am Bull 97:1037–1053

Platt JP, England PC (1993) Convective removal of lithosphere beneath mountain belts: thermal and mechanical consequences. Am J Sci 293:307–336

Powell R (1978) Equilibrium thermodynamics in petrology, an introduction. Harper and Row, New York, 284 pp

Powell R, Holland TJB (1985) An internally consistent dataset with uncertainties and correlations. 1. Methods and a worked example. J Metamorph Geol 3:327–342

Powell R, Holland TJB (1988) An internally consistent dataset with uncertainties and correlations. 3. Applications to geobarometry, worked examples and a computer program. J Metamorph Geol 6:173–204

Powell R, Tim Holland T (2001) Course notes for THERMOCALC workshop 2001: calculating metamorphic phase equilibria: on CD-ROM

Prigogine I (1955) Thermodynamics of irreversible processes. Wiley, New York, 147 pp

Ramberg H (1952) The origin of metamorphic and metasomatic rocks. Univ Chicago Press, Chicago, 317 pp

Robie RA, Hemingway BS, Fisher JR (1978) Thermodynamic properties of minerals and related substances at 298.15 K and 1 bar (10^5 Pascals) pressure and at higher temperatures. US Geol Surv Bull 1452:1–456

Robinson D (1987) Transition from diagenesis to metamorphism in extensional and collision settings. Geology 15:866–869

Rubenach MJ (1992) Proterozoic low-pressure/high-temperature metamorphism and an anti-clockwise P–T–t path for the Hazeldene area, Mount Isa Inlier, Queensland, Australia. J Metamorph Geol 10:333–346

Saxena S, Ganguly J (1987) Mixtures and mineral reactions. Minerals and rocks. Springer, Berlin Heidelberg New York, 260 pp

Sclater JG, Jaupart C, Galson D (1980) The heat flow through oceanic and continental crust and the heat loss of the Earth. Rev Geophys Space Phys 18:269–311

Seyfried WE (1987) Experimental and theoretical constraints on hydrothermal alteration processes at mid-ocean ridges. Annu Rev Earth Planet Sci Lett 15:317–335

Skippen GB, Carmichael DM (1977) Mixed-volatile equilibria. In: Greenwood HJ (ed) Application of thermodynamics to petrology and ore deposits. Short course. Mineralogical Association of Canada, Vancouver, pp 109–125

Skippen GB, Marshall DD (1991) The metamorphism of granulites and devolatilization of the lithosphere. Can Mineral 29:693–706

Sleep NH (1979) A thermal constraint on the duration of folding with reference to Acadian geology, New England, USA. J Geol 87:583–589

Spear FS (1993) Metamorphic phase equilibria and pressure-temperature-time path. Mineralogical Society of America, Washington, DC, 824 pp

Spear FS (1999) Real-time AFM diagrams on your Macintosh. Geol Mater Res 1:1–18

Spear FS, Peacock SM (1989) Metamorphic pressure-temperature-time paths. Short course in geology, vol 7. American Geophysical Union, Washington, DC, 102 pp

Spear FS, Rumble D III, Ferry JM (1982) Linear algebraic manipulation of n-dimensional composition space, In: Ferry JM (ed) Characterization of metamorphism through mineral equilibria. Reviews in mineralogy. Mineralogical Society of America, Washington, DC, pp 53–104

Spear FS, Selverstone J, Hickmont D, Crowley P, Hodges KV (1984) P–T paths from garnet zoning: a new technique for deciphering tectonic processes in crystalline terranes. Geology 12:87–90

Spear FS, Peacock SM, Kohn MJ, Florence FP, Menard T (1991) Computer programs for petrologic P–T–t path calculations. Am Mineral 76:2009–2012

Spiess R, Bell TH (1996) Microstructural controls on sites of metamorphic reaction: a case study of the inter-relationship between deformation and metamorphism. Eur J Mineral 1:165–186

Spooner ETC, Fyfe WS (1973) Sub-sea-floor metamorphism, heat and mass transfer. Contrib Mineral Petrol 42:287–304

Stephenson BJ, Waters DJ, Searle MP (2000) Inverted metamorphism and the Main Central Thrust: field relations and thermobarometric constraints from the Kishtwar Window, NW Indian Himalaya. J Metamorph Geol 18:571–590

Stern T, Smith EGC, Davey FJ, Muirhead KJ (1987) Crustal and upper mantle structure of the northwestern North Island, New Zealand, from seismic refraction data. Geophys J R Astron Soc 91:913–936

Stull DR, Prophet H (1971) JANAF thermochemical tables. Matl Standards Ref Data Ser. National Bureau of Standards, Washington, DC, 1141 pp

Thompson AB (1981) The pressure-temperature (P,T) plane viewed by geophysicists and petrologists. Terra Cognita 1:11–20

Thompson AB, England PC (1984) Pressure-temperature-time paths of regional metamorphism. II. Their inference and interpretation using mineral assemblages in metamorphic rocks. J Petrol 25:929–955

Thompson AB, Ridley JR (1987) Pressure-temperature-time (P–T–t) histories of orogenic belts. Philos Trans R Soc Lond A 321:27–45

Thompson AB, Tracy RJ, Lyttle PT, Thompson JB (1977) Prograde reaction histories deduced from compositional zonation and mineral inclusions in garnet from the Gassetts schist, Vermont. Am J Sci 277:1152–1167

Thompson JB (1982a) Composition space; an algebraic and geometric approach. In: Ferry JM (ed) Characterization of metamorphism through mineral equilibria. Reviews in mineralogy. Mineralogical Society of America, Washington, DC, pp 1–31

Thompson JB (1982b) Reaction space; an algebraic and geometric approach. In: Ferry JM (ed) Characterization of metamorphism through mineral equilibria. Reviews in mineralogy. Mineralogical Society of America, Washington, DC, pp 33–52

Touret JLR (2001) Fluids in metamorphic rocks. Lithos 55:1–25

Vernon RH (1996) Problems with inferring P–T–t paths in low-P granulite facies rocks. J Metamorph Geol 14:143–153

Vigneresse JL (1988) Heat flow, heat production and crustal structure in peri-Atlantic regions. Earth Planet Sci Lett 87:303–312

Vigneresse JL, Cuney M (1991) What can we learn about crustal structure from thermal data? Terra Nova 3:28–34

Walther JV, Orville PM (1982) Rates of metamorphism and volatile production and transport in regional metamorphism. Contrib Mineral Petrol 79:252–257

Whitney DL, Lang HM, Ghent ED (1995) Quantitative determination of metamorphic reaction history: mass balance relations between groundmass and mineral inclusion assemblages in metamorphic rocks. Contrib Mineral Petrol 120:404–411

Wickham S, Oxburgh ER (1985) Continental rifts as a setting for regional metamorphism. Nature 318:330–333

Winslow DM, Bodnar RJ, Tracy RJ (1994) Fluid inclusion evidence for an anticlockwise metamorphic P–T path in central Massachusetts. J Metamorph Geol 12:361–371

Wood BJ, Fraser DG (1976) Elementary thermodynamics for geologists. Oxford Univ Press, Oxford, 303 pp

Wood BJ, Walther JV (1984) Rates of hydrothermal reactions. Science 222:413–415

Zen E-A (1966) Construction of pressure-temperature diagrams for multi-component systems after the method of Schreinemakers – a geometrical approach. US Geol Surv Bull 1225:1–56

Zwart HJ (1962) On the determination of polymetamorphic mineral associations, and its application to the Bosot area (central Pyrénées). Geol Rundsch 52:38–65

CHAPTER 4

Metamorphic Grade

4.1
General Considerations

The intensity of metamorphism and the vigor of metamorphic transformation are expressed by the term **metamorphic grade**. If, for instance, a granitic pluton intrudes a pile of sediments, the resulting contact metamorphic aureole will contain rocks of higher grade close to the contact of the pluton (heat source) and rocks of low grade at greater distance from the pluton where the sediments were only moderately heated. Similarly, in a mountain chain, a metamorphic terrain may have developed and is presently at the erosion surface where rock samples can be collected. They contain mineral assemblages that represent the least hydrated state attained during a metamorphic cycle. The assemblages define a metamorphic field gradient (see Chap. 3). Rocks on this P–T array are said to be of low grade at low P–T and of successively higher grade with higher P–T conditions. The term metamorphic grade is a qualitative indicator of the physical conditions that have been operating on the rocks. Increasing P–T conditions produce metamorphic rocks of increasingly higher grade. It is a useful term for comparison of rocks within a single prograde metamorphic area. When applied to rocks in different regions, its meaning is not always clear, however, and its exact nature should be clearly specified.

According to Turner (1981, p. 85), the term metamorphic grade or grade of metamorphism was originally introduced by Tilley (1924, p. 168) "to signify the degree or state of metamorphism" and, more specifically, "the particular pressure-temperature conditions under which the rocks have arisen" (Tilley 1924, p. 167). Since reliable P–T values were not known for metamorphic rocks at that time, and since temperature was generally accepted as the most important factor of metamorphism (cf. Chap. 3), it became current usage to equate grade rather loosely with temperature. For example, Winkler (1979, p. 7) suggested a broad fourfold division of the P–T field of metamorphism primarily based on temperature, which he named very low-, low-, medium-, and high-grade metamorphism. Even though Winkler noted that information on pressure should be stated as well, a subdivision of metamorphic grade with respect to pressure seemed to be less important. This is very understandable because most of his P–T diagrams were limited to P <12 kbar, whilst the present-day P–T space of metamorphism has to be extended to much higher pressures (Fig. 1.1).

As explained above, metamorphism is currently understood in a geodynamic context. This means that the grade of metamorphic rocks is only a meaningful term for rocks that reached a metamorphic field gradient (or piezo-thermic array) in a single geodynamic process, e.g., a collision event. When, for example, studying a particular shale unit in the Alps, it is useful then to say that a sillimanite-garnet-biotite schist is a high-grade rock and a chlorite-pyrophyllite schist is a low-grade rock. However, comparing the metamorphic grade of an eclogite (25 kbar, 600 °C) with that of an amphibolite (6 kbar, 750 °C) is meaningless in most cases because the two rocks were produced in different geodynamic settings.

In the following sections various concepts to determine metamorphic grade will be discussed.

4.2
Index Minerals and Mineral Zones

Mineral zones were mapped for the first time by Barrow (1893, 1912) in pelitic rocks of the Scottish Highlands. He recognized the systematic entrance of new minerals proceeding upgrade; these minerals were designated as **index minerals**. The following succession of index minerals with increasing metamorphic grade has been distinguished:

chlorite $\Rightarrow$ biotite $\Rightarrow$ almandine-garnet $\Rightarrow$ staurolite $\Rightarrow$ kyanite $\Rightarrow$ sillimanite

The individual minerals are systematically distributed in distinct regional zones in the field, and corresponding **mineral zones** are defined as follows. The low-grade limit is determined by a line on a map joining points of the first appearance of a certain index mineral, after which the zone is named. The high-grade limit is given by a similar line for the following index mineral. A line separating two adjacent mineral zones will be termed a **mineral zone boundary** (and not an isograd, as discussed below). The biotite zone, for instance, is that occurring between the biotite and almandine-garnet mineral zone boundaries. Note that an index mineral generally persists to higher grades than the zone that it characterizes, but is sometimes restricted to a single mineral zone (e.g., kyanite in Fig. 4.1).

The zonal sequences elaborated by Barrow are called **Barrovian zones**. These mineral zones have since been found in many other areas and are characteristic for medium-pressure metapelites. Sequences of mineral zones other than those found by Barrow have been identified in other areas. In the Buchan region of NE Scotland, the **Buchan zones** are defined by the index minerals staurolite, cordierite, andalusite, and sillimanite. The Buchan zones represent a different metamorphic field gradient involving relatively lower pressures than those represented by the Barrovian zones.

Two zonal schemes have been mentioned above in pelites from orogenic metamorphic terrains. It should be made clear, however, that it is, in principle, possible to map mineral zones in many types of rock belonging to any type of metamorphism.

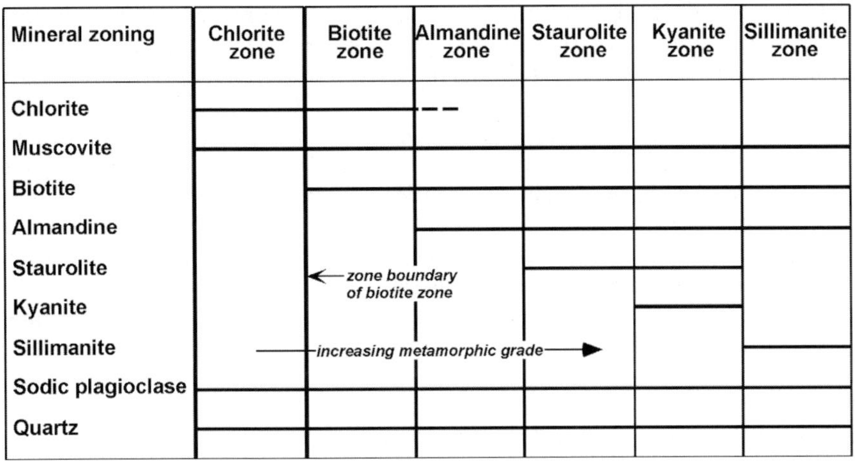

Mineral zoning	Chlorite zone	Biotite zone	Almandine zone	Staurolite zone	Kyanite zone	Sillimanite zone
Chlorite						
Muscovite						
Biotite						
Almandine						
Staurolite						
Kyanite						
Sillimanite						
Sodic plagioclase						
Quartz						

← zone boundary of biotite zone

increasing metamorphic grade →

Fig. 4.1. Distribution of some metamorphic minerals of pelitic rocks from the Barrovian zones of the Scottish Highlands

The mapping of mineral zones in determining metamorphic grade has the great advantage that it is a simple and rapid method, because the distribution of index minerals becomes obvious from inspection of hand specimens and thin sections. When working in an area for the first time, mineral zoning provides a first insight into the pattern of metamorphism. On the other hand, several shortcomings should be mentioned as well. First, by mapping mineral zones one single mineral in each rock is considered instead of a mineral assemblage; the latter, however, contains more petrogenetic information. Second, variations in rock composition are not adequately taken into account, and some index minerals will appear at a higher or lower grade in layers of different composition. Care must be taken to refer the mapped zonal pattern to a specific rock unit, e.g., a specific sedimentary formation. Ideally, this formation should be of uniform composition throughout the mapped area. In order to circumvent these drawbacks, today metamorphic zones are distinguished by associations of two, three or even more minerals rather than by index minerals.

4.3
Metamorphic Facies

4.3.1
Origin of the Facies Concept

The concept of metamorphic facies was introduced by Eskola (1915). He emphasized that mineral assemblages rather than individual minerals are the genetically important characteristics of metamorphic rocks, and a regular relationship between mineral assemblages and rock composition at a given metamorphic grade was proposed. This principle said that in a group of metamorphic rocks that have reached chemical equilibrium under the same definite P–T conditions, the mineral assemblages of the rocks depend only on their bulk chemical composition.

Eskola (1915, p. 115) defined a **metamorphic facies** as follows: "A metamorphic facies includes rocks which... may be supposed to have been metamorphosed under identical conditions. As belonging to a certain facies we regard rocks which, if having an identical chemical composition, are composed of the same minerals". This definition has been a problem ever since Eskola first introduced the idea, as discussed below:

1. A metamorphic facies includes mineral assemblages of a set of associated rocks covering a wide range of composition, all formed under the same broad metamorphic conditions (P and T after Eskola). Therefore, a metamorphic facies in its original meaning does not refer to a single rock-type, even though many facies are named after some characteristic metabasic rocks (e.g., greenschist, amphibolite, eclogite). Furthermore, some mineral assemblages have a large stability range and may occur in several metamorphic facies, whilst other assemblages have a more restricted stability range and may be diagnostic for one facies only. In addition, some rock-types do not show diagnostic assemblages at some particular metamorphic grade, e.g., many metapelites under subgreenschist facies conditions or metacarbonates under eclogite facies conditions. From these considerations it is obvious that we should search for **diagnostic mineral assemblages**, and it may be sufficient to recognize a metamorphic facies with the aid of only one such assemblage. In areas devoid of rock composition suitable for forming these diagnostic assemblages, on the other hand, assignment to a facies cannot be made.

2. Remember that Eskola proposed the idea of metamorphic facies before there were any significant experimental or thermodynamic data on the stability of metamorphic minerals. Under these circumstances, it is understandable that the only factors which this early scheme considered to be variables during metamorphism were temperature and lithostatic pressure. Additional variables, e.g., the composition of fluids present, if any, were not yet recognized, and it was assumed that $P_{H_2O} = P_{total}$. Recent experimental and theoretical studies have shown that minerals and mineral assemblages including, e.g., antigorite, lawsonite, prehnite or zeolites, are stable only in the presence of a very water-rich fluid, i.e., are absent at appreciable concentrations of CO_2. Fluid composition controls the P–T stability of mineral assemblages (Chap. 3). The transition from amphibolite facies to granulite facies rocks depends critically on the partial pressure of H_2O. This means that a more complete representation of metamorphic facies should involve relationships in P–T–X space, but detailed information on fluid composition is often missing.

3. Eskola's definition allows for an unlimited number of metamorphic facies. During his lifetime, Eskola increased the number of facies he recognized from five to eight. As more information on mineral assemblages has become available, more metamorphic facies have been added by other writers and many have been divided into "subfacies" or zones. At some point,

facies classification became impracticable. Some authors proposed the re-
tention of broad divisions, but others advocated for the abolition of metamorphic facies.

Notwithstanding the inherent weaknesses of the concept, the metamorphic
facies scheme continues to be used and is as popular as ever. It is convenient
as a broad genetic classification of metamorphic rocks in terms of two major
variables: lithostatic pressure and temperature. It is especially useful for regional or reconnaissance studies in metamorphic regions, but is too broad
for detailed metamorphic studies. As a good example, the large-scale metamorphic maps of Europe (Zwart 1973) may be mentioned.

4.3.2
Metamorphic Facies Scheme

The metamorphic facies scheme proposed in this volume is shown in
Fig. 4.2, and some diagnostic mineral assemblages are listed in Table 4.1.
Contrary to other schemes, no facies for contact metamorphism are distinguished here, because such facies are only of limited extent and metamorphic
zones are sufficient to designate variations in metamorphic grade. The simple facies scheme of Fig. 4.2 closely follows Eskola's classic treatment. The facies names are linked to characteristic rocks of basic composition. The system is very straightforward. First, we select a standard basic reference rock
composition, MOR basalt is the obvious choice. Then this rock type is placed

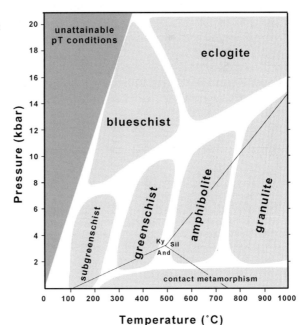

Fig. 4.2. Pressure–temperature regions of metamorphic
facies

Table 4.1. Metamorphic facies and diagnostic assemblages [a]

	Ultramafic rocks	Marbles	Metapelites	Metamarls	Metabasalts	Metagranitiods	Fluids
Protolith	Ol+Opx+Cpx+Spl	Cal+Dol+Qtz+ Kfs±Chl, ±Ab,±"clay"	"clay"+Qtz±Ab±Kfs	Cal+"clay"	Pl+Cpx+Opx+Qtz	Pl+Kfs+Qtz+Hbl± Bt ±Ol+Cpx±Opx	
Subgreenschist	Ol+Opx+Cpx+Grt chrysotil+Brc+Act chrysotil+Tlc+Act± Chl	Cal+Dol+Qtz+ Kfs+Chl+Ms	Kln(Prl)+Chl+ Illite+Qtz	Cal+Kln(Prl)+ Chl+Illite+Qtz	Pl+Cpx+Opx+Ol Zeolites Pmp+Prh+Chl+ Ab±Ep	"clay", illite, zeolites Prh, Stilp, Chl+Kfs	H_2O-CH_4
Greenschist	Atg+Brc+Di+Chl Atg+Tlc+Di+Chl	Cal+Dol+Qtz+Chl Cal+Dol+Qtz+ Kfs+Ab	Prl(Als)+Chl+Ms± Pg±Cld, ±Bt, ±Grt	Cal+Qtz+Mrg+ Chl+Ms±Ep	Ab+Chl+Ep+Act± Phe, ±Bt, ±Stp	Ab+Kfs+Chl+Qtz+ Bt±Act±Ep	H_2O
Amphibolite	Atg+Fo+Di+Chl Atg+Fo+Tr+Chl Tlc+Fo+Tr+Chl Ath+Fo+Tr+Chl En+Fo+Tr+Chl	Cal+Dol+Qtz+Tlc Cal+Dol+Tr+Chl Cal+Dol+Di+Phl Cal+Qtz+Tr+Di+ Phl	St+Chl+Grt+Ms St+Bt+Als+Ms St+Bt+Grt+Ms Crd+Bt+Grt+Ms	Cal+Qtz+Pl± Hbl±Grt±Bt	Pl+Hbl+Ep Pl+Hbl+Grt Pl+Hbl+Cpx±Bt	Pl+Kfs+Qtz±Bt± Ms±Hbl	H_2O-CO_2
Granulite	En+Fo+Hbl+Spl En+Fo+Di+Spl	Cal+Qtz+Di+Phl Cal+Dol+Di+Spl Cal+Dol+Fo+Spl Cal+Qtz+Di+Spl	Bt+Als+Kfs+Grt Opx+Qtz Opx+Crd+Bt+Qtz Opx+Als+Qtz±Spr± Spl	Cal+Qtz+Pl+Cpx± Grt	Pl+Cpx+Grt Pl+Cpx+Opx± Hbl±Bt	Opx+Qtz+Fsp± Ol±Cpx Mesoperthite	No fluid or CO_2
Blueschist	Atg+Fo+Di+Chl	CaCO$_3$+Dol+Qtz+ Phe	Carpholite Phe+Tlc+Grt	Cal+Gln+Ep+ Phe+Pg	Gln+Lws+Chl±Pg Gln+Ep±Grt±Pg± Cld±Tlc±Chl		CO_2
Eclogite	Atg+Fo+Di+Chl En+Fo+Di+Grt		Phe+carpholite Tlc+Ky Jd+Qtz(Coe)+Tlc+ Ky		Omp+Grt±Ky Omp+Grt±Zo±Phe Omp+Grt±Zo± Tlc±Cld	Jd+Qtz±Phe+Ky	H_2O-N_2

[a] "Clay" includes: kaolinite, smectite, montmorillonite, vermiculite, saponite, and many others.

in the P–T space (Fig. 4.2) to all possible pairs of pressure-temperature conditions. A pure H_2O fluid shall be present as well ($P_{H_2O} = P_{total}$). The equilibrium phase assemblage in the meta-basalt at each P–T point is characteristic for the facies fields of Fig. 4.2 and defines the facies. The characteristics of each facies are summarized below.

4.3.2.1
Sub-Blueschist–Greenschist Facies

We continue to use this name after having introduced it in the sixth edition of this volume. Eskola remained unconvinced that certain feebly recrystallized zeolite-bearing rocks contained equilibrium assemblages; hence he chose not to establish a separate metamorphic facies. However, later work by Coombs in New Zealand revealed the ubiquitous, systematic occurrence of such assemblages, and accordingly two very low-grade metamorphic facies have been proposed: the zeolite facies and the prehnite–pumpellyite facies. These two facies can be recognized in meta-basalts and a variety of other rocks. Zeolites occur in hydrothermally altered basalts at very low grade. The distribution of the various zeolite species is systematic and characteristic of P–T conditions (see Chap. 9). Prehnite and pumpellyite occur widespread in meta-basalts that were metamorphosed under conditions between the zeolite and the greenschist facies. Metamorphism has been insufficient to produce the characteristic assemblages of the greenschist and blueschist facies, respectively. In regional orogenic metamorphism, the first greenschist facies assemblages develop in meta-basalts at about 300 °C.

Sub-blueschist-greenschist facies stands for all P–T conditions too weak to produce greenschist or blueschist from basalt. The lower limit of metamorphism has arbitrarily been set at 200 °C (Chap. 1) and diagenetic processes in sediments are normally excluded from the term metamorphism. However, hydrothermal alteration of igneous rocks such as zeolitization of basalt, low-grade alteration of basement gneisses or fissure mineral formation in granite are typical processes at low-grade conditions, often as low as 100 °C. Obviously, metamorphic petrologists are interested in such processes and not sedimentologists. The temperature range of 100–300 °C includes a wealth of metamorphic transformation processes; most of them are, in a broad sense, related to the interaction of (surface) water with rocks. Water–rock interaction (WRI) research is an exceptionally flourishing inter- and multidisciplinary field in geosciences. Almost all WRI research concentrates on problems in the sub-blueschist–greenschist facies.

Because the term sub-blueschist-greenschist facies is rather awkward we suggest the use of the simplified version: sub-greenschist facies (subgreenschist facies).

4.3.2.2
Greenschist Facies

At P–T conditions of the greenschist facies (Fig. 4.2), meta-MORB is transformed to greenschist. Greenschist contains the mineral assemblage actinolite + chlorite + epidote + albite ± quartz. The first three minerals give the rock type its green color. The four-mineral assemblage is mandatory for a greenschist and hence for the greenschist facies. Other bulk rock compositions such as shales (pelites) develop a series of assemblages if metamorphosed under the same conditions. These are, of course, different from the greenschist assemblage of metabasalt. The chlorite, biotite and garnet Barrovian zones in metapelites belong to this facies, and more Al-rich metapelites may contain chloritoid. The metamorphic facies concept implies that a metapelitic schist containing muscovite + chlorite + chloritoid is said to be of greenschist grade. It has been metamorphosed under greenschist facies conditions, i.e., it is not a greenschist itself. A basaltic protolith would be a greenschist if the metapelitic schist contains Ms + Chl + Ctd + Qtz.

This facies roughly covers the temperature range 300–500 °C at low to intermediate pressures. The transition to the amphibolite facies is rather gradual. Above about 450 °C, metabasic rocks gradually develop hornblende and plagioclase. A transitional epidote-amphibolite facies can be found elsewhere in the literature. Here, it is included in the greenschist and amphibolite facies, respectively.

4.3.2.3
Amphibolite Facies

Under the P–T conditions of the amphibolite facies, metabasalt converts to amphibolite containing plagioclase + hornblende ± quartz. In amphibolite, hornblende is modally very abundant (typically >50%). At lower temperature in the amphibolite facies, epidote may still be present in amphibolite; garnet also typically occurs in many amphibolites. In the upper amphibolite facies, clinopyroxene may appear, however, together with garnet only at relatively high pressure. In any case, the diagnostic assemblage in basic rocks is Hbl + Pl. Other rocks such as metapelites contain a large variety of assemblages depending on the rock composition and the precise P–T conditions. The staurolite, kyanite and sillimanite Barrovian zones belong to this facies. Tremolite-bearing marbles are also characteristic. The key point also here, a metapelite or a calcsilicate rock or a meta-lherzolite is said to be of amphibolite facies grade or has been metamorphosed under amphibolite facies conditions if a meta-basalt (if present or not at the same locality) would have developed Hbl + Pl ± Qtz.

4.3.2.4
Granulite Facies

The granulite facies comprises the high-grade rocks formed at maximum temperatures of orogenic metamorphism. The reference rock of the facies concept, MOR-basalt, contains under granulite facies conditions clinopyroxene + plagioclase ± Qtz ± Opx. Note that Cpx + Pl was also the original igneous assemblage of basalt, here we are dealing with a new metamorphic Cpx + Pl pair. Pyroxene has replaced hornblende of the amphibolite facies.

There are several problems with the definition of the granulite facies and, in fact, the terms granulite and granulite facies are used in different ways by metamorphic petrologists. The problem has many different causes. The most obvious is that the pair Cpx + Pl also occurs in rocks that contain Hbl + Pl (i.e., in amphibolites). One could circumvent the problem by requiring that hornblende or any other hydrates such as micas should not be present in the granulite facies. Although this appears to be ingenious, the problem would be that hornblende (and biotite) persists to very high temperature (>950 °C) in metabasalts. Most petrologists would consider this temperature to be well within the granulite facies. In some mafic rocks and under certain conditions within the granulite facies, orthopyroxene forms in addition to Cpx + Pl. A metabasalt that contains Opx + Cpx + Pl is called a two-pyroxene granulite and there is general agreement that the assemblage is diagnostic for granulite facies conditions even if hornblende (and biotite) is present in the 2-Pyx-granulite. Unfortunately, the diagnostic assemblage Opx + Cpx + Pl in metabasalt is restricted to the low-pressure part of the granulite facies. At high pressures in the granulite facies (Fig. 4.2), Opx is replaced by garnet (Opx + Pl = Cpx + Grt). However, the assemblage Grt + Cpx + Pl in rocks that also contain Hbl may occur under P–T conditions of both the amphibolite and the granulite facies, it is not exclusive. Also note that, in the light of the comments made above, the upper limit of the amphibolite facies is difficult to define, since the pair Hbl + Pl occurs well within the granulite facies.

In addition, granulite facies rocks are strongly dehydrated; their formation should be promoted by low water pressure. Conditions of $P_{H_2O} < P_{total}$ have been indeed reported from many granulite facies terrains (however, water saturation is implicit in Fig. 4.2). High CO_2 concentrations in fluid inclusions and large amounts of such inclusions have been reported from many granulites worldwide (see references).

At high water pressure, rocks melt either partially or completely at P–T conditions of the granulite facies. Also, metabasalt melts partially if P_{H_2O} is high and mafic migmatite rocks result from this process. Many rock-forming processes in the granulite facies involve partial melting in the presence of an H_2O-rich vapor phase or in the absence of a free fluid phase.

Few subjects in petrology are as interesting and involved as the amphibolite-granulite facies transition. Perhaps, granulite facies conditions should not be tied to metabasaltic rock compositions because of the lack of truly unique assemblages in such rocks that are unequivocally diagnostic for the granulite facies. In fact, in Chapter 2, we defined the term granulite to include, both

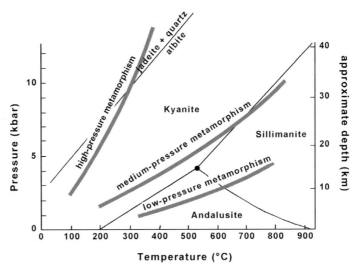

Fig. 4.3. Metamorphic facies series in relation to the Al_2SiO_5 polymorphs and the $Jd + Qtz = Ab$ equilibrium

of course, and result from the fact that each metamorphic belt has been subjected to its own unique set of physical conditions. The metamorphic facies series concept is useful because it links regional orogenic metamorphism to its geodynamic causes. In the broadest sense: metamorphism that follows LP-type facies series is typically related to extension tectonics and heat transfer by igneous intrusions. IP-type metamorphism is the typical collision belt metamorphism (Barrowian metamorphism). HP-type metamorphism normally results from subduction dynamics.

4.4
Isograds and Reaction Isograds

4.4.1
Origin and History of the Isograd Concept

In principle, an isograd is a line on a map connecting points of equal metamorphic grade. The isograd method has been applied by many geologists for several decades, but semantic confusion has arisen because its significance is more complex than was originally assumed. When coining the word "isograd", Tilley (1924, p. 169) referred to the boundaries of Barrow's mineral zones as a "... line joining the points of entry of... (an index mineral) in rocks of the same composition...". Tilley (1924) defined the term **isograd** "as a line joining points of similar P–T values, under which the rocks as now constituted, originated". This definition is inappropriate because in practice it can never be more than an inference; in addition, several arguments exist which indicate that the stipulation of "similar P–T values" may be incorrect (see below).

Later, it was proposed to relate an isograd to a specific metamorphic reaction across the "line". Since a metamorphic reaction depends on temperature, pressure, and fluid composition, an isograd will represent, in general, sets of P–T–X conditions satisfying the reaction equilibrium and not points of equal P–T–X conditions. Therefore, whenever the reaction is known, the term isograd should be replaced by the term reaction-isograd. A **reaction isograd** is a line joining points that are characterized by the equilibrium assemblage of a specific reaction. This line is not necessarily a line of equal or similar grade and therefore it is not an isograd in the proper sense.

4.4.2
Zone Boundaries, Isograds and Reaction Isograds

Tilley's definition of an isograd is identical to what we have called a mineral zone boundary earlier in this chapter, with the additional assumption of similar P–T values existing along this boundary. In order to avoid the latter statement, the terms mineral zone boundary or reaction-isograd will be used in this volume together with the term isograd. The distinction between a mineral zone boundary and a reaction-isograd should be made clear again. Mapping a mineral zone boundary is based on the first appearance of an index mineral, whilst locating a reaction-isograd involves mapping reactants

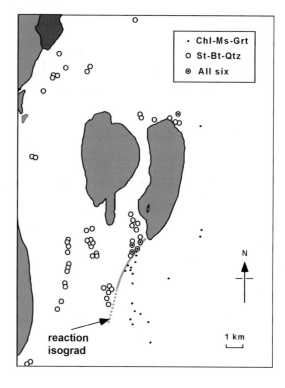

Fig. 4.4. Reaction-isograd based on the reaction Chl+Ms+Grt = St+Bt+Qtz+H$_2$O from the Whetstone Lake area, Ontario. After Carmichael (1970, Fig. 3). *Shaded fields* Rocks other than metapelites

and products of a chemical reaction in rocks (Fig. 4.4). Such reactions may be simple phase transitions such as Ky = Sill or Qtz = Coesite. The mapping geologist will then find an area with kyanite and another area with sillimanite. The line separating the two areas is the reaction-isograd of the Ky-Sill phase transition. Unfortunately, the mapping geologist will commonly find kyanite in sillimanite grade rocks as well. This complication arises from the fact that metastable persistence of reactants or products of a reaction on the "wrong" side of the line is widespread in geologic materials.

This problem is less severe if reaction-isograds involving hydrate minerals are mapped. At very low-grade, kaolinite decomposes to pyrophyllite in the presence of quartz. The corresponding reaction is $Kln + 2 Qtz = Prl + H_2O$. This reaction has been mapped as a reaction-isograd in the Central Alps and it was found that kaolinite never persisted in the pyrophyllite + quartz zone (Frey 1987). Ultramafic rocks are very well suited for reaction-isograd mapping because of simple mineral compositions. For example, in the contact aureole of the Bergell pluton, Trommsdorff and Evans (1972) mapped several reaction-isograds in serpentinites, e.g., the breakdown of antigorite $\Rightarrow$ talc + forsterite + H_2O. Trommsdorff and Evans (1977) mapped a series of reaction-isograds in complex ophicarbonate rocks in the same aureole. In fact, the later study even involved the mapping of isobaric invariant-reaction-isograds. Thus these isograds came very close to true isograds, i.e., lines on a map connecting rocks of the same metamorphic grade.

Reactions that involve minerals of variable composition may be complex and the mass balance coefficients are obtained by applying least-squares regression techniques, e.g., 3.0 Chl + 1.5 Grt + 3.3 Ms + 0.5 Ilm = 1.0 St + 3.1 Bt + 1.5 Pl + 3.3 Qtz + 10.3 H_2O (Lang and Rice 1985 a). The reaction-isograd related to this reaction separates an area on the map with Chl-Grt micaschists from an area with St-Bt schist. Many other excellent field studies have successfully mapped reaction isograds. The mapping of reaction isograds in the field is more time-consuming and more ambitious than the mapping of mineral zone boundaries. On the other hand, a reaction isograd provides more petrogenetic information for two reasons:

1. The position of a reaction-isograd is bracketed by reactants and products, but a mineral zone boundary is limited only towards higher grade and may possibly be displaced towards lower grade as more field data become available. As an example, the chloritoid mineral zone boundary of the Central Alps has been replaced by the reaction-isograd $Prl + Chl \Rightarrow Cld + Qtz + H_2O$, located some 10 km down grade (Frey and Wieland 1975).

2. A reaction-isograd may yield information about conditions of metamorphism provided P–T–X conditions of the corresponding metamorphic reaction are known from experimental or thermodynamic data. This information is dependent on the type of mineral reaction (Chap. 3), its location in P–T–X space, and the distribution of isotherms and isobars in a rock body, as detailed below.

4.4.3
Assessing Isograds, Isobars and Isotherms

Isograds have been used by mapping field geologists to bring information about the intensity of metamorphism onto a geological map. An isograd, a line connecting equal metamorphic grade on a map, is conceptually simple, in principle. However, many geologic features make it almost impossible to find these lines in the field in practice. The first problem is that rocks are collected at today's erosion surface of the earth and this sampling surface is unrelated to the metamorphic structure that was established many million years ago. The second point is that the intersection of an isobaric and an isothermal surface in 3D space is the true line connecting equal P–T conditions at a given instant in time. When later cut by the erosion surface, this line appears as a point on a map rather than a line! Most metamorphic rocks conserve the least hydrated state attained during their metamorphic evolution. This state is characterized by a unique P–T pair for each rock. Rocks with equal P–T conditions at the least hydrated state are on a line in 3D and a single point on the erosion surface. So, true isograds are, except for some special cases, related to contact aureoles, not mappable features in the field.

Because one would like to relate isograds to field studies the concept must work in the field. Now, consider a P–T diagram; here isotherms are perpendicular to isobars. If a metamorphic reaction was dependent on temperature only, the reaction-isograd would parallel an isothermal surface and it could be mapped in the field as a line connecting rocks that experienced the same temperature. Nothing could be said about the pressure during metamorphism. Similarly, a reaction dependent on pressure only would produce a reaction-isograd parallel to an isobaric surface. However, metamorphic reactions are, in general, dependent on both P and T and their equilibria have general slopes on a P–T diagram. Yet, many dehydration reactions are nearly isothermal over some pressure range, and this is also true for some polybaric traces of isobaric invariant assemblages in mixed volatile systems (see, e.g., tremolite-in curve in Fig. 6.6). We may conclude that, although most mapped reaction isograds are lines with variable P and T, this variation may be relatively small for many dehydration reaction based isograds.

Further insight into the significance of reaction-isograds with respect to the P–T distribution during orogenic metamorphism is obtained by considering the relative position of isotherms and isobars in a cross-section, termed a **P–T profile** by Thompson (1976). Let us assume that isobars are parallel to the earth's surface during metamorphism. In a stable craton or during burial metamorphism, isotherms are also parallel to the earth's surface, and, in this limiting case, P and T are not independent parameters; at a given pressure, the temperature is fixed, and reaction-isograds corresponding to mineral equilibria are isothermal and isobaric (Chinner 1966). During orogenic metamorphism, however, isotherms are rarely expected to be parallel to isobars. Furthermore, the temperature gradient – the increase in temperature perpendicular to isotherms; not to be confused with the geothermal gradient, the increase of temperature with depth, both expressed in °C/km – is not likely

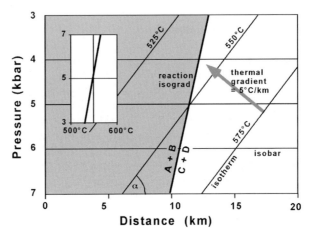

Fig. 4.5. Geometrical relations between a hypothetical mineral reaction A+B = C+D, isotherms, and isobars in a P–T profile. The mineral reaction has an assumed dP/dT slope of 8 °C/kbar as shown in the inset P–T diagram. Isobars are horizontal with a pressure gradient of 0.286 kbar/km (corresponding to a mean rock density of 2.8 g/cm^3). Isotherms are inclined with respect to isobars at an angle a of 60 ° with a thermal gradient of 5 °C/km. Note the relatively large angle between the mineral reaction and isotherms in the P–T profile

to be constant. Thompson (1976) and Bhattacharyya (1981) have analyzed the geometrical relations among reaction-isograds, the angle a between isotherms and isobars, and the temperature gradient. It was shown that the temperature gradient and the angle a are important in determining P–T distributions from reaction-isograd patterns besides, of course, the slope of the corresponding mineral reaction on a P–T diagram. As an example, for a given metamorphic reaction with a low dT/dP slope of 8 °C/kbar, if temperature gradients are low the angle between such reaction-isograds and isotherms can be large (Fig. 4.5).

From the preceding discussion it is clear that along a reaction isograd, both P and T will generally vary, and that "similar P–T values" as suggested by Tilley either require the special precondition of isotherms being parallel to isobars, or may be true for nearby localities only. Reaction-isograds intersecting at high angles, as described by Carmichael (1970), provide further convincing evidence for changing P–T–X conditions along such metamorphic boundaries.

Reaction-isograds mentioned so far refer to univariant or discontinuous mineral reactions. In the field, such reaction-isograds will define relatively sharp lines which may be mapped within some 10 to 100 m (e. g., Fig 4.4). Many metamorphic reactions are, however, continuous in nature due to extensive solid solution in minerals, e. g., Mg–Fe minerals in metapelites. Corresponding reaction-isograds are not sharp mappable lines in the field; instead, rocks containing the reactant or the product assemblage of the reaction will occur together for several 100 m or even km, because of the range of P–T conditions over which the continuous reaction occurs in rocks with different Mg/Fe ratios.

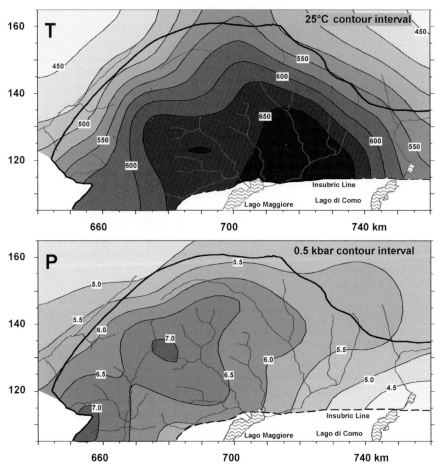

Fig. 4.6. Isotherm and isobar map of the Tertiary metamorphism of the Ticino area, Central Alps (Todd and Engi 1997). *Numbers on axes* refer to the Swiss coordinate network in km. The metamorphic pattern is typical of "Barrowian" regional orogenic metamorphism. It is the result of the late stage continent collision during formation of the Alps and has been modified by even younger processes particularly along the Insubric line (the major suture zone of the Alps). It is important to understand that the isotherms and the isobars are diachronous. The temperature at location xy is about maximum T during the last thermal event and the pressure at the same location is assumed to be P at T_{max}. The P–T data were derived from a statistical analysis of MET solutions (see Chap. 4.7.7) from a large number of samples

A reaction isograd line recorded on a map is only a two-dimensional representation of a reaction-isograd, which is a surface when viewed in three dimensions, as already noted by Tilley (1924). The shape of the reaction-isograd surfaces in orogenic belts is of great interest from a geodynamic point of view; however, unfortunately in many areas, such information is difficult to obtain because of poor relief (in addition to poor exposure or scarcity of rock horizons of suitable composition). A few examples are known, however, from the Alps where the shapes of reaction-isograds have been determined with some confidence (e.g., Bearth 1958; Fox 1975; Thompson 1976). But even for areas with a long-lasting erosion history and where the vertical dimension is not exposed, Thompson (1976) has described a method to reconstruct the distribution of isotherms in a portion of the earth's crust at the time of metamorphism from the mapped pattern of reaction-isograds.

The zone boundaries can be mapped by hand specimen examination directly in the field; reaction-isograd mapping normally requires thin sections and a microscope. If, in a regional-scale study mineral composition data are available from a large number of rocks of similar composition, then P–T data can be derived for each sample by the methods of thermobarometry (as will be discussed below). The only studies to date that have systematically analyzed P–T data from thermobarometry of metapelitic rocks on a regional scale are those by Engi et al. (1995) and Todd and Engi (1997) from the Tertiary orogenic collision type metamorphic structure of the Ticino area of the central Alps (Fig. 4.6). Engi et al. were able to map the metamorphic structure as intersections of 25-Ma-old isobaric and isothermal surfaces with the present-day erosion surface (iso-T and iso-P contours). It can be seen from this figure that in this area the location of the highest temperature does not coincide with the highest pressure. As an extremely instructive exercise you may display the P–T data along a number of chosen profiles on a P–T diagram and make your conclusions about the meaning of metamorphic field gradients. In this figure it is obvious that the intersection of, e.g., the 600 °C isotherm with, e.g., the 6 kbar isobar is a point on a map. This point is the "isograd" condition of equal metamorphic grade, all 600 °C–6 kbar pairs are on a 3D line that intersects the erosion surface at that point on the map.

4.5
Bathozones and Bathograds

This concept, proposed by Carmichael (1978), relies on the following considerations. On a P–T diagram, any invariant point in any model system separates a lower-pressure mineral assemblage from a higher-pressure assemblage. An invariant model reaction relates these two critical assemblages, which have no phase in common (i.e., these assemblages are, at least in part, different from divariant assemblages derived from a Schreinemakers' analysis). A **bathograd** is then defined "as a mappable line that separates occurrences of the higher-P assemblage from occurrences of the lower-P assemblage" (Carmichael 1978, p. 771). The field between two neighboring bathograds is termed a **bathozone**.

Carmichael considered five invariant points in a model pelitic system: Four are based on intersections of the Al_2SiO_5 phase boundaries with three curves referring to the upper thermal stabilities of St + Ms + Qtz and Ms + Qtz ± Na-feldspar, one invariant point corresponds to the Al_2SiO_5 triple point (Fig. 4.7). Carmichael's invariant reactions are as follows (with P increasing from left to right and in sequential order, and numbers referring to bathozones to be defined below):

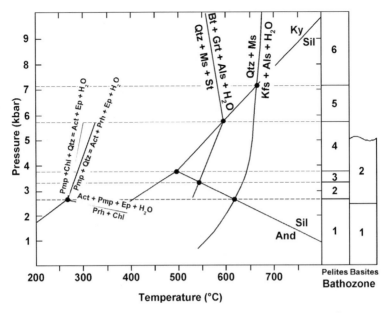

Fig. 4.7. P–T phase diagram for pelitic rocks divided into bathozones. (After Carmichael 1978, Fig. 2; Archibald et al. 1984, Fig. 2)

Bathograd	Invariant model reaction
1/2	$Kfs + And + H_2O = Qtz + Ms + Sil$
2/3	$Bt + Grt + And + H_2O = Qtz + Ms + St + Sil$
3/4	$And = Ky + Sil$
4/5	$Qtz + Ms + St + Sil = Bt + Grt + Ky + H_2O$
5/6	$Qtz + Ab + Ms + Sil = Kfs + Ky + granitic\ liquid$

Reactants and products of these reactions define diagnostic mineral assemblages of six bathozones. With the exception of bathozones 1 and 6, which are "open-ended" and therefore characterized by a single diagnostic assemblage, each bathozone is characterized by two diagnostic assemblages, one to constrain its lower-P boundary and the other to constrain its higher-P boundary:

Bathozone	Diagnostic assemblage		Pressure range
	Lower limit	Upper limit	
1		Kfs + And	P<2.2 kbar
2	Qtz + Ms + Sil	Bt + Grt + And	P = 2.2–3.5 kbar
3	Qtz + Ms + St + Sil	And	P = 3.5–3.8 kbar
4	Ky + Sil	Qtz + Ms + St + Sil	P = 3.8–5.5 kbar
5	Bt + Grt + Ky	Qtz + Ab + Ms + Sil	P = 5.5–7.1 kbar
6	Kfs + Ky (+ neosomes)		P>7.1 kbar

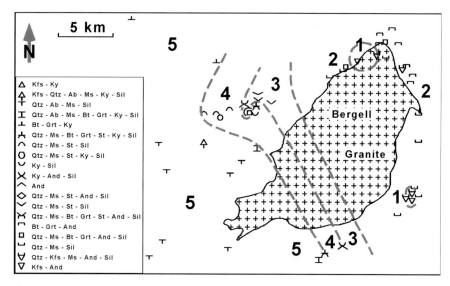

Fig. 4.8. Bathozones in metapelites from the Bergell Alps of Switzerland and Italy. (After Carmichael 1978, Fig. 6)

These bathozones are shown in Fig. 4.7. Note that the apparently precise pressure calibration is somewhat misleading; no errors in the equilibrium curves have been taken into account, and several of the reactions are continuous in natural rocks and also depend on the activity of water. In fact, the equilibrium $Qtz + Ms + St = Bt + Grt + Als + H_2O$ was located by Carmichael (1978, p. 793) to be "consistent with a huge body of field data" and later slightly adjusted by Archibald et al. (1984, Fig. 2), but is inconsistent with available experimental or thermodynamic data (see Chap. 7).

Nevertheless, Carmichael's scheme provides an elegant and simple way for reconnaissance geobarometry in metapelites of the amphibolite facies, requiring only thin-section observations. Figure 4.8 illustrates such an example for mapping of bathozones and bathograds. The Bergell pluton intruded in the Tertiary the nappe stack of the central Alps. The erosion surface intersects the structure in the eastern portion of the pluton at high levels in the crust (bathozones 1 and 2). A well-developed contact metamorphic aureole is present in the east and pressures range from 2–3 kbar. Towards the west, pressure rapidly increases and deeper parts of the pluton are exposed. Bathozones increase progressively from 1 through 5 from east to west. Bathozone 5 defines a minimum pressure of about 6 kbar along the western contact of the pluton. Consequently, the narrow contact metamorphic aureole widens towards the west and grades into the regional orogenic metamorphic terrain of the central Alps in the Ticino area.

Powell et al. (1993) introduced a bathograd and two bathozones for low-grade metabasites. The assemblages $Prh + Chl$ and $Act + Pmp + Ep + Qtz$ provide low-pressure and high-pressure constraints, respectively, on a bathograd that corresponds to a pseudo-invariant point at which chlorite, epidote, prehnite, pumpellyite, actinolite, quartz and water coexist. P–T coordinates of the invariant point determined for metabasites of the Abitibi greenstone belt, Canada, suggest that this sub-greenschist bathograd corresponds approximately to the Sil-Ms-Qtz bathograd, separating bathozones 1 and 2 of Carmichael (1978), for amphibolite facies metapelites. Accordingly, the lower-pressure bathozone, as defined in the low-grade metabasites, is referred to as bathozone 1 and the higher-pressure bathozone as bathozone 2 (Fig. 4.7).

4.6
Petrogenetic Grid

P–T curves of different mineral equilibria generally have different slopes and will therefore intersect. Such intersecting curves will thus cut a P–T diagram up into a grid which Bowen (1940, p. 274) called a "petrogenetic grid". His idea was to construct a P–T grid with univariant reaction curves bounding all conceivable divariant mineral assemblages for a given bulk chemical composition. Each mineral assemblage would lie in a unique P–T pigeonhole and inform the investigator immediately as to the conditions of metamorphism. When proposing the concept, Bowen regarded it rather as a vision than a tool, realizing that "the determinations necessary for the production of such a grid constitute a task of colossal magnitude..." (Bowen 1940, p. 274). In general, a **petrogenetic grid** is a diagram whose coordinates are intensive parameters characterizing the rock-forming conditions (e.g., P, T) on which may be plotted equilibrium curves delimiting the stability fields of specific minerals and mineral assemblages.

Early petrogenetic grids consisted of a few experimentally determined mineral equilibria belonging to different chemical systems, and thus the geometrical rules of Schreinemakers could not be applied. This resulted in rather vague conclusions with regard to the conditions of formation of associated rocks. Today, univariant curves delineating stability limits of many end member minerals and mineral assemblages have been determined by experimental studies. Furthermore, internally consistent sets of thermodynamic data are available that include many rock-forming minerals, making it possible to calculate petrogenetic grids with all stable reaction curves of some model systems. Such an approach will be extensively used in Part II of this volume.

However, even after five decades of extensive experimental and theoretical effort, the goal of a wholly comprehensive petrogenetic grid has still not been achieved. The main limitation is due to widespread solid solution in minerals like chlorite, mica, garnet, pyroxene, and amphibole, and one of the main thrusts in future studies will be the evaluation of activity-composition relations for such phases. Nevertheless, available petrogenetic grids for simple end-member systems provide important constraints on the P–T conditions of metamorphism in chemically complex systems. A compilation of some published petrogenetic grids is listed in Table 4.2.

Table 4.2. Information on selected petrogenetic grids

System components/rock type	Phase components (excess phases are underlined)	P–T range	Substitutions considered[a]	Reference
CFMASH/ ultramafics	Amph, Brc, Cal, <u>Chl</u>, Cpx, <u>Opx</u>, Ol, <u>Dol</u>, Mgs, <u>Qtz</u>, Srp, Tlc, <u>Fluid</u> (H_2O-CO_2)	2 kbar/ 440–600 °C	$FeMg_{-1}$ Tsch	Will et al. (1990 a)
CMSH-CO_2/ carbonates	Atg, Ath, Brc, Cal, Di, Dol, En, Fo, Mgs, Per, Qtz, Tlc, Tr, Fluid (H_2O-CO_2)	0–10 kbar/ 400–800 °C		Carmichael (1991)
CMSH-CO_2/ carbonates	Atg, Cal, Di, Dol, En, Mgs, Qtz, Tlc, Tr, ±Fluid (H_2O-CO_2)	1.5–14 kbar/ 465–715 °C		Connolly and Trommsdorff (1991)
CMASH-CO_2/ carbonates	Cal, Chl, Di, Dol, Fo, Qtz, Spl, Tr, Fluid (H_2O-CO_2)	1 kbar/ 400–640 °C; 4 kbar/ 630–700 °C	Fixed Tr activity	Chernosky and Berman (1988)
CFMASH-CO_2/ carbonates and basites	Amph, An, And, Cal, Chl, Cld, Czo, Dol, Grt, Mrg, Qtz, Tlc, <u>Fluid</u> (H_2O-CO_2)	2 and 5 kbar/ 400–500 °C	$FeMg_{-1}$/Tsch	Will et al. (1990b)
NCMASH/basites	Ab, An, Chl, Czo, Di, Gln, Jd, Lws, Pg, Pmp, Prp, Qtz, Tr, Fluid (H_2O)	2–20 kbar/ 250–600 °C	Activity-corrected/curves	Evans (1990)
NCMASH/basites	Ab, An, Cln, Gln, Grs, Hul, <u>Jd</u>, Lmt, Lws, Pg, Prh, Pmp, Qtz, Stb, Tr, Wr, Zo, <u>Fluid</u> (H_2O)	0.1–10 kbar/ 0–500 °C	Activity-corrected/curves	Frey et al. (1991)
FMASH/pelites	Car, Chl, Cld, Kln, Ky, Prl, Qtz, Sud, Fluid (H_2O)	3–15 kbar/ 150–500 °C	Activity-corrected/curves	Vidal et al. (1992)
KFMASH/pelites	And, Bt, Chl, Cld, Crd, Grt, Ky, <u>Ms</u>, <u>Qtz</u>, Sil, St, <u>Fluid</u> (H_2O)	0–18 kbar/ 460–720 °C	$FeMg_{-1}$/Tsch	Powell and Holland (1990)
NFMASH/pelites	Ab, Car, Chl, Cld, Gln, Grt, Jd, Ky, Pg, <u>Qtz/Coe</u>, Tcl, Fluid (<u>H_2O</u>)	5–50 kbar/ 300–850 °C	$FeMg_{-1}$/Tsch	Guiraud et al. (1990)
MnNKFMASH/ pelites	Ab, And, Bt, Chl, Cld, Crd, Grt, Kfs, Ky, Ms, Prl, Qtz, Sil, St, <u>Fluid</u> (H_2O)	3,5,7 kbar/ 400–650 °C		Symmes and Ferry (1992)
KFMASH/pelites	Bt, Crd, Grt, Kfs, Opx, Osm, Qtz, Sil, Melt	5–12.5 kbar/ 840–1000 °C	Fixed XMg	Carrington and Harley (1995)
MnKFMASH	And, Bt, Chl, Cld, Crd, Grt, Ky, Ms, Qtz, Sil, FeMn	0–20 kbar/ 450–700 °C	FeMn, Mn	Mahar et al. (1997)

Table 4.2 (continued)

System compo-nents/rock type	Phase components (excess phases are underlined)	P–T range	Substitutions considered[a]	Reference
CMASH/ ultramafics	Am, An, Chl, Cpx, Fo, Grt, Opx, Qtz	3–22 kbar/ 700–950 °C	Ca	Schmädicke (2000)
NCMASH/ ultramafics	Am, Chl, Cpx, Fo, Grt, Opx, Pl	3–22 kbar/ 700–950 °C		
NCKFMASH/ pelites	Bt, Chl, Grt, Ky, Mrg, Ms, Pg, Pl, Qtz, St, Zo	2–12 kbar/ 525–700 °C	CaFe	Worley and Powell (1998 a,b, 2000)
NCFMASH/ basites	Am, Chl, Cld, Cp, Grt, Ky, Lws, Mrg, Omp, Pg, Pl, Prl, Qtz, Sil, Tlc	4–23 kbar/ 380–620 °C	Ca	Will et al. (1998)

[a] Tsch = Tschermak exchange, $MgSiAl_{-1}Al_{-1}$.

4.6.1
Polymorphic Transitions

Examples of polymorphic transitions in metamorphic rocks include andalusite-kyanite-sillimanite, calcite-aragonite, quartz-coesite, and diamond-graphite. Theoretically, these simple conversions correspond to univariant reactions which are relatively well known in P–T space (Fig. 4.9). Due to the small energy changes involved in these reactions, however, such equilibria may become divariant through the preferential incorporation of minor element contents, by order-disorder relations, by kinetic factors, or by the influence of strain effects. Neglecting these complications for the moment, polymorphic transitions are easy to apply to rocks because the mere presence of an appropriate phase (or pseudomorphs thereafter) is sufficient to yield some P–T information. A disadvantage is that only P–T limits are usually provided, and absolute P–T values can be determined only in rare cases where univariant assemblages are preserved. In principle, using polymorphic transitions as P–T indicators is related to the petrogenetic grids rather than to thermobarometry proper.

4.6.1.1
Andalusite–Kyanite–Sillimanite

Because of their common occurrence in peraluminous rocks, the three Al_2SiO_5 polymorphs have been widely used as index minerals, for the definition of facies series (Sect. 4.3), and for estimating P–T in metamorphic rocks (e.g., Sect. 4.5). The system Al_2SiO_5 has an eventful history of experimentation (for a review, see, e.g., Kerrick 1990). In the 1970s and 1980s, most petrologists referred to the Al_2SiO_5 triple point of Richardson et al. (1969), 5.5 kbar and 620 °C, or that of Holdaway (1971), 3.8 kbar and 500 °C. Both determinations were obtained in a hydrostatic pressure apparatus. The main difference between these two studies concerns the And = Sil equilibrium. Richardson et al. used fibrolitic sillimanite whilst Holdaway experimented with coarse-grained sillimanite, and Salje (1986) suggested that the different thermodynamic properties of these materials account for the experimental discrepancies.

Bohlen et al. (1991) performed the most recent phase equilibrium experiments for the system Al_2SiO_5. Combining the dP/dT slopes of the equilibria Ky = Sil and Ky = And, these authors obtained a revised triple point at 4.2 ± 0.3 kbar and $530 \pm 20\,°C$. Considering the low angle of intersection of the two equilibria mentioned above, the small uncertainty appears rather optimistic. Pattison (1992) combined field data from a contact aureole with available experimental and thermodynamic data in order to constrain the triple point. Firstly, the P–T position of the And = Sil equilibrium in the aureole was estimated. Secondly, this equilibrium was extrapolated, using entropy and volume data, to intersect the Ky = And and Ky = Sil equilibria, giving an estimate of the triple point of 4.5 ± 0.5 kbar and $550 \pm 35\,°C$.

Both experimental and field evidence indicate that Fe^{3+} and Mn^{3+} have an effect on the stability relations of the Al_2SiO_5 polymorphs (see Kerrick 1990 for details). Grambling and Williams (1985) have shown that these transition metals stabilized the And-Ky-Sil assemblage, in apparent chemical equilibrium, across a P–T interval of 500–540 °C, 3.8–4.6 kbar.

In summary, it is clear that a single Al_2SiO_5 triple point exists on petrogenetic grids only, but, depending on the degree of "fibrolitization" of sillimanite and the transition metal content of the aluminosilicate, some variation in the location of the triple point is to be expected in nature. Nevertheless, the 4.5 kbar and 550 °C triple point is consistent with field data worldwide and the P–T coordinates are accepted by the petrologic community.

4.6.1.2
Calcite–Aragonite

At ambient P–T, the stable polymorph is calcite; at high pressure, however, aragonite is the stable Ca-carbonate (Fig. 4.9). Consequently, the presence of aragonite may be a valuable pressure indicator in rocks of the blueschist facies. Interestingly, aragonite is widespread at the earth's surface in marine animal shells and spring sinters, etc. This aragonite is metastable relative to calcite; it forms, e.g., by biological activity.

The large number of experimental attempts to define the stable polymorphic transition Cal = Arg are thoroughly reviewed by Newton and Fyfe (1976); see also Carlson (1983). Only small discrepancies exist between results obtained in the pure $CaCO_3$ system by different workers in the late 1960s and early 1970s. The effects of solid solution were evaluated for Mg and Sr. Common Mg and Sr contents in metamorphic carbonates resulted in small displacements of the equilibrium boundary, not exceeding a few hundreds of bars.

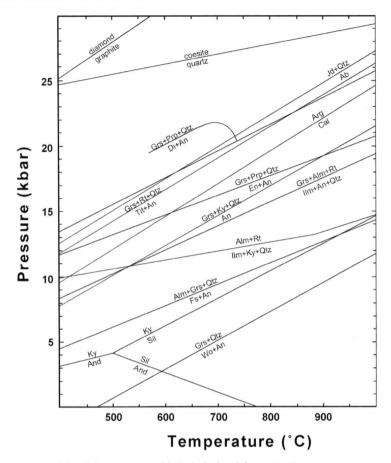

Fig. 4.9. Some solid–solid reaction equilibria (calculated from HP98)

Application of the simple equilibrium reaction Cal = Arg to metamorphic rocks is, however, complicated by several factors. First, aragonite is a widely distributed phase in some high-pressure metamorphic rocks of the Franciscan formation in California, but is uncommon in other blueschist terranes. These field observations have been explained by high reaction rates, meaning that aragonite-bearing rocks en route to the surface must enter the calcite stability field between 125 and 175 °C (Carlson and Rosenfeld 1981), and this aragonite will be transformed into calcite if it crosses the equilibrium curve at higher temperatures. Secondly, the metastable growth of aragonite in modern and ancient oceans and its subsequent conversion to calcite is a well-known phenomenon, and one wonders whether metamorphic aragonite might form outside its stability field as well. It has been shown that aragonite may form instead of calcite at much lower pressures if it is precipitated from fluids with other dissolved ions, rather than from solution in pure water. The frequent occurrence of aragonite in the Franciscan formation as coarse crystals in late-stage veins suggests that it may often form in a solution-precipitation process. The question remains open whether such metastable aragonite grows at lower pressure outside its stability field. Thirdly, severely deformed calcite will accumulate strain energy, and Newton et al. (1969) were able to grow aragonite at the expense of calcite strained by prolonged mortar grinding at pressures several kilobars below the aragonite stability limit. This suggests that deformational energy can be a significant parameter in Cal-Arg stability considerations.

In summary, more experimental and fieldwork will be needed to establish definitely the circumstances under which aragonite forms in nature. Where associated with other high-pressure minerals (e.g., jadeite), it is likely that aragonite occurring in metamorphic rocks is an indicator of high pressure.

4.6.1.3
Quartz–Coesite

Coesite, the high-pressure polymorph of SiO_2, has been found at many locations worldwide. Its presence in eclogites and subducted supracrustal rocks (for references, see Reinecke 1991 and Schmädicke 1991) has generated great interest in the quartz-coesite transition. The Qtz–Cs equilibrium has been well located experimentally as a function of P and T (Bohlen and Lindsley 1987, Fig. 1), and it was observed in the experiments that conversion of one polymorph to the other was rapid. It is thus astonishing that coesite survived the long return trip from great depth to the surface of the earth. Under static conditions, the experimental results imply unusually high metamorphic pressures of 25–30 kbar for stable coesite to form. Depending on temperature, this pressure corresponds to depths in excess of 75 km. However, it is crucial to know whether coesite formed under static conditions or not, because it has been shown experimentally that coesite can grow from highly strained quartz as much as 10 kbar below the transition determined under static conditions. From the description of several coesite localities it appears that strain energy was considered to be an irrelevant factor, and that the presence of coesite actually indicated very high pressures of formation.

4.7
Geothermobarometry

The pressure and temperature at which rocks formed can be accessed by various methods. Finding the pigeonhole for a characteristic mineral assemblage on a petrogenetic grid is one popular possibility (see Sect. 4.6). An even more powerful technique to gather P–T data for metamorphic rocks is geologic thermometry and barometry or, in short, **geo-thermo-barometry (GTB)**. The P–T conditions of the least hydrated state of a rock along a metamorphic evolution path hold much geologic information and are of great geodynamic significance. These conditions can be inferred from mineral equilibria (stable or metastable) together with the chemical composition of minerals in a specific sample, especially from the distribution of elements between coexisting minerals in that sample.

During the last three decades, laboratory experiments, thermodynamic modeling and calculations, and mineral analyses with the aid of modern electron microprobe instruments have led to a much better understanding of P–T conditions prevailing during metamorphism. Geothermobarometry made enormous progress in the 1980s and has evolved into a standard technique in metamorphic petrology. General reviews on this topic were provided by Essene (1982), Bohlen and Lindsley (1987), Essene (1989) and Yardley and Schumacher (1991). More specific reviews refer to the geobarometry of granulites (Bohlen et al. 1983; Newton 1983) and the geothermobarometry of eclogites (Newton 1986; Carswell and Harley 1989).

Today, several sophisticated software packages greatly facilitate the use of geothermobarometry and make the technique a very general tool. Notably, the program "Thermobarometry" by Spear et al. (1999) is extremely user-friendly and the program's continuous updates incorporate latest developments in the field (we use GTB V.2.1 2001). The program can be downloaded from Prof. F. Spear's website and a comprehensive manual and tutorial are provided by the authors. The web address is given below (you may also find it via our software link at: minpet.uni-freiburg.de): http://ees2.geo.rpi.edu/MetaPetaRen/Frame_research.html.

4.7.1
Concept and General Principle

The basic idea of geothermobarometry is simple: (1) find a suitable equilibrium phase assemblage in a rock that allows you to formulate two reactions among the phase components, and (2) solve for the equilibrium pressure and temperature conditions of the two reactions. The equilibrium conditions of the two reactions are two lines on a P–T diagram; the P–T intersection indicates the conditions of simultaneous equilibrium of the two reactions in one rock and, hence, the values of P–T at which the assemblage equilibrated.

The two reactions must be independent on the composition of a now vanished fluid phase. Consequently, no dehydration reactions or other reactions that involve volatile species can be used in thermobarometry. There are no exceptions to this rule. This does not mean that only anhydrous minerals are suitable for thermobarometry. It means that if hydrates participate in the reactions, OH groups of hydrous minerals must be conserved among the solid phases.

The most commonly used GTB (geothermobarometry) applications utilize one exchange reaction and one net-transfer reaction to determine the P–T intersection and so the condition of "rock formation". In Chapter 3, the following reaction was used as an example of a net-transfer reaction:

Grs + Qtz = An + 2 Wo

The reaction has some relevance for calcsilicate rocks and is used in the following to illustrate the concept of GTB. The equilibrium conditions of this

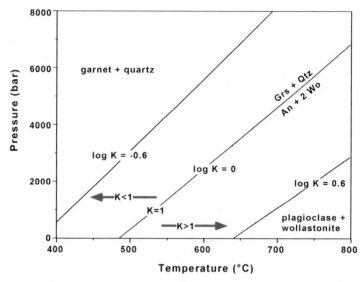

Fig. 4.10. Equilibrium conditions of the Grs+Qtz = An+2 Wo (GWAQ) reaction for three different values of log K. The log K = –0.6 contour corresponds to An$_{20}$ and Grs$_{80}$ composition of Pl and Grt using the solution model a = X (thermodynamic data of phase components from Berman 1988)

reaction can be calculated from thermodynamic data and are shown in Fig. 4.10. At low temperature grossular + quartz is stable, at high temperature the products of the reaction, anorthite + wollastonite. In contact aureoles, calcsilicate rocks may contain all four minerals, Grs-rich garnet + plagioclase + wollastonite + quartz. Assuming equilibrium among the phase components of the reaction given above, we may write an expression for the equilibrium constant:

$$\log K_{PT} = \log a_{An} - \log a_{Grs}$$

The microprobe analysis of coexisting plagioclase and garnet in a sample has been recalculated to, for instance, $X_{An} = 0.2$ and $X_{Grs} = 0.8$. If we find a function a = f(X), called a solution model, the required activities of the phase components and the equilibrium constant can be calculated. The simplest function is a = X. It serves the purpose of illustrating the principle here and the resulting log K = –0.6. The solution model also implies $a_{Wo} = a_{Qtz} = 1$ and the corresponding log a_i of Wo and Qtz are equal to zero.

Then follows the second step of thermobarometry, solving for P and T. For this purpose, remember the equation from Chapter 3:

$$\log K_{PT} = -0.6 = -\Delta G_r^\circ / RT.$$

The problem is to find all values of P and T, where ΔG_r° has values that satisfy this equation. The equation represents the equilibrium condition of the re-

Fig. 4.11. ln K contours for the equilibrium conditions of the GRAIL reaction

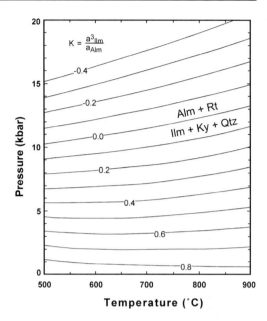

$$K = \frac{a^3_{Ilm}}{a_{Alm}}$$

action. In other words, we need an equation that permits us to calculate the right-hand side of the equation at any values of P and T. The most popular form of such an equation is:

$$\log K_{PT} = -(A/T) + B + (C/T)P$$

In this equation the coefficients A, B, and C are constants, specific to a thermometer or barometer, that have been determined experimentally or were calculated from thermodynamic data. The equation describes a straight line in a P–T diagram with the slope (B/C) and the intercept –(A/C) for a given value of log K (Fig. 4.10). The equation represents a family of lines for specific values of log K, one of which is log K=0, the equilibrium among end-member phase components An and Grs. Another line stands for log K=–0.6, the equilibrium between An_{20} and Grs_{80} in the presence of Qtz + Wo. There is, of course, an infinite number of lines for all possible values of log K. For practical purposes, one can contour a P–T frame by a series of lines with constant log K. Such a figure permits direct graphic application of a thermobarometer (e.g., Fig. 4.11).

From the above, it follows that an equilibrium with a steep slope in P–T space depends mostly on T, one with a flat slope mainly on P. The first type of reaction is a preferred thermometer, the second type an excellent barometer. The slope (B/C) is, in fact, the so-called Clausius-Clapeyron slope given by $\Delta S_r/\Delta V_r$. An excellent thermometer has a large entropy change and a small volume change of reaction; a good barometer is one with a large volume change. From Chapter 3 we may recall that exchange reactions are typically associated with small ΔV_r and make good thermometers. Net-transfer reactions are the prime candidates for barometers.

After having solved the above thermobarometer for all possible P–T solutions, we know that the equilibrium conditions of the GWAK reaction in our rock can be anywhere along a line in P–T space. In order to find a P–T point that we then may associate with the least hydrated state or loosely speaking with the "peak conditions" of metamorphism, we need a second geothermobarometric equilibrium in our rock sample. The equilibrium conditions of this second reaction represent a second line in P–T space.

In general, two (straight) lines intersect somewhere on a P–T diagram and great care and geological common sense must be used in the interpretation of such P–T data. Never think the numbers are "true" because they are derived from a "clever" computer program. Easy cases are intersections at negative pressure or negative temperature. They indicate calculation errors or very severe disequilibrium. More difficult to detect are erroneous P–T solutions in plausible P–T portions of P–T space. Some of the pitfalls of GTB are briefly discussed below.

4.7.2
Assumptions and Precautions

Successful geothermobarometry relies on several assumptions and many pitfalls have to be avoided; some of these are mentioned briefly below. Recent studies dealing with errors and difficulties in geothermobarometry include Carson and Powell (1997) and Holdaway (2001).

4.7.2.1
Question of Equilibrium

The basic assumption of thermobarometry is that a mineral assemblage formed at equilibrium, but it is impossible to prove that minerals in an individual rock ever achieved equilibrium (Chaps. 2 and 3). It is highly desirable, on the other hand, to find different generations of sub-assemblages in the same rock, each hopefully preserved in a state of small-scale domain equilibrium. It might then be possible to infer the pressures and temperatures at various stages in the history of a rock, and to reconstruct its pressure-temperature-time path. As an example, St-Onge (1987) found zoned poikiloblastic garnets with inclusions of biotite, plagioclase, quartz, and less commonly Al_2SiO_5. This enabled simultaneous determination of successive metamorphic P and T conditions from core to rim of a single garnet grain, using the garnet-biotite (Fe-Mg) exchange reaction and the garnet–Al_2SiO_5–plagioclase–quartz solid–solid reaction. Over the years, many studies have appeared in the literature that successfully deduced P–T paths taken by a single sample from thermobarometric data (see references).

4.7.2.2
Retrograde Effects

If an equilibrium assemblage has remained unchanged, determined P–T values presumably refer to peak temperature conditions (= least hydrated state; see Chap. 3). Such situations may be nearly valid for some quickly cooled or low-temperature rocks, but are not realized in slowly cooled rocks formed at high temperatures (higher amphibolite facies and granulite facies). In some cases, retrogression is obvious from thin-section examination, e. g. chloritization of biotite and garnet, pinitization of cordierite, or exsolution in feldspars, carbonates, oxides, and sulfides. In other cases, however, retrograde diffusion may be detected only by careful microchemical analysis. The following two examples serve to demonstrate how such problems may be overcome. Bohlen and Essene (1977) used feldspar and iron–titanium–oxide equilibria to determine metamorphic temperatures in a granulite facies terrain. Exsolution of albite in alkali feldspar and of ilmenite in magnetite was observed, and geologically reasonable temperatures were obtained only after reintegration of exsolved components. Edwards and Essene (1988) applied two-feldspar and garnet–biotite thermometry to an amphibolite–granulite facies transition zone, whereby biotite-garnet pairs gave erratic temperatures compared with two-feldspar temperatures. In detail, it was found that some garnets were homogeneous in composition, and some were zoned with rapidly changing composition near the rim but with sillimanite included. Biotites had different compositions depending on whether they were included in garnet, touching but not included in garnet, or not touching garnet. In addition, included and touching biotites from the same sample had highly variable Fe/Mg ratios, whereas biotites not touching garnet tended to have the same composition. These observations were explained by the retrograde exchange of Fe and Mg between biotite and garnet and by the retrograde hydration reaction $Grt + Kfs + H_2O = Bt + Sil + Qtz$. It is clear that the use of biotite not touching garnet and garnet cores should have resulted in peak metamorphic temperatures, but this was not the case with the calibrations applied.

4.7.2.3
Quality of Calibration

The basic reaction or equilibrium must be well calibrated, either through experimental or thermodynamic data. The experiments that are used to calibrate a given system should be well reversed and not rely on synthesis runs only. Furthermore, starting materials and run products should be carefully characterized.

4.7.2.4
Long Extrapolations in P–T

Some geothermobarometers were calibrated at much higher P–T conditions than those of metamorphism, generating large errors if long extrapolations are required. As an example, the reaction Grs + 2 Ky + Qtz = 3 An has been determined experimentally over the temperature range 900–1600 °C, but can be applied at temperatures as low as 500 °C. This problem may be diminished if an "accurate" slope calculation is feasible, but calculated uncertainty limits may turn out to be disappointingly large (see, e.g., Hodges and McKenna 1987, Fig. 4).

4.7.2.5
Sensitivity of Thermobarometry

Some systems are sensitive over a restricted P–T range only. Many solvus thermometers, e.g., are useful at relatively high temperatures, but less so at relatively low temperatures where mineral compositions are located on the steep limbs of the solvus. In such cases, even small errors involved in the location of the solvus as well as in the mineral analyses will produce large errors in inferred temperatures. In contrast to solvus thermometry, fractionations of oxygen isotopes between coexisting minerals are most sensitive at low temperatures, because fractionations decrease to almost zero at high temperatures (800–1000 °C).

4.7.2.6
Variable Structural State

Many minerals have variably ordered cation distributions, but the extent and temperature dependence of the ordering process are presently not well known for most phases. In an ideal case, the synthetic products should have the same structural state as the metamorphic minerals had at the time of metamorphism; of course, the structural state of disordered metamorphic minerals may reset upon cooling. In using feldspar thermometry, for instance, one must decide whether the metamorphic feldspars were ordered, partially disordered or completely disordered.

4.7.2.7
Effect of Other Components

Most geothermobarometers are based on mineral equilibria that were calibrated using simple mineral compositions. Most minerals in rocks contain additional components and form complex solid solutions. Thermodynamic solution models can account for these compositional complexities. This requires the accurate knowledge of activity–composition relations. The activities of all phase components of a reaction equilibrium must be calculated from mineral composition data that have been analyzed with an electron microprobe. The activities are inserted into an equilibrium constant expression that can be solved for P and T. The activity–composition function, $a = f(x)$, is commonly known as the solution model. Unfortunately, the connection between activity and composition is often complex and only poorly known for most minerals. As a consequence, there is often disagreement among geologists about the most suitable solution model for a given mineral species. This has led to the widespread practice of reporting multiple P–T estimates reflecting the effects of different solution models. For example, for the garnet-biotite thermometer (see below), at least seven calibrations based on the experimental study of Ferry and Spear (1978) are currently available. "In some cases, P–T estimates calculated using different solution models can vary widely, implying that uncertainties in solution behavior constitute a major source of error in thermobarometry. In general, this error cannot be quantified because the assumptions inherent in different models are often mutually exclusive" (Hodges and McKenna 1987, p. 672).

If the composition of a metamorphic mineral (e.g., the garnet in your sample) deviates strongly from that used in the calibration experiment for the geothermobarometer, then the effect of "other" components will lead to large errors of extrapolation. For instance, the garnet-Al_2SiO_5–quartz–plagioclase thermobarometer is commonly applied to metapelites containing garnets with only 5–10% grossular component and plagioclase with only 10–30% anorthite component. However, the experimental calibration is available for the end-member equilibrium Grs + 2 Als + Qtz = 3 An.

4.7.2.8
Estimation of Fe^{2+}/Fe^{3+} in Mineral Analysis

The electron microprobe renders information on total iron content, and the ferric and ferrous iron contents must be calculated assuming charge balance and some site occupancy model for a mineral. For relatively simple minerals with low ferric iron (e.g., pelitic garnets, some pyroxenes), such calculations can be done in principle. For other, more complex minerals with partially vacant cation sites and variable H_2O content (e.g., amphiboles and micas), such calculations are often questionable or even meaningless. Meaningful calculation of ferric iron in garnet and pyroxene (and spinel, etc.) requires extremely carefully produced microprobe data of outstanding quality. Routine data are not good enough! In a given sample, calculated Fe^{3+}/Fe^{2+} ratios in Grt and Pyx, for example, should not vary by more than ~5%. The redox conditions do not normally vary from grain to grain or within a grain of a given sample. A larger variation is not permitted because we claim that the sample reached an equilibrium state. This is a simple, but very restrictive, quality test. If the variation is larger than that, then you should use the best guess $Fe_{total} = Fe^{2+}$. Spinels are notoriously difficult because of the common presence of vacancies and of Fe^{3+}; few spinels are simple hercynite–spinel solutions, assuming $Fe_{total} = Fe^{2+}$ is recommended for amphiboles, micas and other complex minerals. However, if excellent garnet and pyroxene data from a sample indicate the presence of some ferric iron, one might consider a fixed Fe^{3+}/Fe^{2+} in coexisting amphibole and mica. In any case, Fe^{3+}/Fe^{2+} calculations from probe data need to be conducted with great care and the data should be used thoughtfully (see also Schumacher 1991).

4.7.3
Exchange Reactions

Exchange equilibria involve interchange of two similar atoms between different sites in one mineral (intracrystalline exchange) or between two minerals (intercrystalline exchange). The atoms may be elements of equal charge and similar ionic radius or isotopes of the same element. Because the volume changes involved are very small and the entropy changes are relatively large, exchange reactions are largely independent of pressure and have a good potential as thermometers. Unfortunately, retrograde exchange by diffusion below peak metamorphic conditions occurs very easily in many cases and generally leaves no textural evidence. This is especially true for intracrystalline cation distributions, for the diffusion paths are but a fraction of a unit cell.

Table 4.3. Some Fe–Mg exchange reactions used for geothermometry

No.	Mineral pair	Exchange reaction	Range of application [a]	Reference [b]
1	Ol-Spl	Fo+Hc = Fa+Spl	GRE, AMP, GRA	Engi (1983)
2	Ol-Opx	Fo+Fs = Fa+En	GA, ECL	Docka et al. (1986); Carswell and Harley (1989)
3	Ilm-Opx	Gei+Fs = Ilm+En	GRA	Docka et al. (1986)
4	Ilm-Cpx	Gei+Hd = Ilm+Di	GRA, AMP	Docka et al. (1986)
5	Ilm-Ol	Gei+Fa = Ilm+Fo	GRA	Docka et al. (1986)
6	Opx-Cpx	En+Hd = Fs+Di	GRA	Stephenson (1984); Docka et al. (1986)
7	Opx-Bt	En+Ann = Fs+Phl	GRA	Fonarev and Konilov (1986); Sengupta et al. (1990)
8	Crd-Spl	Mg-Crd+Hc = Fe-Crd+Spl	GRA	Vielzeuf (1983)
9	Grt-Ol	Prp+Fa = Alm+Fo	GRA, ECL	Carswell and Harley (1989)
10	Grt-Opx	Prp+Fs = Alm+En	GRA, ECL	Carswell and Harley (1989); Bhattacharya et al. (1991)
11	Grt-Cpx	Prp+Hd = Alm+Di	GRA, ECL, ±AMP	Carswell and Harley (1989); Pattison and Newton (1989)
12	Grt-Crd	Prp+Fe-Crd = Alm+Mg-Crd	GRA	Bhattacharya et al. (1988)
13	Grt-Bt	Prp+Ann = Alm+Phl	AMP, GRE, ±GRA	Thoenen (1989)
14	Grt-Phe	Prp+ Fe-Cel = Alm+Mg-Cel	ECL, BLU	Carswell and Harley (1989)
15	Grt-Hbl[c]	Prp+Fprg = Alm+Prg	AMP, ±GRA	Graham and Powell (1984)
16	Grt-Chl	Prp+Fe-Chl = Alm+Mg-Chl	GRE, BLU, ±AMP	Laird (1988); Grambling (1990)

[a] Abbreviations for metamorphic facies: *AMP* amphibolite; *BLU* blueschist; *ECL* eclogite; granulite; *GRE* greenschist.
[b] One or two recent references for each reaction are given only.
[c] Empirical calibrations; all other calibrations were determined by experiment or calculation.

Therefore, intracrystalline exchange reactions are unsuitable for peak metamorphic thermometry and will not be considered further.

The most widely applied exchange thermometers in metamorphic rocks involve Fe^{2+} and Mg, including the following minerals: olivine, garnet, clinopyroxene, orthopyroxene, spinel, ilmenite, cordierite, biotite, phengite, chlorite and hornblende. For many pairs in this list calibrations exist for some end-member mineral compositions, either from experimental work or from thermodynamic calculations. In a few cases, experimental calibrations are also available for complex "natural" systems. Based on these calibrations, many reformulations exist, using different solution models to account for the effects of the impurities. Furthermore, some exchange thermometers were calibrated empirically against natural assemblages. Data for some 18 systems are available at present, and some of these are listed in Table 4.3. Most of the experiments on the anhydrous pairs were obtained at T >800 °C and many were not compositionally reversed. If not reset by retrograde exchange, their best application is in high-temperature rocks from the granulite facies and the mantle, and extrapolation of these data down to relatively low temperatures is often unjustified. The potential problem with Mg–Fe^{2+} thermometry in using microprobe analyses was already mentioned earlier.

In the following, the two most commonly used Fe–Mg exchange thermometers are briefly reviewed.

4.7.3.1
Garnet–Clinopyroxene

The Grt-Cpx thermometer has been considered by many workers because of its potential importance for garnet-granulites, garnet-amphibolites, garnet-peridotites, and eclogites. At least ten calibrations of this thermometer exist, relying on a large body of experimental work. Experiments were conducted in the temperature range 600–1500 °C, but predominantly at T >900 °C. Only three calibrations will be mentioned here, while references to other studies can be found in Pattison and Newton (1989).

The most often used calibration of the Grt-Cpx thermometer is that derived by Ellis and Green (1979). These authors first experimentally evaluated the effect of grossular substitution (X_{Ca}^{Grt}) on exchange reaction (11), see Table 4.3, and derived the following equation:

$$T\,(^{\circ}C) = (3030 + 10.86\,P\,(kbar) + 3104\,(X_{Ca}^{GRT}))/(\ln K_D + 1.9034) - 273$$

where K_D is the distribution ratio $(Fe^{2+}/Mg)^{Grt}/(Fe^{2+}/Mg)^{Cpx}$. More recent experimental studies have shown that the Ellis and Green geothermometer overestimates temperatures when applied to granulites formed at about 10 kbar. According to Green and Adam (1991, p. 347), "this overestimate could be of the order of 50–150 °C". However, no new thermometric equation was presented.

Krogh (1988), using earlier experimental data, derived a new expression for this thermometer with a curvilinear relationship between $\ln K_D$ and X_{Ca}^{Grt}

$$T\,(^{\circ}C) = \{1879 + 10\,P\,(kbar) - 6173\,(X_{Ca}^{Grt})^2 + 6731\,X_{Ca}^{Grt}\}/(\ln K_D + 1.393) - 273\;.$$

Pattison and Newton (1989) presented a large set of new experimental data. They found that the Grt-Cpx Fe/Mg distribution curves were asymmetric, implying nonideal solid solution behavior, in contrast to those of previous experimental studies. The Grt-Cpx thermometer as formulated by Pattison and Newton (1989) seems to work fine for upper amphibolite facies, granulite facies and high-temperature eclogite facies rocks. Application to low- and medium-temperature crustal eclogites is uncertain because their clinopyroxene is ordered omphacite of high jadeite content, very different to those of the simpler system investigated by Pattison and Newton. In almost all recent studies, the Ellis and Green (1979) calibration has been applied to eclogites. In many cases a long extrapolation in temperature is involved, and, in this respect, the lowest temperature estimates obtained are critical, i.e., in the range 400–500 °C. As an example, Schliestedt (1986) determined Grt-Cpx temperatures of 471 ± 31 °C (range = 392–512 °C, 16 samples) on eclogites and blueschists from Sifnos, based on a preliminary version of the Krogh (1988) calibration. Schliestedt concluded that these temperatures were in accordance with those estimated by other methods (calcite-dolomite solvus and oxygen isotope fractionation). Note that the Ellis and Green (1979) thermometer gave equilibrium temperatures ca. 50 °C too high for these rocks (Krogh 1988). These limited data seem to indicate that reasonable results can be obtained with the Krogh formulation, even in low-temperature eclogites.

The Grt-Cpx thermometer calibrations mentioned above do not include any possible effects of chemical variations in Cpx. Such effects were noted for variations in the jadeite content, with K_D tending to decrease with an increase in X_{Jd}^{Cpx} of the sodic omphacite, and especially at $X_{Jd}^{Cpx} > 0.6$ (e. g., Koons 1984; Heinrich 1986; Benciolini et al. 1988). However, X_{Jd}^{Cpx} will in most cases be lower than 0.6 and will exert only a minor effect on K_D. On the other hand, no definite compositional dependence of K_D's on the Fe^{3+} content in clinopyroxene (aegirine component) was observed by Benciolini et al. (1988), but it should be remembered that the problem of the determination of ferrous/ferric iron in the clinopyroxene from microprobe data is not solved. Additional chemical effects on the Fe–Mg partition coefficient K_D are discussed by Krogh (1988).

In conclusion, the present status of Grt-Cpx Fe–Mg exchange geothermometry is certainly far from being ideal, but nevertheless it is a valuable thermometer for granulites and eclogites. The calibrations by Krogh (1988) and Pattison and Newton (1989) are recommended for low-and medium-temperature eclogites and for high-temperature eclogites plus granulites, respectively.

4.7.3.2
Garnet–Biotite

The garnet-biotite exchange thermometer is the most popular of all geothermometers because of its wide application in a large variety of rocks covering a broad range of metamorphic grade. At least 18 versions of this thermometer are available: 3 have been calibrated by field observations, 2 by laboratory experimentation, and 13 were derived from the experimental calibrations by using different non-ideal mixing models for garnet and biotite.

Ferry and Spear (1978) published experimental data on the Fe–Mg exchange between synthetic annite-phlogopite and almandine-pyrope in systems with Fe/(Fe + Mg) held at 0.9. The following equation was derived:

$$T\,(^{\circ}C) = \{[2089 + 9.56\,P\,(kbar)]/(0.782 - \ln K_D)\} - 273$$

with $K_D = (Fe/Mg)^{Bt}/(Fe/Mg)^{Grt}$. As it is likely that solid solution of additional components affects this thermometer, Ferry and Spear (1978) suggested that it should be restricted for usage with garnet low in Ca and Mn, with $(Ca + Mn)/(Ca + Mn + Fe + Mg) = 0.2$, and with biotite low in Al^{VI} and Ti, with $(Al^{VI} + Ti)/(Al^{VI} + Ti + Fe + Mg) \leq 0.15$.

Perchuk and Lavrent'eva (1983) performed the second experimental study on the Fe–Mg exchange between garnet and biotite. For most experiments, natural minerals served as starting materials in systems that covered the range 0.3–0.7 Fe/(Fe + Mg). As pointed out by Thoenen (1989), the thermometric equation given by Perchuk and Lavrent'eva is unlikely because it gives negative Clausius-Clapeyron slopes. A corrected version of the equation is

$$T\,(^{\circ}C) = \{[3890 + 9.56\,P\,(kbar)]/(2.868 - \ln K_D)\} - 273\,.$$

Temperatures obtained from this calibration are ~30 °C higher in the 500 °C temperature range, but ~60 °C lower in the 700 °C temperature range compared with the Ferry and Spear calibration.

Several modifications of the Ferry and Spear calibration have been made, based on the effects of Ca and Mn in garnet and of Al and Ti in biotite. Despite exhaustive research efforts, the magnitude of the effects of impurities on the Fe–Mg exchange between garnet and biotite is not sufficiently well known at present (for a discussion, see Bhattacharya et al. 1992). This makes it difficult to choose between the many versions of this geothermometer. Some authors have tried to evaluate the quality of different calibrations by comparing the calculated temperatures with temperature estimates based on some other mineral equilibria or by comparing the scatter of data within a given metamorphic zone. Still another approach was taken by Chipera and Perkins (1988), who applied a trend surface analysis to determine how well the temperatures fitted a regional temperature surface across their study area. Eight calibrations of the garnet-biotite thermometer were compared and it was found that the Perchuk and Lavrent'eva (1983) thermometer yielded the most precise results.

In general, the Grt-Bt thermometer seems to work fairly well in the high-temperature greenschist facies and the amphibolite facies. For several areas the Grt-Bt temperatures indicate a regular increase in temperature with increasing metamorphic grade over a range of some 100–150 °C (e.g., Ferry 1980; Lang and Rice 1985 b; Holdaway et al. 1988). In the upper amphibolite facies and granulite facies, however, retrograde Fe-Mg exchange generally occurs, resulting in anomalously low Grt-Bt temperatures if garnet rims and biotites in contact with garnets are analyzed. Nevertheless, reasonable prograde temperatures can still be obtained from garnets that have undergone retrograde reactions. As suggested by Tracy et al. (1976), in biotite-rich rocks retrograde reactions will cause negligible changes in matrix biotite composition while the garnet-rim composition changes greatly. In biotite-rich rocks, therefore, reasonable estimates of maximum prograde temperatures can be made using garnet-core composition and matrix biotite composition (provided that these two mineral compositions once were in chemical equilibrium).

The GTB program reported by Spear et al. (1999) offers a choice of 11 different Bt-Grt thermometers, including recent calibrations by: Patiño Douce et al. (1993), Holdaway et al. (1997), Gessmann et al. (1997), and Kleemann and Reinhardt (1994). We tested the thermometers with various Bt and Grt data from amphibolite facies metapelites. The Ferry and Spear (1978), Hodges and Spear (1982), Perchuk and Lavrent'eva (1984) and Patiño Douce et al. (1993) formulations gave consistent and plausible temperatures.

4.7.3.3
Isotopic Exchange Thermometry

Most exchange thermometers involve cations. Isotopic thermometers are an important exception and are based on the temperature dependence of the equilibrium partition of light stable isotopes (typically of the elements C, O, S) between two coexisting phases. Stable isotopes are chosen because their ratios will not vary with time due to radioactive decay. Heavy isotopes are not useful for thermometry because they do not fractionate as much as light isotopes. The fractionation between light isotopes is dependent on temperature, mineral chemical composition and crystal structure. Fractionation between two coexisting minerals is largest at low temperatures and decreases to almost zero at high temperatures (800–1000 °C). Fractionation tends to be largest between phases of widely different compositions and structures, so pairs like magnetite-quartz, ilmenite-quartz, rutile-quartz, and calcite-quartz are most useful. Isotopic exchange has virtually no volume change, which means that this thermometer is effectively independent of pressure.

Isotopic thermometry is potentially a very powerful technique, and it would appear to be the ideal thermometer for metamorphic rocks. However, several problems do exist. Many different experimentally determined calibration curves are in use that are in general not internally consistent. Retrograde isotopic exchange will occur during slow cooling from high temperatures, yielding discordant temperature values for different mineral pairs from the same rock. The set of paleotemperatures obtained for a rock will, in general, give neither the mineral closure temperatures nor the formation or crystallization temperatures. On the other hand, the cooling rate of the rock may be derived from the data (Giletti 1986). Lack of equilibrium may be a problem, especially during contact metamorphism and at low temperatures. For further information, the reader is referred to Valley (1986), Hoefs (1987), and Savin and Lee (1988).

4.7.4
Net-Transfer

The prime equilibria in geobarometry are net-transfer reactions in which one or more of the involved minerals shows variable composition. Due to their generally small positive slope in P–T space, solid–solid transfer reactions are primarily used as pressure indicators and the temperature can be retrieved from exchange or other thermometers. Net-transfer reactions are continuous or sliding equilibria that are multivariant in P–T–X space and, as a result, the entire thermobarometric assemblage can coexist over a wide range of P and T values. Application of these reactions requires that (1) the appropriate univariant end-member equilibrium has been well calibrated, (2) the chemical compositions of the phases are known, and (3) activity-composition relations are available for those minerals with appreciable solid solution. The net-transfer reaction $Grs + Qtz = 2 Wo + An$ was used in Section 4.7.1 to illustrate the general principles of thermobarometry.

As a further example, consider the assemblage $Grt + Rt + Als + Ilm + Qtz$ (GRAIL), typical of some metapelitic rocks. The following end-member reaction can be formulated:

$$Fe_3Al_2Si_3O_{12} + 3\,TiO_2 = 3\,FeTiO_3 + Al_2SiO_5 + 2\,SiO_2$$
$$\text{Alm} \qquad\quad \text{Rt} \qquad\ \text{Ilm} \qquad \text{Als} \qquad \text{Qtz}$$

Equilibrium of the reaction is displayed in Fig. 4.11. The effect of solid solution in garnet and ilmenite leads to isopleths of $\log K_{PT} = 3 \log a_{Ilm} - \log a_{Alm}$ contoured on a P–T diagram. If chemical analyses of coexisting Ilm and Grt can be converted to activities of the respective Fe end-member phase components by using an appropriate solution model, then log K for the specific assemblage in your rock can be calculated. This specific log K value corresponds to one P–T line or a nearby line on Fig. 4.11.

As mentioned above, thermodynamic models of activity-composition relations are a prerequisite for application of thermobarometers involving solid solutions. Unfortunately, solution models for the minerals most commonly employed in thermobarometry are poorly constrained (see, e.g., Essene 1989, p. 2), and this represents one of the main obstacles to accurate geothermobarometry. For this reason, the use of any thermodynamic model at large dilutions, i.e., where the actual mineral compositions deviate largely from those of the calibrated end-member reaction, should be avoided or at least regarded with scepticism, except where experimental a-X data are available.

Table 4.4. Some net-transfer reactions used in geothermobarometry

No.	Mineral assemblage	net-transfer reaction	Range of application[a]	Reference[b]
1	Grt-Opx	1 Prp=1 En+1 MgTs	ECL	Carswell and Harley (1989)
2	Opx-Ol-Qtz	1 Fs=1 Fa+1 Qtz	GRA	Bohlen and Boettcher (1981); Newton (1983)
3	Cpx-Pl-Qtz	1 CaTs+1 Qtz=1 An	GRA	Newton (1983), Gasparik (1984)
4	Cpx-Pl-Qtz	1 Jd+1 Qtz=1 Ab	BLU, ECL, ±GRA	Liou et al. (1987); Carswell and Harley (1989)
5	Grt-Pl-Ol	1 Grs+2 Prp=3 An+3 Fo	GRA	Johnson and Essene (1982)
6	Grt-Pl-Ol (GAF)[a]	1 Grs+2 Alm=3 An+3 Fa	GRA	Bohlen et al. (1983a)
7	Sp-Po-Py	—	GRE, AMP, GRA, ±BLU	Jamieson and Craw (1987) Bryndzia et al. (1988, 1990)
8	Wo-Pl-Grt-Qtz (WAGS)[c]	1 Grs+1 Qtz=1 An+2 Wo	AMP, GRA	Huckenholz et al. (1981)
9	Grt-Spl-Sil-Qtz	1 Alm+2 Sil=3 Hc+5 Qtz	GRA	Bohlen et al. (1983b)
10	Grt-Spl-Sil-Crn	1 Alm+5 Crn=3 Hc+3 Sil	GRA	Shulters and Bohlen (1989)
11	Grt-Pl-Opx-Qtz (GAES)[c]	1 Grs+2 Prp+3 Qtz=3 An+3 En	GRA	Eckert et al. (1991)
12	Grt-Pl-Opx-Qtz (GAFS)[c]	1 Grs+2 Alm+3 Qtz=3 An+3 Fs	GRA	Faulhaber and Raith (1991)
13	Grt-Pl-Cpx-Qtz (GADS)[c]	2 Grs+1 Prp+3 Qtz=3 An+3 Di	GRA	Eckert et al. (1991)
14	Grt-Pl-Cpx-Qtz (GAHS)[c]	2 Grs+1 Alm+3 Qtz=3 An+3 Hd	GRA	Moecher et al. (1988)
15	Grt-Als-Qtz-Pl (GRIPS)[c]	1 Grs+2 Als+1 Qtz=3 An	AMP, GRA	Koziol and Newton (1988); McKenna and Hodges (1988)
16	Grt-Hbl-Pl-Qtz[d]	2 Grs+1 Prp+3 Prg+18 Qtz=6 An+3 Ab+ 3 Tr	AMP, GRA	Kohn and Spear (1989)
17	Grt-Hbl-Pl-Qtz[d]	2 Grs+1 Alm+3 Fprg+18 Qtz=6 An+3 Ab+ 3 Fac	AMP, GRA	Kohn and Spear (1989)
18	Gr-Ms-Pl-Bt[d]	1 Grs+1 Prp+1 Ms=3 An+1 Phl	AMP	Powell and Holland (1988)
19	Grt-Ms-Pl-Bt[d]	1 Grs+1 Alm+1 Ms=3 An+1 Ann	AMP	Powell and Holland (1988)
20	Ttn-Ky-Pl-Rt	1 Ttn+1 Ky=1 An+1 Rt	ECL	Manning and Bohlen (1991)
21	Grt-Opx-Cpx-Pl-Qtz	1 Prp+1 Di+1 Qtz=2 En+1 An	GRA	Paria et al. (1988)
22	Grt-Opx-Cpx-Pl-Qtz	1 Alm+1 Hd+1 Qtz=2 Fs+1 An	GRA	Paria et al. (1988)
23	Grt-Rt-Als-Ilm-Qtz (GRAIL)[c]	1 Alm+3 Rt=3 Ilm+1 Als+2 Qtz	AMP, GRA	Bohlen et al. (1983b)
24	Grt-Rt-Pl-Ilm-Qtz (GRIPS)[c]	1 Grs+2 Alm+6 Rt=3 An+6 Ilm+3 Qtz	AMP, GRA	Bohlen and Liotta (1986); Anovitz and Essene (1987b)

[a] Abbreviations for metamorphic facies: AMP amphibolite; BLU blueschist; ECL eclogite; GRA granulite; GRE greenschist.
[b] One or two recent references for each reaction are given only.
[c] Abbreviations according to Essene (1989).
[d] Empirical calibrations; all other calibrations were determined by experiment or calculation.

Some of the more commonly used net-transfer reactions together with some potentially useful solid–solid end-member reactions in thermobarometry are listed in Table 4.4, together with their range of applicability in terms of metamorphic facies. Note that these reactions are listed with increasing number of phases involved, ranging from two up to five. Before some specific reactions are considered, a few general comments concerning Table 4.4 are given.

1. No equilibria involving cordierite or sapphirine have been considered because the effects of various fluid species (contained in cordierite), order/disorder, and nonideal a-X relations are not known well enough (Essene 1989).
2. There are two calibrated reactions available for six of the mineral assemblages listed, one each for the Fe and Mg end members, respectively.
3. Most of these thermobarometers have a rather restricted range of applicability, often limited to one or two metamorphic facies.
4. There are at least 19 thermobarometers based on solid–solid reactions available for the granulite facies and still 9 for the amphibolite facies, but the situation is quite unfavorable for the other facies.

4.7.4.1
Garnet–Aluminosilicate–Quartz–Plagioclase (GASP)

The equilibrium $Grs + 2Als + Qtz = 3An$ (Table 4.4) has become the most widely used barometer for amphibolite and granulite facies rocks. The advantage of this tool is the widespread occurrence of the GASP assemblage in metapelites, due to extensive solid solution in garnet and plagioclase. An expression of the popularity of this geobarometer is the existence of at least eight different calibrations. However, the large uncertainties involved (see below) are cause for concern for the successful application of this barometer.

The end member reaction has been calibrated in several studies (for references, see, e.g., Koziol and Newton 1988) in the temperature range 900–1400 °C. Application to rocks that equilibrated between 500–800 °C requires a rather stretched extrapolation and this contributes to the large uncertainty in pressure estimates from this barometer. Based on 27 brackets from 5 different experimental studies, McKenna and Hodges (1988) give the following equation for use in thermobarometry and with kyanite as the aluminosilicate polymorph: $P(bar) = (22.0 \pm 1.5)T(°K) - (6200 \pm 3000)$ where the uncertainties are at the 95% confidence level. According to McKenna and Hodges (1988), these inaccuracies propagate into paleopressure uncertainties of ±2.5 kbar.

An additional problem with this barometer is the generally small content of grossular component in metapelitic garnets. This leads to relatively large errors in analyzing small concentrations of Ca in garnet. Moreover, this entails large extrapolations in composition as well because the experimental data refer to pure Grs componens. Furthermore, the activity of grossular in garnet at such extreme dilution is unknown which results in large additional uncertainties in applying solution models. Our recommendation: use it with great care for high-pressure amphibolite and granulite facies rocks (see also Todd 1998; Holdaway 2001). It works best with high-pressure mafic rocks containing An-rich plagioclase and fairly Grs-rich garnet (P = 6–10 kbar).

4.7.4.2
Garnet–Rutile–Aluminosilicate–Ilmenite–Quartz (GRAIL)

The GRAIL reaction (Table 4.4) has been tightly reversed by Bohlen et al. (1983 b) in the temperature range 750–1100 °C. This reaction has many advantages as a barometer for metapelites from the amphibolite and the granulite facies:
1. It was calibrated within or in the vicinity of the P–T range for which it will be used and, therefore, long extrapolations are unnecessary.
2. Its location is quite insensitive to temperature.
3. Only two minerals in the requisite assemblage, garnet and ilmenite, have appreciable solid solution in ordinary metamorphic rocks.
4. Analyses of garnet and ilmenite are for their major components, and long compositional extrapolations are not required.

Furthermore, in the absence of rutile, the GRAIL barometer may be used to estimate an upper pressure limit. Application of this barometer in numerous terrains yields geologically reasonable pressures that agree well with other barometers.

4.7.4.3
Clinopyroxene–Plagioclase–Quartz

$Jd + Qtz = Ab$ (Table 4.4) provides the basis of a widely used thermobarometer. Because of its petrological significance, this reaction has been evaluated in several experimental studies (for references, see, e.g., Essene 1982, p. 176; Berman 1988, Fig. 38). All these experiments were performed at T >500 °C, using high albite as starting material. Many petrologists have used the calibration by Holland (1980) for P–T estimates, and this procedure is also recommended by Carswell and Harley (1989, p. 93) for the eclogite facies. However, Liou et al. (1987, p. 91) found that Holland's data are not consistent with some observed mineral assemblages and mineral composition data from the blueschist facies. For this reason, at low temperatures, Liou et al. recommended the use of the Jd–Ab–Qtz stability relations as determined by Popp and Gilbert (1972).

Application of the jadeitic pyroxene equilibrium to natural assemblages is complicated by the difficulty of accurately measuring sodium in clinopyroxene, by the order-disorder transitions in albite and clinopyroxene, and by the limited $a = f(X)$ data for Ca-Na pyroxenes. For a recent discussion of some of these questions, the reader is referred to Carswell and Harley (1989) and Meyre et al. (1997). Also note that eclogite facies assemblages are, by definition, devoid of plagioclase and many eclogites do not contain quartz as well. Therefore, the jadeite content of clinopyroxene in such rocks will yield only a minimum pressure of formation.

4.7.4.4
Sphalerite–Pyrrhotite–Pyrite

The iron content of sphalerite in equilibrium with hexagonal pyrrhotite and pyrite continuously decreases as a function of pressure, and provides a barometer in the temperature range 300–750 °C (e.g., Bryndzia et al. 1988, 1990). Applications of this barometer to metamorphic rocks have yielded excellent to poor results. At T <300 °C, extensive resetting of sulfide compositions and transformation of hexagonal to monoclinic pyrrhotite occurs, resulting in decreased FeS contents in sphalerite and high apparent pressures. It is important to use unexsolved, coexisting three-phase assemblages following the textural criteria proposed by Hutchison and Scott (1980), and we caution against the indiscriminate use of this sphalerite geobarometer. Sufide barometers for mantle samples are described in Eggler and Lorand (1993).

4.7.5
Miscibility Gaps and Solvus Thermometry

Some mineral pairs with similar structure but different composition show a mutual solubility that increases with temperature. The immiscible region is called a miscibility gap. Well-known examples include the pairs K-feldspar-albite, calcite-dolomite, and many others. At some temperature, e.g., 400 °C, Kfs can dissolve a certain amount of Ab, and vice versa. With increasing temperature, the composition of Kfs becomes more sodic and the coexisting Nafs contains an increasing amount of K. At some critical temperature, the two feldspars have an identical composition and the miscibility gap has closed. At higher temperature, Na-K-feldspar forms a continuous solid solution. In the temperature region where a miscibility gap exists, the composition of the two coexisting minerals can be used for thermometry. Most gaps are not very pressure-dependent, however; some are and therefore the pressure dependence should not be ignored. A miscibility gap between isostructural phases is called a solvus. The term solvus is often also used for all miscibility gaps and also for the curve that separates the two-phase from the one-phase region as well (Fig. 4.12). Although this is not completely correct, we will follow this practice, too.

Because of the general shape of miscibility gaps (Fig. 4.12), small variations in mineral composition correspond to a large temperature difference at the steep limbs at low T, but relate to a moderate temperature difference at higher T. Thus geothermometers become more sensitive temperature indicators as temperature increases. If disorder occurs in one or both minerals, then there will be a continuum of solvi with gradually increasing partial disorder.

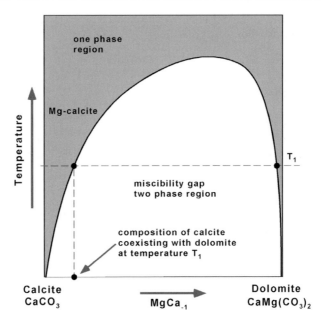

Fig. 4.12. Schematic T–X diagram showing the asymmetric miscibility gap between calcite and dolomite [compositional variable is $MgCa_{-1}$, it is 0 in $CaCO_3$ and 1 in $CaMg(CO_3)_2$]. A given Mg content in calcite coexisting with dolomite corresponds to an equilibrium temperature of the Ca-Mg exchange equilibrium between Cal and Dol. A given analytical error translates to a large T error at low T and a smaller T error at high T

The following mineral pairs are often used for solvus geothermometry: orthopyroxene-clinopyroxene, plagioclase-alkali feldspar, calcite-dolomite, and muscovite-paragonite, and these will be discussed briefly below.

4.7.5.1
Orthopyroxene–Clinopyroxene

A large body of experimental and theoretical data exists on the enstatite-diopside solvus, and it has been shown that, at high temperatures, above about 800 °C, the diopside limb of the asymmetrical solvus is strongly temperature-dependent. Because natural pyroxenes rarely belong to the binary join $Mg_2Si_2O_6$–$CaMgSi_2O_6$, the major problem with this thermometer lies in correction for other components. This has led to the formulation of several multicomponent two-pyroxene thermometers. If applied to ultramafic rocks, where pyroxenes closely approach the enstatite-diopside binary join, satisfactory results have been obtained (e.g., Carswell and Gibb 1987), but much less so for iron-rich pyroxenes from granulites.

A different, and apparently more successful, approach to Opx-Cpx thermometry was taken by Lindsley (1983). This author determined phase equilibria for pure Ca-Mg-Fe pyroxenes over the temperature range 800–1200 °C and for pressures up to 15 kbar. Solvus isotherms for coexisting Opx–Cpx were displayed graphically in the quadrilateral pyroxene system composed of Di–Hd–En–Fs. The effect of pressure was found to be $= 8\,°C/kbar$, and graphs are presented for solvus relations at 1 atm and at 5, 10, and 15 kbar, along with approximate formulas for interpolating between these pressures. Pyroxenes with appreciable contents of nonquadrilateral components require special projections onto the Di–Hd–En–Fs pyroxene quadrilateral; corresponding methods have been presented by Lindsley and Anderson (1983) and Lindsley (1983). This correction scheme is largely empirical and Lindsley (1983) suggests that application be limited to pyroxenes with relatively low (<10%) amounts of nonquadrilateral components. An updated version of Cpx-Opx thermometry is implemented in the code QUILF (Andersen et al. 1993). QUILF is briefly described in Section 4.8.1 on ilmenite–magnetite oxygenbarometry.

Slowly cooled pyroxenes often show exsolution lamellae, and reintegration of exsolved material in the microprobe analyses is necessary to obtain meaningful thermometric results. Also, an incorrect calculation of ferric iron will affect this geothermometer.

4.7.5.2
Plagioclase–Alkali Feldspar

Most feldspars consist of the three components, Ab, Or, An, with extensive solid solution between Ab and An (the plagioclase series) and between Ab and Or (the alkali feldspar series). The composition of coexisting feldspars contains important thermometric information. The majority of formulations of the two-feldspar thermometer concentrate on two feldspar binaries, Ab-An and Ab-Or, and obtain temperatures from the partitioning of albite components between coexisting plagioclase and alkali feldspar (e.g., Stormer 1975; Haselton et al. 1983). This approach may be valid at lower temperatures where feldspars have only trivial amounts of the third component. At higher temperatures, however, the effects of ternary solution become increasingly important. Thus, the most recently developed feldspar thermometers are based on ternary solution models that account for the effects of K in plagioclase and Ca in alkali feldspar (e.g., Fuhrman and Lindsley 1988; Elkins and Grove 1990). A great advantage of the ternary approach is that it allows three temperatures to be calculated, one for each component, providing a valuable test of equilibrium. The more sophisticated thermometers cannot be published as simple thermometric equations; however, in many cases, a computer program can be downloaded from the websites of the authors. A very nice program for ternary feldspar thermometry is SOLVCALC (Wen and Nekvasil 1994). SOLVCALC is a Windows-based interactive graphics program package for calculating the ternary feldspar solvus and for two-feldspar geothermometry. It is available via: http://www.ndsu.nodak.edu/instruct/sainieid/software/software_list.shtml.

Two-feldspar thermometry has been applied with some success to high-grade gneisses from the granulite facies (e.g., Bohlen et al. 1985; see also Fuhrman and Lindsley 1988). In such rocks, exsolution of Ab-rich plagioclase from alkali feldspar and of K-rich feldspar from plagioclase upon cooling is a common phenomenon, and it is essential that exsolved grains are reintegrated to obtain the compositions of the original feldspars (for analytical details, see Bohlen et al. 1985). Applications of feldspar geothermometry to rocks from the amphibolite and greenschist facies have been less successful, often yielding unreasonable low-temperature values. Evidently, alkali exchange with a fluid phase below peak metamorphic conditions must have taken place.

4.7.5.3
Calcite–Dolomite

In the system $CaCO_3$–$CaMg(CO_3)_2$ a miscibility gap exists and the two phases on either side of the gap are calcite and dolomite, respectively. The amount of $MgCO_3$ in calcite in equilibrium with dolomite can be used to estimate metamorphic temperatures. This strongly asymmetric solvus appears to be well determined and the effect of pressure on temperature estimates is small. Most natural carbonates, however, contain additional components like $FeCO_3$ or $MnCO_3$, and the effect of $FeCO_3$ on calcite–dolomite thermometry has been evaluated, e.g., by Powell et al. (1984) and Anovitz and Essene (1987a).

Application of the two-carbonate thermometer to metamorphic rocks is complicated by retrograde resetting. Magnesian calcite may exsolve dolomite that must be reintegrated during microprobe analysis. Occasionally, the originally high magnesium content of calcite is obliterated through diffusion, yielding temperatures, e.g., of 300–400 °C for high-grade marbles (e.g., Essene 1983). Therefore, this thermometer is more useful in low-grade orogenic metamorphic rocks and in contact metamorphic rocks where cooling rates are high.

Recently, Ferry (2001) has shown that a high Mg content of calcite can be preserved by calcite inclusions in forsterite of high-grade marble. Small calcite crystals were trapped by growing forsterite in dolomite marble and did not exsolve Dol or lose Mg during cooling. The computed Cal-Dol solvus temperatures (600–700 °C) were high and reflected the temperature at the time of Fo growth.

4.7.5.4
Muscovite–Paragonite

The binary solvus between coexisting muscovite and paragonite has been repeatedly investigated experimentally (for references, see Essene 1989, p. 8); a detailed study being that of Flux and Chatterjee (1986). However, even their solvus is not well constrained relying on three experimental brackets only, and with one bracket determined with 1M micas run products (instead of the stable $2M_1$ polytype). The solvus calculated by Chatterjee and Flux (1986) shows a considerable pressure dependence.

The use of the muscovite-paragonite solvus geothermometer has not been successful in the past (see, e.g., Guidotti 1984, p. 409). Nevertheless, the muscovite limb of the solvus determined by Chatterjee and Flux (1986, Fig. 6) may be provisionally used to obtain geothermometric data for coexisting muscovite-paragonite pairs, provided (1) the equilibrium pressure is independently known and (2) the chemical composition of the micas is close to the ideal muscovite-paragonite join. This latter requirement will be rarely fulfilled for phengitic muscovites formed at high pressures and/or low temperatures, and use of the solvus should be restricted to muscovites with Si <3.05 atoms per formula unit.

4.7.6
Uncertainties in Thermobarometry

The development of geothermobarometry has been one of the most exciting advances in metamorphic petrology in the past decades. This has allowed petrologists to quantify the conditions of metamorphism. But how accurate are such data? Ideally, application of geothermobarometry should be accompanied by a statement regarding the uncertainties of temperature and pressure. In practice, however, uncertainties are rarely reported because it is difficult to quantify some sources of errors.

Kohn and Spear (1991) performed an evaluation of the uncertainty in pressure when a barometer is applied to an assemblage in a rock. Four barometers were considered, including GASP and GRAIL discussed above. Sources of error considered included (with estimated 1σ uncertainties in pressure given in brackets): accuracy of the experimentally determined, barometric end-member reaction (±300 to ±400 bar); volume measurement errors (±2.5 to ±10 bar); analytical imprecision of mineral analyses using an electron microprobe (±55 to ±185 bar); thermometer calibration errors (±250 to ±1000 bar); variation in garnet and plagioclase activity models (±60 to ±1500 bar); and compositional heterogeneity of natural minerals (±150 to ±500 bar). Collectively, the accuracy of a barometer may typically range from ±600 to ±3250 bar (1σ), with the most significant sources of error being uncertainty in thermometer calibration and poorly constrained activity models. However, continued experimental and empirical work should substantially reduce these uncertainties in the future.

The situation is much better, however, if we are interested in comparative thermobarometry, which involves applying a single pair of thermobarometers to different samples in order to calculate differences in P–T conditions. In this case, the systematic errors associated with experimental calibrations and solution modeling are eliminated, leaving only the effects of analytical uncertainties. It is then possible to confidently resolve P–T differences of as little as a few hundred bars and a few tens of degrees.

Fig. 4.13. P–T diagram showing the results of a TWEEQU calculation for an amphibolite sample containing the mineral assemblage Grt–Cpx–Hbl-Pl-Qtz–Ilm–Rt. 14 equilibria among (Grs–Alm–Prp)–(Di–Hd)–An–Qtz–Ilm–Rt, excluding hornblende, were considered. Note that only 3 of the 14 equilibria are linearly independent. (Lieberman and Petrakakis 1991, Fig. 8b)

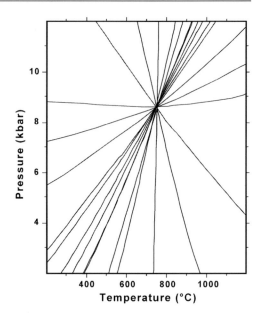

4.7.7
Thermobarometry Using Multi-equilibrium Calculations (MET)

Above, we used two intersecting equilibria to determine possible P–T conditions of rock formation (the P–T conditions of the least hydrated state). Many rocks, however, contain mineral assemblages that allow for the formulation of more than two reactions among phase components of the assemblage. Often a large number of reactions can be written down for the assemblage of a metamorphic rock. If the rock achieved a state of equilibrium at one point in P–T space all reactions must be simultaneously in equilibrium at this point. A set of equations, one for each reaction, can be solved for these conditions using (1) a consistent set of thermodynamic data for all material involved in the reactions, (2) appropriate equations of state to describe thermodynamic properties of all phase component at P and T, (3) appropriate solution models to describe the thermodynamics of solid and fluid solutions for all minerals and fluids involved, and (4) an efficient computer code that performs the desired computation of the equilibrium conditions. The equilibrium conditions of all reactions will appear as lines on a P–T diagram that intersect at one point, the 'frozen in equilibrium' conditions of P and T for that metamorphic rock (Fig. 4.13).

All four requisites for successful MET applications are available today and the tools can be downloaded from websites. The technique is widely used, sometimes a little unmindfully. Thermodynamic data sets particularly useful for metamorphic rocks have been published by Berman (1988) and Holland and Powell (1990). The HP database is regularly maintained and expanded as new experimental data become available or the science makes progress in

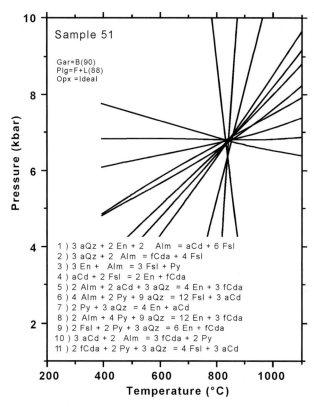

Fig. 4.14. MET solution of a granulite sample from the Bamble area in southern Norway (unpublished data Bucher). Assemblage: Qtz+Crd+Grt+Opx, 3 independent equilibria, 11 equilibria total. Software: PTAX (Berman et al. 1987), GridLoc (Lieberman 1991)

some other way. The latest version is HP98 (Holland and Powell 1998). It is the most widely used geologic data set today. The codes THERMOCALC (Powell and Holland 1988) and TWEEQU (Berman 1991; Lieberman and Petrakakis 1991) have been mentioned earlier (see Chap. 3). THERMOCALC is regularly maintained and updated; it can be downloaded from the authors excellent website at the University of Cambridge. The current web address is: http://www.esc.cam.ac.uk/astaff/holland/index.html.

Other, highly recommended programs for MET-type calculations include PERPLEX (Connolly and Kerrick 1987) and GIBBS (Spear and Menard 1989; Spear et al. 1999). We recommend surfing the web; many sites provide and maintain geologic software links. A good start is http://www.minpet.uni-freiburg.de, where you will find links to the homes of a large variety of geologic and petrologic software. The web address of Prof. J. Connolly's site is: http:/www.perplex.ethz.ch and you will find GIBBS under the same address as for GTB given above.

Fig. 4.15. MET solution of a cordierite gneiss from the Thor Range, Antarctica (mineral data from Bucher and Otha 1993). Assemblage: Qtz+Pl+Grt+Crd+Sil+Bt+Kf-s. Software: PTAX (Berman et al. 1987), GridLoc Lieberman (1991)

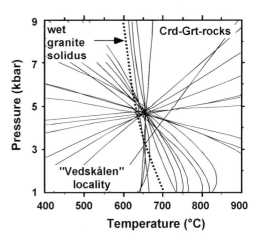

If you run your own rocks through the computer mill, please don't stop thinking, be judicious and observant. The perfectly well-behaved sample does not exist in practice. Multiequilibria rarely intersect at one P–T point (except for linear-dependent equilibria which always intersect at one point and are therefore no proof of MET success). Significant deviations from this ideal single point case often occur. When results for several rocks are compared with each other and with petrographic observations, it may become possible to conclude whether inaccurate thermodynamic data, compositional data, or disequilibrium may have caused discrepant results. In some samples, subsets of the assemblage may intersect at one P–T point and another subset may define a second intersection. Some examples of MET calculations can be found in Vidal et al. (2001) and examples from metapelitic rocks from the Bamble granulite area in southern Norway and from the Thor range in Antarctica are shown in Figs. 4.14 and 4.15, respectively.

4.8
Gibbs Method

In a classic benchmark paper, Spear et al. (1982) introduced "the Gibbs method" for applications in petrology. The method is based on differential thermodynamics and permits the computation of successive pressure-temperature differences (P–T paths) recorded by chemically zoned minerals in rocks. The method essentially solves a set of partial differential equations that express the thermodynamic consequences of chemically zoned minerals. It has been mostly used to model P–T paths from chemical zonation in pelitic garnets. It has been very successful in resolving remarkably detailed paths taken by garnet-bearing metapelitic schists in the pressure-temperature space. The differential calculations are tied to a starting point in P–T space where an assemblage and the associated mineral compositions are well known. From there the movements of the garnet-bearing sample can be reconstructed with high resolution of a few °C and bars (e. g., Selverstone et al. 1984). The software for GIBBS method calculations is available for download from the website of Prof. F. Spear, given in Section 4.7.

The entire package includes the programs GIBBS, GTB (Geothermobarometry), plus the tutorial notes (pdf format) for running the software. The texts were prepared originally for a short course presented at the Northeast Section Meeting of the Geological Society of America at the University of Vermont, Burlington, Vermont, in 2001. An excellent account of differential thermobarometry using zoned minerals can be found in Spear (1993); Chapter 17: "The origin and interpretation of zoned metamorphic minerals".

4.9
Assemblage Stability Diagrams

Assemblage stability diagrams, or **pseudosections**, are diagrams that display the stable mineral assemblages for a specific bulk rock composition over a range of P-T conditions. They can be P-T, T-X or P-X diagrams. Pseudosections are relatives of the petrogenetic grid. The great advantage of pseudosections is that they predict the stable mineral assemblage and the composition of minerals at P-T-X conditions for the specific rock of interest.

Pseudosections can be computed by very different techniques and, consequently, different computer programs. One popular method of computation is utilized by the codes GIBBS and THERMOCALC. It employs the methods of integrated thermodynamics the same way as to construct P-T curves, grids and contour diagrams. Using the divariant contour routines (also perplex provides this option), one can draw 0.0 mol contours for a phase. The contours must be truncated at the absolute stability limit of the assemblage defined by univariant reactions. In the program GIBBS, a shortcut has been implemented for the KFMASH system using the petrogenetic grid routines. In this mode, the program calculates the contours (as before) and compares the P-T conditions of the contour with the stability field for the assemblage as defined by the grid, plotting only those curves where there is an intersection. Using this mode, a complete P-T pseudosection for a bulk composition in the KFMASH system can be calculated in around an hour (see comment on Spear's web page). A pseudosection for average pelite compositions is shown in Fig. 7.8. The P-T diagram has been computed using GIBBS and is a download from the website cited above.

THERMOCALC uses a similar approach to pseudosections and the software and tutorials can be downloaded from: http://www.esc.cam.ac.uk/astaff/holland/index.html.

This website (maintained by Prof. T. Holland) offers many attractive software programs, including THERMOCALC, aX, and a hornblende–plagioclase thermobarometer. THERMOCALC is a very versatile program that can be used for conventional phase diagram calculations, petrogenetic grid creation and, as mentioned, for pseudosection construction. aX is a utility for the computation of phase component activities from mineral analyses utilizing a number of solution models. THERMOCALC is a very sophisticated problem solver in metamorphic petrology and its versatility and complexity necessarily makes it intricate to use. In addition, the code is well suited for the "ultimate" petrologist with a higher level of skill than that imparted by this volume.

Finally, the most elegant and mature concept in the field of assemblage stability diagram computation is perhaps the program package THERIAK-DOMINO by de Capitani. It is based on a Gibbs free energy minimization algorithm and solves directly for the assemblage with the lowest G at a specified P–T for a given bulk rock composition in the multidimensional composition space and utilizing a general thermodynamic description of multicomponent solid solutions (de Capitani and Brown 1987). The general and unique approach is an offspring of an ingenious early program EQUILAS by Brown and Skinner (1974). An example of a modified EQUILAS computation is shown in Fig. 3.22B. The DOMINO part of the package calculates and draws x-y equilibrium phase diagrams (x,y = P, T, a_i, ln a_i, log a_i or a pseudobinary or -ternary system). The diagram shows either areas of equal assemblages, isopleths or the amount of a phase. The software package can be downloaded together with examples and a tutorial from: http://therion. minpet.unibas.ch/minpet/groups/theriak/theruser.html.

The same can be said for this package as what has been said about THERMOCALC, perhaps even more so; it is ideal for the experienced specialist with a sound knowledge of thermodynamics, particularly thermodynamics of solid solutions. Nevertheless, we encourage readers of this volume to explore the advances of the front-end of our science by, for instance, having a look at the THERIAK-DOMINO-generated assemblage stability diagrams in Meyre et al. (1997).

4.10
More P–T Tools

4.10.1
Reactions Involving Fluid Species

Univariant equilibria involving volatile components cannot be directly applied to thermobarometry because their position in P–T space depends on properties of a now vanished fluid phase (Chap. 3). Dehydration reactions are often used in constructing petrogenetic grids assuming $P_{H_2O} = P_S$ or $a_{H_2O} = 1$, but it is now well known that CH_4 is a significant fluid species in many low-temperature rocks whilst CO_2 is abundant in many high-temperature rocks. For this reason, dehydration reactions are, with one exception, not further considered here.

4.10.1.1
Phengite–Biotite–K-Feldspar–Quartz

The white mica phengite is an intermediate member of the muscovite-celadonite solid-solution series. The two end members are related by the exchange of the Tschermak component, $\{Mg,Fe^{2+}Si(AlAl)_{-1}\}$, and the silica content of phengite is the preferred measure of this substitution. Field observations and data have shown that phengite increases in Si from nearly 3 per formula unit to almost 4 with increasing pressure. In the KMASH assemblage phengite + biotite + K-feldspar + quartz, the composition of phengite is invariant at fixed P, T and a_{H_2O}. The reaction:

$$3\,KMgAlSi_4O_{10}(OH)_2 = KMg_3\,AlSi_3O_{10}(OH)_2 + 2\,KAlSi_3O_8 + 3\,SiO_2 + 2H_2O$$
$$\text{celdonite} \qquad\qquad \text{phlogopite} \qquad \text{K-feldspar} \quad \text{quartz}$$

has been investigated experimentally by Massonne and Schreyer (1987) for varying Si content in phengite. Si isopleths have a small positive P–T slope and the Si content of phengite increases strongly and linearly with pressure. In practice, in rocks with the assemblage Phe–Bt–Kfs–Qtz, only phengite is chemically analyzed and the Si content of phengite used for a pressure estimate.

This is a potentially powerful geobarometer, especially in low- to medium-grade rocks, where calibrated barometers are scarce. However, several drawbacks exist: (1) the isopleths determined by Massonne and Schreyer (1987) are based on synthetic experiments that produced mainly 1M and Md K-white mica polytypes; (2) since the above reaction involves dehydration, estimates of a_{H_2O} are crucial for correct application; water activities below unity shift the Si isopleths of phengite towards higher pressure; and (3) experiments were conducted in an iron-free system whilst natural micas may contain considerable amounts of Fe^{2+} and Fe^{3+}. Since the ratio Fe^{2+}/Mg is generally higher in biotite than coexisting muscovite, the introduction of Fe will stabilize biotite + K-feldspar and thus reduce the celadonite content of the K-white mica, displacing the Si-isopleths to higher pressure (Evans and Patrick 1987). Application of the Massonne and Schreyer phengite geobarometer yielded reasonable results for metagranites from the eclogite facies (Massonne and Chopin 1989), but gave pressures up to 4 kbar too high for greenschist and amphibolite facies metamorphic terranes (Patrick and Ghent 1988). It may be most useful for relative pressure estimates across a field area.

4.10.1.2
Magnetite–Ilmenite Equilibria and the "QUILF" System

Two Fe-Ti oxides coexist in some metamorphic rocks. Magnetite shows solid solution with ulvöspinel (Fe_2TiO_4) whilst ilmenite shows solid solution with hematite. Chemical equilibrium between titanian magnetite and ferrian ilmenite can be described by a temperature-dependent Fe-Ti exchange reaction:

$$Fe_3O_4 + FeTiO_3 = Fe_2TiO_4 + Fe_2O_3$$

and a redox reaction:

$$4\,Fe_3O_4 + O_2 = 6\,Fe_2O_3$$

Buddington and Lindsley (1964) presented a graphic form of this thermometer and oxybarometer. They contoured isopleths of ilmenite (X_{FeTiO_3}) and ulvöspinel ($X_{Fe_2TiO_4}$) on a log f_{O_2}-T diagram and showed that intersection of these isopleths yields f_{O_2} and T. Since then, there have been a number of thermodynamic treatments.

Application of the Mag-Ilm thermometer is limited because these oxides are easily reset during retrogression. Upon cooling, Ti-bearing magnetite will be oxidized with the appearance of ilmenite lamellae within magnetite; conversely, Fe_2O_3-rich ilmenite will be reduced, yielding lamellae of Ti-magnetite within ilmenite. In order to obtain peak metamorphic temperatures, these "exsolution" lamellae must be reintegrated (Bohlen and Essene 1977).

However, spinel-ilmenite equilibria can be combined with equilibria involving silicate minerals in high-grade rocks. In this way, Lindsley and Frost (1992) and Frost and Lindsley (1992) have developed an excellent system of thermobarometers. The system and the code QUILF (Andersen et al. 1993) can be used for $P - T - f_{O_2}$ estimates among Cpx–Opx–pigeonite–Ol–SiO$_2$–ilmenite–spinel. The excellent program can be downloaded from the address below. It is particularly useful for igneous rocks but can be applied to granulite facies mafic and felsic rocks as well. Notably, charnockite and other Opx-bearing meta-granitoid rocks are very well suited for QUILF-o-metry. Prof. D. Lindsley's website can be found at: http://pbisotopes.ess.sunysb.edu/geo/general/faculty/quilf96.zip.

4.10.2
P–T Tools for Very Low-Grade Rocks

None of the many geothermobarometers mentioned above is applicable to sub-blueschist-greenschist facies conditions. Fortunately, a few other methods are available to estimate T, and sometimes also P, at very low metamorphic grade.

4.10.2.1
Fluid Inclusions

Fluid inclusions from metamorphic rocks generally yield information about the retrograde P–T path, but at very low metamorphic grade it is possible to obtain thermobarometric data about peak (or near-peak) conditions of metamorphism (e. g., Mullis 1987). This author developed a method to determine approximate P and T values of fluid trapping based on fluid inclusion studies from fissure quartz. This microthermometric method takes advantage of immiscibility in H_2O–CH_4 fluids. As a result, the homogenization temperature of water-rich inclusions saturated with CH_4 can be interpreted as temperature of formation. Methane-rich inclusions, on the other hand, allow the measurement of the density of CH_4. With these data the pressure of inclusion formation can be determined from known experimental P–V–T–X properties. Application of this method recorded temperatures of up to 270 °C and pressures of up to about 3 kbar.

4.10.2.2
Vitrinite Reflectance

Vitrinite is a predominant constituent in many coals and in finely dispersed organic matter in sedimentary rocks. As organic material is converted from complex hydrocarbons to graphite, the optical reflectance of vitrinite increases and indicates the degree of transformation. Various models exist relating vitrinite reflectance and physical variables. While temperature is regarded uniformly as the main controlling variable, the role played by the time factor is controversial and ranges from "important" to "negligible". Presumably, the best model available at present is that by Sweeney and Burnham (1990), called EASY%R_0, based on a chemical kinetic model of vitrinite maturation by Burnham and Sweeney (1989). In this Arrhenius reaction model, equations were integrated over temperature and time to account for the elimination of water, carbon dioxide, methane, and higher hydrocarbons from vitrinite. The model can be implemented on a spreadsheet or in a small computer program on a personal computer, but results are also shown graphically in nomograms of vitrinite reflectance versus exposure time and maximum temperature (Sweeney and Burnham 1990, Fig. 5). EASY%R_0 can be used for vitrinite reflectance values of 0.3 to 4.5%, and for heating rates ranging from igneous intrusions (1 °C/day), and geothermal systems (10 °C/ 100 years) to burial diagenesis (1 °C/10 Ma).

4.10.2.3
Conodont Color Alteration Index (CAI)

Conodonts are apatitic marine microfossils that contain trace amounts of organic material that colors them, occurring from the Cambrian to the Triassic, mainly in carbonate rocks. In the range of diagenesis and incipient metamorphism, five stages of CAI are distinguished according to color changes from pale yellow to brown to black. Epstein et al. (1977) and Rejebian et al. (1987) reproduced these color changes by heating in the laboratory and found a time and temperature dependence. Experimental data obtained between 300 and 950 °C and measured up to 10^3 h were extrapolated to geologic times. A graph of reciprocal absolute temperature versus a logarithmic time scale then forms the basis of this geothermometer, which has a resolution of possibly 50 °C. An advantage of the conodont method is that it can be used in carbonate rocks in which vitrinite is rarely present.

References

Anderson DJ, Lindsley DL (1988) Internally consistent solution models for Fe–Mg–Mn–Ti oxides: Fe–Ti oxides. Am Mineral 73:714–726

Andersen DJ, Lindsley DH, Davidson PM (1993) QUILF: a PASCAL program to assess equilibria among Fe–Mg–Mn–Ti oxides, pyroxenes, olivine, and quartz. Comput Geosci 19:1333–1350

Anovitz IM, Essene EJ (1987a) Phase equilibria in the system $CaCO_3$–$MgCO_3$–$FeCO_3$. J Petrol 28:389–414

Anovitz IM, Essene EJ (1987b) Compatibility of geobarometers in the system CaO–FeO–$Al_2O_3SiO_2$–TiO_2 (CFAST): implications for garnet mixing models. J Geol 95:633–645

Aranovich LY, Podlesskii KK (1989) Geothermobarometry of high-grade metapelites: simultaneously operating reactions. In: Daly S, Cliff RA, Yardley BWD (eds) Evolution of metamorphic belts. Geological Society Special Publication, Blackwell, Oxford, pp 45–62

Archibald DA, Krogh TE, Armstrong RL, Farrar E (1984) Geochronology and tectonic implications of magmatism and metamorphism, southern Kootenay Arc and neighbouring regions, southeastern British Columbia. II. Mid-Cretaceous to Eocene. Can J Earth Sci 21:567–583

Barrow G (1893) On an intrusion of muscovite biotite gneiss in the SE Highlands of Scotland and its accompanying metamorphism. Q J Geol Soc Lond 49:330–358

Barrow G (1912) On the geology of lower Deeside and the southern Highland border. Proc Geol Assoc 23:268–284

Bearth P (1958) Über einen Wechsel der Mineralfazies in der Wurzelzone des Penninikums. Schweiz Mineral Petrogr Mitt 38:363–373

Bégin NJ (1992) Contrasting mineral isograd sequences in metabasites of the Cape Smith Belt, northern Québec, Canada: three new bathograds for mafic rocks. J Metamorph Geol 10:685–704

Benciolini L, Lombardo B, Martin S (1988) Mineral chemistry and Fe/Mg exchange geothermometry of ferrogabbro-derived eclogites from the Northwestern Alps. Neues Jahrb Mineral Abh 159:199–222

Berman RG (1988) Internally-consistent thermodynamic data for minerals in the system K_2O–Na_2O–CaO–MgO–FeO–Fe_2O_3–Al_2O_3–SiO_2–TiO_2–H_2O–CO_2. J Petrol 29:445–522

Berman RG (1991) Thermobarometry using multi-equilibrium calculations: a new technique, with petrological applications. Can Mineral 29:833–856

Berman RG, Brown TH, Perkins EH (1987) GEØ-CALC: software for calculation and display of P–T–X phase diagrams. Am Mineral 72:861–862

Bhattacharya A, Krishnakumar KR, Raith M, Sen SK (1991) An improved set of a-X parameters for Fe–Mg–Ca garnets and refinements of the orthopyroxene-garnet thermometer and the orthopyroxene-garnet-plagioclase-quartz barometer. J Petrol 32:629–656

Bhattacharya A, Mazumdar AC, Sen SK (1988) Fe-Mg mixing in cordierite: constraints from natural data and implications for cordierite-garnet geothermometry in granulites. Am Mineral 73:338–344

Bhattacharya A, Mohanty L, Maji A, Sen SK, Raith M (1992) Non-ideal mixing in the phlogopite-annite binary: constraints from experimental data on Mg-Fe partitioning and a reformulation of the biotite-garnet geothermometer. Contrib Mineral Petrol 111:87–93

Bhattacharyya DS (1981) Geometry of isograds in metamorphic terrains. Tectonophysics 73:385–395

Bohlen SR, Boettcher AL (1981) Experimental investigations and geological applications of orthopyroxene geobarometry. Am Mineral 66:951–964

Bohlen SR, Essene EJ (1977) Feldspar and oxide thermometry of granulites in the Adirondack Highlands. Contrib Mineral Petrol 62:153–169

Bohlen SR, Lindsley DH (1987) Thermometry and barometry of igneous and metamorphic rocks. Annu Rev Earth Planet Sci 15:397–420

Bohlen SR, Liotta JJ (1986) A barometer for garnet amphibolites and garnet granulites. J Petrol 27:1025–1056

Bohlen SR, Wall VJ, Boettcher AL (1983a) Experimental investigation and application of garnet granulite equilibria. Contrib Mineral Petrol 83:52–61

Bohlen SR, Wall VJ, Boettcher AL (1983b) Experimental investigations and geologic applications of equilibria in the system FeO–TiO_2–Al_2O_3–SiO_2–H_2O. Am Mineral 68:1049–1058

Bohlen SR, Valley JW, Essene EJ (1985) Metamorphism in the Adirondacks. 1. Pressure and temperature. J Petrol 26:971–992

Bohlen SR, Dollase WA, Wall VJ (1986) Calibration and application of spinel equilibria in the system FeO–Al$_2$O$_3$–SiO$_2$. J Petrol 27:1143–1156

Bohlen SR, Montana A, Kerrick DM (1991) Precise determinations of the equilibria kyanite = sillimanite and kyanite = andalusite and a revised triple point for Al$_2$SiO$_5$ polymorphs. Am Mineral 76:677–680

Bowen NL (1940) Progressive metamorphism of siliceous limestone and dolomites. J Geol 48:225–274

Brown TH, Skinner BJ (1974) Theoretical prediction of equilibrium phase assemblages in multicomponent systems. Am J Sci 274:961–986

Bryndzia LT, Scott SD, Spry PG (1988) Sphalerite and hexagonal pyrrhotite geobarometer: experimental calibration and application to the metamorphosed sulfide ores of Broken Hill, Australia. Econ Geol 83:1193–1204

Bryndzia LT, Scott SD, Spry PG (1990) Sphalerite and hexagonal pyrrhotite geobarometer: correction in calibration and application. Econ Geol 85:408–411

Bucher K, Ohta Y (1993) Granulites and garnet-cordierite gneisses from Dronning Maud Land, Antarctica. J Metamorph Geol 11:691–703

Buddington AF, Lindsley DH (1964) Iron-titanium oxide minerals and their synthetic equivalents. J Petrol 5:310–357

Burnham AK, Sweeney JJ (1989) A chemical kinetic model of vitrinite maturation and reflectance. Geochim Cosmochim Acta 53:2649–2657

Carlson WD (1983) The polymorphs of CaCO$_3$ and the aragonite-calcite transformation. In: Reeder RJ (ed) Carbonates: mineralogy and chemistry. Reviews in Mineralogy, vol. 11. Mineralogical Society of America, Washington, DC, pp 191–225

Carlson WD, Rosenfeld JL (1981) Optical determination of topotactic aragonite-calcite growth kinetics: metamorphic implications. J Geol 89:615–638

Carmichael DM (1970) Intersecting isograds in the Whetstone Lake area, Ontario. J Petrol 11:147–181

Carmichael DM (1978) Metamorphic bathozones and bathograds: a measure of the depth of post-metamorphic uplift and erosion on the regional scale. Am J Sci 278:769–797

Carmichael DM (1991) Univariant mixed-volatile reactions: pressure-temperature phase diagrams and reaction isograds. Can Mineral 29:741–754

Carrington DP, Harley SL (1995) The stability of osumilite in metapelitic granulites. J Metamorph Geol 13:613–625

Carson CJ, Powell R (1997) Garnet-orthopyroxene geothermometry and geobarometry; error propagation and equilibration effects. J Metamorph Geol 15:679–686

Carswell DA, Gibb FGF (1987) Evaluation of mineral thermometers and barometers applicable to garnet lherzolite assemblages. Contrib Mineral Petrol 95:499–511

Carswell DA, Harley SL (1989) Mineral barometry and thermometry. In: Carswell DA (ed) Eclogites and related rocks. Blackie, Glasgow, pp 83–110

Chatterjee ND, Flux S (1986) Thermodynamic mixing properties of muscovite-paragonite crystalline solutions at high temperatures and pressures, and their geological applications. J Petrol 27:677–693

Chernosky JV, Berman RG (1988) The stability of Mg-chlorite in supercritical H$_2$O–CO$_2$ fluids. Am J Sci 288A:393–420

Chinner GA (1966) The distribution of pressure and temperature during Dalradian metamorphism. Q J Geol Soc Lond 122:159–186

Chipera SJ, Perkins D (1988) Evaluation of biotite-garnet geothermometers: application to the English River subprovince, Ontario. Contrib Mineral Petrol 98:40–48

Connolly JAD, Kerrick DM (1987) An algorithm and computer program for calculating computer phase diagrams. CALPHAD 11:1–55

Connolly JAD, Trommsdorff V (1991) Petrogenetic grids for metacarbonate rocks: pressure-temperature phase-diagram projection for mixed-volatile systems. Contrib Mineral Petrol 108:93–105

Cooke RA, O'Brien PJ, Carswell DA (2000) Garnet zoning and the identification of equilibrium mineral compositions in high-pressure-temperature granulites from the Moldanubian Zone, Austria. J Metamorph Geol 18:551–569

Coombs DS, Ellis AJ, Fyfe WS, Taylor AM (1959) The zeolite facies, with comments on the interpretation of hydrothermal syntheses. Geochim Cosmochim Acta 17:53–107

Currie KL, Van Staal CR (1999) The assemblage stilpnomelane-chlorite-phengitic mica: a geothermobarometer for blueschist and associated greenschist terranes. J Metamorph Geol 17:613–620

Dachs E (1994) Uncertainties in the activities of garnets and their propagation into geothermobarometry. Eur J Mineral 6:291–295
de Capitani C, Brown TH (1987) The computation of chemical equilibrium in complex systems containing non-ideal solutions. Geochim Cosmochim Acta 51:2639–2652
Docka JA, Berg JH, Klewin K (1986) Geothermometry in the Kiglapait aureole. II. Evaluation of exchange thermometry in a well-constrained setting. J Petrol 27:605–626
Eckert JO, Newton RC, Kleppa OJ (1991) The DH of reaction and recalibration of garnet-
   pyroxene-plagioclase-quartz geobarometers in the CMAS system by solution calorimetry.
   Am Mineral 76:148–160
Edwards RL, Essene EJ (1988) Pressure, temperature and C-O-H fluid fugacities across the
   amphibolite-granulite facies transition, NW Adirondack Mtns, NY. J Petrol 29:39–72
Eggler DH, Lorand JP (1993) Mantle sulfide geobarometry. Geochim Cosmochim Acta
   57:2213–2222
Elkins LT, Grove TL (1990) Ternary feldspar experiments and thermodynamic models. Am
   Mineral 75:544–559
Ellis DJ, Green DH (1979) An experimental study of the effect of Ca upon garnet-clinopyr-
   oxene Fe-Mg exchange equilibria. Contrib Mineral Petrol 71:13–22
El-Shazly AK, Liou JG (1991) Glaucophane chloritoid-bearing assemblages from NE Oman:
   petrologic significance and a petrogenetic grid for high P metapelites. Contrib Mineral
   Petrol 107:180–201
Engi M (1983) Equilibria involving Al-Cr spinel. I. Mg-Fe exchange with olivine; experi-
   ments, thermodynamic analysis, and consequences for geothermometry. In: Greenwood
   HJ (ed) Orville volume; studies in metamorphism and metasomatism. Am J Sci 283A:
   29–71
Engi M, Todd CS, Schmatz DR (1995) Tertiary metamorphic conditions in the eastern
   Lepontine Alps. Schweiz Mineral Petrogr Mitt 75:347–369
Epstein AG, Epstein JB, Harris LD (1977) Conodont color alteration – an index to organic
   metamorphism. US Geol Surv Prof Paper, 955 pp
Ernst WG (1976) Petrologic phase equilibria. Freeman, San Francisco, 333 pp
Eskola P (1915) On the relations between the chemical and mineralogical composition in
   the metamorphic rocks of the Orijarvi region. Bull Comm Geol Finlande 44
Essene EJ (1982) Geologic thermometry and barometry. In: Ferry JM (ed) Characterization
   of metamorphism through mineral equilibria. Reviews in mineralogy, vol 10. Mineralogical Society of America, Washington, DC, pp 153–205
Essene EJ (1983) Solid solutions and solvi among metamorphic carbonates with applica-
   tions to geologic thermometry. In: Reeder RJ (ed) Carbonates: mineralogy and chemis-
   try. Reviews in mineralogy, vol 11. Mineralogical Society of America, Washington, DC,
   pp 77–96
Essene EJ (1989) The current status of thermobarometry in metamorphic rocks. In: Daly St,
   Cliff RA, Yardley BWD (eds) Evolution of metamorphic belts. Geological Society Special
   Publication. Blackwell, Oxford, pp 1–44
Evans BW (1990) Phase relations of epidote-blueschists. Lithos 25:3–23
Evans BW, Guidotti CV (1966) The sillimanite-potash feldspar isograd in western Maine,
   USA. Contrib Mineral Petrol 12:25–62
Evans BW, Patrick BE (1987) Phengite-3T in high-pressure metamorphosed granitic ortho-
   gneisses, Seward Peninsula, Alaska. Can Mineral 25:141–158
Faulhaber S, Raith M (1991) Geothermometry and geobarometry of high-grade rocks: a case
   study on garnet-pyroxene granulites in southern Sri Lanka. Mineral Mag 55:33–56
Ferry JM (1980) A comparative study of geothermometers and geobarometers in pelitic
   schists from southern-central Maine. Am Mineral 65:720–732
Ferry JM (2000) Patterns of mineral occurence in metamorphic rocks. Am Mineral 85:1573–
   1588
Ferry JM (2001) Calcite inclusions in forsterite. Am Mineral 86:773–779
Ferry JM, Spear FS (1978) Experimental calibration of the partitioning of Fe and Mg between biotite and garnet. Contrib Mineral Petrol 66:113–117
Flux S, Chatterjee ND (1986) Experimental reversal of the Na-K exchange reaction between
   muscovite-paragonite crystalline solutions and a 2 molal aqueous (Na,K)Cl fluid. J Petrol
   27:665–676
Fonarev VI, Konilov AN (1986) Experimental study of Fe-Mg distribution between biotite
   and orthopyroxene at P = 490 MPa. Contrib Mineral Petrol 93:227–235
Fox JS (1975) Three-dimensional isograds from the Lukmanier Pass, Switzerland, and their
   tectonic significance. Geol Mag 112:547–564

Frey M (1987) The reaction-isograd kaolinite+quartz = pyrophyllite+H$_2$O, Helvetic Alps, Switzerland. Schweiz Mineral Petrogr Mitt 67:1–11

Frey M, Wieland B (1975) Chloritoid in autochthon-parautochthonen Sedimenten des Aarmassivs. Schweiz Mineral Petrograph Mitt 55:407–418

Frey M, De Capitani C, Liou JG (1991) A new petrogenetic grid for low-grade metabasites. J Metamorph Geol 9:497–509

Frost BR, Chacko T (1989) The granulite uncertainty principle: limitations on thermobarometry in granulites. J Geol 97:435–450

Frost BR, Lindsley DH (1992) Equilibria among Fe-Ti oxides, pyroxenes, olivine, and quartz. II. Application. Am Mineral 77:1004–1020

Fuhrman ML, Lindsley DH (1988) Ternary feldspar modeling and thermometry. Am Mineral 73:201–215

Fyfe WS, Turner FJ, Verhoogen J (1958) Metamorpic reactions and metamorphic facies. Geol Soc Am Bull 73:25–34

Gasparik T (1984) Experimental study of subsolidus phase relations and mixing properties of pyroxene in the system CaO–Al$_2$O$_3$–SiO$_2$. Geochim Cosmochim Acta 48:2537–2546

Gessmann CK, Spiering B, Raith M (1997) Experimental study of the Fe-Mg exchange between garnet and biotite: constrains on the mixing behaviour and analysis of the cation exchange mechanisms. Am Mineral 82:1225–1240

Ghent ED, Gordon TM (2000) Application of INVEQ to the geothermobarometry of metamorphic rocks near a kyanite-sillimantite isograd, Mica Creek, British Colombia. Am Mineral 85:9–13

Giletti BJ (1986) Diffusion effects on oxygen isotope temperatures of slowly cooled igneous and metamorphic rocks. Earth Planet Sci Lett 77:218–228

Graham CM, Powell R (1984) A garnet-hornblende geothermometer and application to the Peloma Schist, southern California. J Metamorph Geol 2:13–32

Grambling JA (1990) Internally-consistent geothermometry and H$_2$O barometry in metamorphic rocks: the example garnet-chlorite-quartz. Contrib Mineral Petrol 105:617–628

Grambling JA, Williams ML (1985) The effects of Fe^{3+} and Mn^{3+} on aluminum silicate phase relations in north-central New Mexico, USA. J Petrol 26:324–354

Green TH, Adam J (1991) Assessment of the garnet-clinopyroxene Fe-Mg exchange thermometer using new experimental data. J Metamorph Geol 9:341–347

Guidotti CV (1984) Micas in metamorphic rocks. In: Bailey SW (ed) Micas. Reviews in mineralogy, vol 13. Mineralogical Society of America, Washington, DC, pp 357–467

Guidotti CV, Dyar MB (1991) Ferric iron in metamorphic biotite and its petrologic and crystallochemical implications. Am Mineral 76:161–175

Guidotti CV, Sassi FP, Blencoe JG, Selverstone J (1994) The paragonite-muscovite solvus. I. P-T-X limits derived from the Na-K compositions of natural, quasibinary paragonite-muscovite pairs. Geochim Cosmochim Acta 58:2269–2275

Guiraud M, Holland T, Powell R (1990) Calculated mineral equilibria in the greenschist-blueschist-eclogite facies in Na$_2$O–FeO–MgO–Al$_2$O$_3$–SiO$_2$–H$_2$O. Contrib Mineral Petrol 104:85–98

Haselton HT, Hovis GL, Hemingway BS, Robie RA (1983) Calorimetric investigation of the excess entropy of mixing in analbite-sanidine solid solutions: lack of evidence for Na, K short range order and implications for two-feldspar thermometry. Am Mineral 68:398–413

Heinrich CA (1986) Eclogite facies regional metamorphism of hydrous mafic rocks in the Central Alpine Adula Nappe. J Petrol 27:123–154

Hodges KV, McKenna LW (1987) Realistic propagation of uncertainties in geologic thermobarometry. Am Mineral 72:671–680

Hodges KV, Spear FS (1982) Geothermometry, geobarometry and the Al$_2$SiO$_5$ triple point at Mt. Moosilauke, New Hampshire. Am Mineral 67:1118–1134

Hoefs J (1987) Stable isotope geochemistry, 3rd edn. Springer, Berlin Heidelberg New York, 241 pp

Hokada T (2001) Feldspar thermometry in ultrahigh-temperature metamorphic rocks: evidence of crustal metamorphism attaining ~1100 °C in the archean Napier Complex, East Antarctica. Am Mineral 86:932–938

Holdaway MJ (1971) Stability of andalusite and the aluminum silicate phase diagram. Am J Sci 271:97–131

Holdaway MJ (2000) Application of new experimental and garnet Margules data to the garnet-biotite geothermometer. Am Mineral 85:881–892

Holdaway MJ (2001) Recalibration of the GASP geobarometer in light of recent garnet and plagioclase activity models and versions of the garnet-biotite geothermometer. Am Mineral 86:1117–1129

Holdaway MJ, Mukhopadhyay B (1993) Geothermobarometry in pelitic schists: a rapidly evolving field. Am Mineral 78:681–693

Holdaway MJ, Dutrow BL, Hinton RW (1988) Devonian and Carboniferous metamorphism in west-central Maine: the muscovite-almandine geobarometer and the staurolite problem revisited. Am Mineral 73:20–47

Holdaway MJ, Mukhopadhyay B, Dyar MD, Guidotti CV, Dutrow BL (1997) Garnet-biotite geothermometry revised; new Margules parameters and a natural specimen data set from Maine. Am Mineral 82:582–595

Holland TJB (1980) The reaction albite=jadeite+quartz determined experimentally in the range 600–1200 °C. Am Mineral 65:129–134

Holland TJB, Powell R (1990) An enlarged and updated internally consistent thermodynamic dataset with uncertainties and correlations: the system $K_2O-Na_2O-CaO-MgO-MnO-FeO-Fe_2O_3-Al_2O_3-TiO_2-SiO_2-C-H_2-O_2$. J Metamorph Geol 8:89–124

Holland TJB, Powell R (1998) An internally-consistent thermodynamic dataset for phases of petrological interest. J Metamorph Geol 16:309–343

Huckenholz HG, Lindhuber W, Fehr KT (1981) Stability of grossular+quartz+wollastonite+anorthite: the effect of andradite and albite. Neues Jahrb Mineral Abh 142:223–247

Hutchison MN, Scott SD (1980) Sphalerite geobarometry applied to metamorphosed sulfide ores of the Swedish Caledonides and U.S. Appalachians. Norges Geol Undersøkelse 360:59–71

Jamieson RA, Craw D (1987) Sphalerite geobarometry in metamorphic terranes: an appraisal with implications for metamorphic pressure in the Otago Schist. J Metamorph Geol 5:87–99

Johnson CA, Essene EJ (1982) The formation of garnet in olivine-bearing metagabbros from the Adirondacks. Contrib Mineral Petrol 81:240–251

Kerrick DM (ed) (1990) The Al_2SiO_5 polymorphs. Reviews in mineralogy, vol 22. Mineralogical Society of America, Washington, DC, 406 pp

Kirschner DL, Sharp ZD, Masson H (1995) Oxygen isotope thermometry of quartz-calcite veins: unraveling the thermal-tectonic history of the subgreenschist facies Morcles nappe (Swiss Alps). Geol Soc Am Bull 107:1145–1156

Kitchen NE, Valley JW (1995) Carbon isotope thermometry in marbles of the Adirondack Mountains, New York. J Metamorph Geol 13:577–594

Kleemann U, Reinhardt J (1994) Garnet-biotite thermometry revisited. I. The effect of Al^{VI} and Ti in biotite. Eur J Mineral 6:925–941

Kohn MJ, Spear FS (1989) Empirical calibration of geobarometers for the assemblage garnet+hornblende+plagioclase+quartz. Am Mineral 74:77–84

Kohn MJ, Spear FS (1991) Error propagation for barometers. 2. Application to rocks. Am Mineral 76:138–147

Koons PO (1984) Implications to garnet-clinopyroxene geothermometry of non-ideal solid solution in jadeitic pyroxenes. Contrib Mineral Petrol 88:340–347

Koziol AM, Newton RC (1988) Redetermination of the garnet breakdown reaction and improvement of the plagiclase-garnet-Al_2SiO_5-quartz geobarometer. Am Mineral 73:216–223

Kretz R (1991) A note on transfer reactions. Can Mineral 29:823–832

Krogh EJ (1988) The garnet-clinopyroxene Fe-Mg geothermometer – a reinterpretation of existing experimental data. Contrib Mineral Petrol 99:44–48

Kroll H, Evangelakakis C, Voll C (1993) Two-feldspar geothermometry, a review and revision for slowly cooled rocks. Contrib Mineral Petrol 114:510–518

Laird J (1988) Chlorites: metamorphic petrology. In: Bailey SW (ed) Hydrous phyllosilicates. Reviews in mineralogy, vol 19. Mineralogical Society of America, Washington, DC, pp 404–454

Lal RK (1997) Internally consistent calibrations for geothermobarometry of high-grad Mg-Al rich rocks in the system $MgO-Al_2O_3-SiO_2$ and their application to sapphirine-spinel granulites of Eastern Ghats, India and Enderby Land, Antarctica. Earth Planet Sci Lett 106:91–113

Lang HM, Rice JM (1985a) Regression modelling of metamorphic reactions in metapelites, Snow Peak, northern Idaho. J Petrol 26:857–887

Lang HM, Rice JM (1985b) Geothermometry, geobarometry and T–X (Fe-Mg) relations in metapelites, Snow Peak, northern Idaho. J Petrol 26:889–924

Letargo CMR, Lamb WM, Park J-S (1995) Comparsion of calcite+dolomite thermometry and carbonate+silicate equilibria: constraints on the conditions of metamorphism of the Llano uplift, central Texas, USA. Am Mineral 80:131–143

Lieberman J (1991) GridLoc v.1.3 a phase diagram plotter. (unpublished)

Lieberman J, Petrakakis K (1991) TWEEQU thermobarometry: analysis of uncertainties and applications to granulites from western Alaska and Austria. Can Mineral 29:857–887

Lindsley DH (1983) Pyroxene thermometry. Am Mineral 68:477–493

Lindsley DH, Andersen DJ (1983) A two-pyroxene thermometer. Proceedings of the 13th Lunar and Planetary Science Conference, part 2. J Geophys Res 88 [Suppl]:A887–A906

Lindsley DH, Frost BR (1992) Equilibria among Fe-Ti oxides, pyroxenes, olivine, and quartz. I. Theory. Am Mineral 77:987–1003

Liogys VA, Jenkins DM (2000) Hornblende geothermometry of amphibolite layers of the Popple Hill gneiss, north-west Adirondack Lowlands, New York, USA. J Metamorph Geol 18:513–530

Liou JG, Maruyama S, Cho M (1987) Very low-grade metamorphism of volcanic and volcaniclastic rocks – mineral assemblages and mineral facies. In: Frey M (ed) Low temperature metamorphism. Blackie, Glasgow, pp 59–114

Mahar EM, Baker JM, Powell R, Holland TJB (1997) The effect of Mn on mineral stability in metapelites. J Metamorph Geol 15:223–238

Manning CE, Bohlen SR (1991) The reaction titanite+kyanite = anorthite+rutile and titanite-rutile barometry in eclogites. Contrib Mineral Petrol 109:1–9

Massonne HJ, Chopin C (1989) P–T history of the Gran Paradiso (western Alps) metagranites based on phengite geobarometry. In: Daly JS, Cliff RA, Yardley BWD (eds) Evolution of metamorphic belts. Geol Soc Spec Publ 43:545–549

Massonne HJ, Schreyer W (1987) Phengite geobarometry based on the limiting assemblage with K-feldspar, phlogopite, and quartz. Contrib Mineral Petrol 96:212–224

McKenna LW, Hodges KV (1988) Accuracy versus precision in locating reaction boundaries: implications for the garnet-plagioclase-aluminum silicate-quartz geobarometer. Am Mineral 73:1205–1208

Meyre C, De Capitani C, Partsch JH (1997) A ternary solid solution model for omphacite and its application to geothermobarometry of eclogites from the Middle Adula nappe (Central Alps, Switzerland). J Metamorph Geol 15:687–700

Miyashiro A (1961) Evolution of metamorphic belts. J Petrol 2:277–318

Miyashiro A (1972) Pressure and temperature conditions and tectonic significance of regional and ocean floor metamorphism. Tectonophysics 13:141–159

Moecher DP, Essene EJ, Anovitz LM (1988) Calculation of clinopyroxene-garnet-plagioclase-quartz geobarometers and application to high grade metamorphic rocks. Contrib Mineral Petrol 100:92–106

Mullis J (1987) Fluid inclusion studies during very low-grade metamorphism. In: Frey M (ed) Low temperature metamorphism. Blackie, Glasgow, pp 162–199

Newton RC (1983) Geobarometry of high-grade metamorphic rocks. Am J Sci 283-A:1–28

Newton RC (1986) Metamorphic temperatures and pressures of Group B and C, eclogites. Geol Soc Am Spec Pap 164:17–30

Newton RC, Fyfe WS (1976) High pressure metamorphism. In: Bailey DK, MacDonald R (eds) The evolution of the crystalline rocks. Academic Press, London, pp 100–186

Newton RC, Smith JV (1967) Investigations concerning the breakdown of albite at depth in the earth. J Geol 75:268–286

Newton RC, Goldsmith JR, Smith JV (1969) Aragonite crystallization from strained calcite at reduced pressures and its bearing on aragonite in low-grade metamorphism. Contrib Mineral Petrol 22:335–348

Paria P, Bhattacharya A, Sen A (1988) The reaction garnet+clinopyroxene+quartz = 2 orthopyroxene+anorthite: a potential geobarometer for granulites. Contrib Mineral Petrol 199:126–133

Patiño Douce AE, Johnston AD, Rice JM (1993) Octahedral excess mixing properties in biotite: a working model with applications to geobarometry and geothermometry. Am Mineral 78:113–131

Patrick BE, Ghent ED (1988) Empirical calibration of the K-white mica+feldspar+biotite geobarometer. Mineral Soc Canada, program with abstracts 13:A95

Pattison DRM (1992) Stability of andalusite and sillimanite and the Al_2SiO_5 triple point: constraints from the Ballachulish aureole, Scotland. J Geol 100:423–446

Pattison DRM, Newton RC (1989) Reversed experimental calibration of the garnet-clinopyroxene Fe-Mg exchange thermometer. Contrib Mineral Petrol 101:87–103

Pattison DRM, Spear FS, Cheney JT (1999) Polymetamorphic origin of muscovite+cordier-ite+staurolite+biotite assemblages: implications for the metapelitic petrogenetic grids and for P–T paths. J Metamorph Geol 17:685–703

Pattison DRM, Spear FS, DeBuhr CL, Cheney JT, Guidotti CV (2002) Natural assemblage and thermodynamic analysis of the reaction muscovite+cordierite = Al_2SiO_5+biotite+ quartz+H_2O: implications for the metapelitic petrogenetic grid. J Metamorph Geol (in press)

Perchuk LL, Lavrent'eva IV (1983) Experimental investigation of exchange equilibria in the system cordierite-garnet-biotite. In: Saxena SK (ed) Kinetics and equilibrium in mineral reactions. Advances in physical geochemistry, vol 3. Springer, Berlin Heidelberg New York, pp 199–239

Perkins EH, Brown TH, Berman RG (1986) PT-system, TX-system, PX-system: three pro-grams which calculate pressure-temperature-composition phase diagrams. Comput Geo-sci 12:749–755

Popp RK, Gilbert MC (1972) Stability of acmite-jadeite pyroxenes at low pressure. Am Mineral 57:1210–1231

Powell R, Holland TJB (1988) An internally consistent dataset with uncertainties and corre-lations. 3. Applications to geobarometry, worked examples and a computer program. J Metamorph Geol 6:173–204

Powell R, Holland TJB (1990) Calculated mineral equilibria in the pelite system, KFMASH (K_2O–FeO–MgO–Al_2O_3–SiO_2–H_2O). Am Mineral 75:367–380

Powell R, Holland T (1995) Optimal geothermometry and geobarometry. Am Mineral 79:120–133

Powell R, Holland T (2001) Course notes for "THERMOCALC workshop 2001: calculating metamorphic phase equilibria" on CD-ROM (available from the authors)

Powell R, Condliffe DM, Condliffe E (1984) Calcite-dolomite geothermometry in the system $CaCO_3$–$MgCO_3$–$FeCO_3$: an experimental study. J Metamorph Geol 2:33–41

Powell WG, Carmichael DM, Hodgson CJ (1993) Thermobarometry in a sub-greenschist to greenschist transition in metabasites of the Abitibi greenstone belt, Superior Province, Canada. J Metamorph Geol 11:165–178

Ramberg H (1952) The origin of metamorphic and metasomatic rocks. Univ Chicago Press, Chicago, 317 pp

Reinecke T (1991) Very-high-pressure metamorphism and uplift of coesite-bearing meta-sediments from the Zermatt-Saas zone, Western Alps. Eur J Mineral 3:7–17

Rejebian VA, Harris AG, Huebner JS (1987) Conodont color and textural alteration: an index to regional metamorphism, contact metamorphism, and hydrothermal alteration. Geol Soc Am Bull 99:471–479

Richardson SW, Gilbert MC, Bell PM (1969) Experimental determination of kyanite-anda-lusite and andalusite-sillimanite equilibrium; the aluminium silicate triple point. Am J Sci 267:259–272

Salje E (1986) Heat capacities and entropies of andalusite and sillimanite: the influence of fi-brolization on the phase diagram of the Al_2SiO_5 polymorphs. Am Mineral 71:1366–1371

Savin SM, Lee M (1988) Isotopic studies of phyllosilicates. In: Bailey SW (ed) Hydrous phyl-losilicates. Reviews in mineralogy, vol 19. Mineralogical Society of America, Washington, DC, pp 189–223

Schliestedt M (1986) Eclogite-blueschist relationships as evidenced by mineral equilibria in the high-pressure metabasic rocks of Sifnos (Cycladic Islands), Greece. J Petrol 27:1437–1459

Schmädicke E (1991) Quartz pseudomorphs after coesite in eclogites from the Saxonian Erzgebirge. Eur J Mineral 3:231–238

Shulters JC, Bohlen SR (1989) The stability of hercynite and hercynite-gahnite spinels in corundum- or quartz-bearing assemblages. J Petrol 30:1017–1031

Schumacher JC (1991) Empirical ferric iron corrections: necessity, assumptions and effects on selected geothermobarometers. Mineral Mag 55:3–18

Selverstone J, Spear FS, Franz G, Morteani G (1984) High-pressure metamorphism in the SW Tauern Window, Austria: P–T paths from hornblende–kyanite–staurolite schists. J Petrol 25:501–531

Sengupta P, Dasgupta S, Bhattacharya PK, Mukherjee M (1990) An orthopyroxene-bitote geothermometer and its application in crustal granulites and mantle-derived rocks. J Metamorph Geol 8:191–198

Simpson GDH, Thompson AB, Connolly JAD (2000) Phase relations, singularities and thermobarometry of metamorphic assemblages containing phengite, chlorite, biotite, K-feldspar, quartz and H_2O. Contrib Mineral Petrol 139:555–569

Soto JI (1993) PTMAFIC: software for thermobarometry and activity calculations with mafic and ultramafic assemblages. Am Mineral 78:840–844

Spear FS (1989) Relative thermobarometry and metamorphic P–T paths. In: Daly S, Cliff RA, Yardley BWD (eds) Evolution of metamorphic belts. Geological Society Special Publication, Blackwell, Oxford, pp 63–82

Spear FS (1992) On the interpretation of peak metamorphic temperatures in light of garnet diffusion during cooling. J Metamorph Geol 9:379–388

Spear FS (1993) Metamorphic phase equilibra and pressure-temperature-time path. Mineralogical Society of America, Washington, DC, 824 pp

Spear FS, Menard T (1989) Program GIBBS: a generalized Gibbs method algorithm. Am Mineral 74:942–943

Spear FS, Ferry JM, Rumble D III (1982) Analytical formulation of phase equilibria: the Gibbs' method. In: Ferry JM (ed) Characterization of metamorphism through mineral equilibria. Reviews in Mineralogy, vol 10. Mineralogical Society of America, Washington, DC, pp 105–152

Spear FS, Selverstone J, Hickmont D, Crowley P, Hodges KV (1984) P–T paths from garnet zoning: a new technique for deciphering tectonic processes in crystalline terranes. Geology 12:87–90

Spear FS, Peacock SM, Kohn MJ, Florence FP, Menard T (1991) Computer programs for petrologic P–T–t path calculations. Am Mineral 76:2009–2012

Spear FS, Kohn MJ, Cheney JT (1999) P–T paths from anatectic pelites. Contrib Mineral Petrol 134:17–32

Stephenson NCN (1984) Two-pyroxene thermometry of Precambrian granulites from Cape Riche, Albany-Fraser Province, Western Australia. J Metamorph Geol 2:297–314

St-Onge MR (1987) Zoned poikiloblastic garnets: P–T paths and syn-metamorphic uplift through 30 km of structural depth, Wopmay Orogen, Canada. J Petrol 28:1–21

Stormer JC (1975) A practical two-feldspar geothermometer. Am Mineral 60:667–674

Streckeisen A, Wenk E (1974) On steep isogradic surfaces in the Simplon area. Contrib Mineral Petrol 47:81–95

Stüwe K, Sandiford M (1994) Contribution of deviatoric stresses to metamorphic P–T paths: an example appropriate to low-P, high-T metamorphism. J Metamorph Geol 12:445–454

Sweeney JJ, Burnham AK (1990) Evaluation of a simple model of vitrinite reflectance based on chemical kinetics. Am Assoc Petroleum Geologists Bull 74:1559–1570

Symmes GH, Ferry JM (1992) The effect of whole-rock MnO content on the stability of garnet in pelitic schists during metamorphism. J Metamorph Geol 10:221–237

Thoenen T (1989) A comparative study of garnet-biotite geothermometers. Doctoral Dissertation, University of Basel, 118 pp

Thompson PH (1976) Isograd patterns and pressure-temperature distributions during regional metamorphism. Contrib Mineral Petrol 57:277–295

Tilley CE (1924) The facies classification of metamorphic rocks. Geol Mag 61:167–171

Tilley CE (1925) Metamorphic zones in the southern Highlands of Scotland. Q J Geol Soc 81:100–112

Todd CS (1998) Limits on the precision of geobarometry at low grossular and anorthite content. Am Mineral 83:1161–1167

Todd CS, Engi M (1997) Metamorphic field gradients in the Central Alps. J Metamorph Geol 15:513–530

Tracy RJ, Robinson P, Thompson AB (1976) Garnet composition and zoning in the determination of temperature and pressure of metamorphism, Central Massachusetts. Am Mineral 61:762–775

Triboulet C, Thiéblemont D, Audren C (1992) The (Na-Ca) amphibole-albite-chlorite-epidote-quartz geothermobarometer in the system S-A-F-M-C-N-H_2O. 2. Applications to metabasic rocks in different metamorphic settings. J Metamorph Geol 10:557–566

Trommsdorff V, Evans BW (1972) Progressive metamorphism of antigorite schist in the Bergell tonalite aureole (Italy). Am J Sci 272:487–509

Trommsdorff V, Evans BW (1977) Antigorite–ophicarbonates: contact metamorphism in Valmalenco, Italy. Contrib Mineral Petrol 62:301–312

Turner FJ (1981) Metamorphic petrology – mineralogical, field and tectonic aspects, 2nd edn, McGraw-Hill, New York

Valley JW (1986) Stable isotope geochemistry of metamorphic rocks. In: Valley JW, Taylor
    HP Jr, O'Neil JR (eds) Stable isotopes in high temperature geological processes, Reviews
    in Mineralogy, vol 16. Mineralogical Society of America, Washington, DC, pp 445–489
Vidal O, Goffé B, Theye T (1992) Experimental study of the stability of sudoite and magne-
    siocarpholite and calculation of a new petrogenetic grid for the system FeO–MgO–
    Al_2O_3–SiO_2–H_2O. J Metamorph Geol 10:603–614
Vidal O, Parra T, Trotet F (2001) A thermodynamic model for Fe-Mg aluminous chlorite
    using data from phase equilibrium experiments and natural pelitic assemblages in the
    100° to 600°C, 1 to 15 kb range. Am J Sci 301:557–592
Vielzeuf D (1983) The spinel and quartz associations in high grade xenoliths from Tallante
    (SE Spain) and their potential use in geothermometry and barometry. Contrib Mineral
    Petrol 82:301–311
Wen S, Nekvasil H (1994) SOLVCALC: an interactive graphics program package for calculat-
    ing the ternary feldspar solvus and for two-feldspar geothermometry. Comput Geosci
    20:1025–1040
Will TM, Powell R, Holland TJB (1990a) A calculated petrogenetic grid for ultramafic rocks
    in the system CaO–FeO–MgO–Al_2O_3–SiO_2–CO_2–H_2O at low pressures. Contrib Mineral
    Petrol 105:347–358
Will TM, Powell R, Holland T, Guiraud M (1990b) Calculated greenschist facies mineral
    equilibria in the system CaO–FeO–MgO–Al_2O_3–SiO_2–CO_2–H_2O. Contrib Mineral Petrol
    104:353–368
Will T, Okrusch M, Schmaedicke E, Chen G (1998) Phase relations in the greenschist-blue-
    schist-amphibolite-eclogite facies in the system Na_2O–CaO–FeO–MgO–Al_2O_3–SiO_2–H_2O
    (NCFMASH), with application to metamorphic rocks from Samos, Greece. Contrib
    Mineral Petrol 132:85–102
Winkler HGF (1979) Petrogenesis of metamorphic rocks. Springer, Berlin Heidelberg New
    York, 348 pp
Worley B, Powell R (1998a) Making movies; phase diagrams changing in pressure, tempera-
    ture, composition and time. In: Treloar PJ, O'Brien PJ (eds) What drives metamorphism
    and metamorphic relations? Geological Society Special Publications, vol 138. pp 269–280
Worley B, Powell R (1998b) Singularities in NCKFMASH (Na_2O–CaO–K_2O–FeO–MgO–
    Al_2O_3–SiO_2–H_2O). J Metamorph Geol 16:169–188
Worley B, Powell R (2000) High-precision relative thermobarometry; theory and a worked
    example. J Metamorph Geol 18:91–101
Yardley BWD, Schumacher JC (eds) (1991) Geothermometry and geobarometry. Mineral
    Mag (special volume on GTB) 55:378 pp
Yui T.-F, Huang E, Xu J (1996) Raman spectrum of carbonaceous material: a possible meta-
    morphic grade indicator for low-grade metamorphic rocks. J Metamorph Geol 14:115–
    124
Zwart HJ (1973) Metamorphic map of Europe. Leiden/UNESCO, Paris

Part II

CHAPTER 5

Metamorphism of Ultramafic Rocks

5.1
Introduction

The earth's mantle consists predominantly of ultramafic rocks. The mantle is, with the exception of some small anomalous regions, in a solid state. The ultramafic rocks undergo continuous recrystallization due to large-scale convection in the sublithosphere mantle and as a result of tectonic processes in the lithosphere. The bulk of the mantle rocks, therefore, meet the criteria of metamorphic rocks. Metamorphic ultramafic rocks represent the largest volume of rocks of the planet.

Most of the ultramafic rocks found in the earth's crust and consequently found in surface outcrops have been tectonically implanted from the mantle during orogenesis. Such ultramafic mantle fragments in the crust are often referred to as Alpine-type peridotites. Ultramafic rocks of igneous origin do occur in the crust but are relatively rare. Most of the igneous ultramafic rocks in the crust formed from fractional crystallization from basic (gabbroic, basaltic) magmas in crustal magma chambers (olivine-saturated cumulate layers).

There are two basic types of mantle fragments in the crust: (1) mantle fragments from beneath oceanic crust. They constitute integral members of ophiolite sequences and are often lherzolitic in bulk composition (see below); (2) mantle fragments from subcontinental mantle occur in rock associations typical of continental crust. Such ultramafites are usually of harzburgitic (or dunitic) composition.

5.2
Metamorphic Ultramafic Rocks

Ultramafic mantle rock fragments experience strong mineralogical and structural modifications during emplacement in the crust and subsequent crustal deformation and metamorphism. Two completely different situations can be conceived.

1. Mantle fragments may retain parts of the original mineralogy and structure. Equilibration may be incomplete, because of limited access of water or slow reaction kinetics at low temperature. The ultramafics may not (or only partially) display mineral assemblages that equilibrated at the same conditions as the surrounding crustal rocks. Such ultramafic rocks may be

termed **allofacial**. Allofacial ultramafics are particularly typical members
of ophiolite complexes, metamorphosed under relatively low-grade conditions.

2. Ultramafic rocks may completely equilibrate at the same P–T conditions as
the surrounding crustal rocks. They may show rare relics of original
mineralogy and structure (e.g., cores of Cr-spinel in metamorphic magnetite). The ultramafic rocks register identical P–T histories as their crustal
envelope. Ultramafic rocks showing such characteristics may be designated **isofacial**. Most ultramafics in higher-grade terrains conserve very
little of their mantle heritage. On the other hand, transitions between the
two extremes, as usual, do occur (or are probably the rule). For example,
the assemblage Fo + Atg + Di (see below) found in antigorite schists may
be isofacial but structures and relict minerals often permit a fairly confi-
dent reconstruction of the pre-crustal state of the ultramafics (chlorite
exsolution from high-Al mantle Cpx, Cr-spinel, polygonization of mantle
olivine megacrysts, etc.).

5.2.1
Rock Types

A few examples of metamorphic ultramafic rocks are listed below. **Serpentinites** are massive or schistose rocks containing abundant minerals of the serpentine group. Serpentinites represent hydrated low-temperature versions of
mantle lherzolites or harzburgites. **Peridotite** is often used for olivine-bearing
ultramafic mantle rocks. Partially serpentinized allofacial mantle peridotite
may still be called peridotite; a recrystallized isofacial olivine-bearing anti-
gorite-schist is described as serpentinite, but not as peridotite. Rocks containing metamorphic olivine and enstatite are named En + Fo **felses**. Carbonate-bearing serpentinites are often designated as **ophicarbonate rocks**
(ophicalcite, etc.). Carbonate-bearing talc schist (**talc fels**) is known as **soapstone**. The name **sagvandite**, from the type locality Sagelv vatnet in northern
Norway, is often used for carbonate-bearing enstatite fels. In the Scandinavian Caledonides, sagvandite is very widespread.

The very characteristic yellow-brownish or even reddish weathering rind
of ultramafic rocks make peridotite outcrops very special. Such places have
fascinated and attracted people at all times. Knobs and hills with peridotite
outcrops have been favorite cult sites in many cultures. This is often still dis-
cernable from geographical place names.

Soapstone has been a sought-after material in all ancient cultures. It has
been used to build stoves and utensils for cooking and of course as material
for cult objects. Many place names on modern topographic maps still indicate the presence or the former presence of a soapstone occurrence.

5.2.2
Chemical Composition

Ultramafic rocks consist predominantly of ferro-magnesian silicates. Anhydrous ultramafic rocks contain the three minerals olivine, orthopyroxene and calcic clinopyroxene in various proportions. Olivine, Opx and Cpx together dominate the modal composition of anhydrous ultramafites. Consequently, the system components SiO_2, FeO, MgO, and CaO constitute >95 wt% of almost all anhydrous ultramafites. Fe represents an important component in most ultramafic rocks. Much of it is normally attached to a spinel phase (magnetite, spinel, chromite) and, consequently, Fe occurs in two different oxidation states. In the silicates, Fe-Mg substitution in olivine and Opx seldom exceeds 5–15 mol% (X_{Mg} ~0.85–0.95). Thus, phase relationships in ultramafic rocks can be discussed in Fe-free systems as a first approximation. We will discuss possible effects of additional components such as iron where necessary and appropriate. Most ultramafic rocks contain hydrates (amphiboles, sheet silicates, etc.) and very often also carbonates. Therefore, H_2O and CO_2 must be added to the set of components in order to describe phase relationships in partially or fully hydrated (and/or carbonated) versions of Ol + Opx + Cpx rocks.

The system SiO_2–MgO–CaO–H_2O–CO_2 (CMS–HC system) or subsystems thereof are adequate for a discussion of the metamorphism of ultramafic rocks (note: the same system will be used to describe reactions and assemblages in sedimentary carbonate rocks, Chap. 6). Figure 5.1 shows the chemography of the CMS-HC system projected from H_2O and CO_2 onto the SiO_2-CaO-MgO plane. The composition of mantle rocks with the anhydrous mineralogy Ol + Opx + Cpx is restricted to the shaded area defined by forsterite (Ol), enstatite (Opx) and diopside (Cpx). Rocks with compositions inside

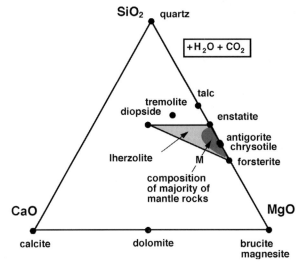

Fig. 5.1. Chemography of the system CMS-HC projected from CO_2 and H_2O onto the plane CaO–MgO–SiO_2 showing some rock and mineral compositions typical of ultramafic rocks

Table 5.1. Reactions in ultramafic rocks

Composition of phase components:

Chrysotile	Ctl	$Mg_3Si_2O_5(OH)_4$
Antigorite	Atg	$Mg_{48}Si_{34}O_{85}(OH)_{62}$
Brucite	Brc	$Mg(OH)_2$
Talc	Tlc	$Mg_3Si_4O_{10}(OH)_4$
Forsterite	Fo	Mg_2SiO_4
Enstatite	En	$Mg_2Si_2O_6$
Anthophyllite	Ath	$Mg_7Si_8O_{22}(OH)_2$
Diopside	Di	$CaMgSi_2O_6$
Tremolite	Tr	$Ca_2Mg_5Si_8O_{22}(OH)_2$
Chlorite	Chl	$Mg_5Al_2Si_3O_{10}(OH)_8$
Anorthite	An	$CaAl_2Si_2O_8$
Pyrope	Py	$Mg_3Al_2Si_3O_{12}$

MSH	(1)	$15\,Ctl + Tlc \Rightarrow Atg$	First antigorite
	(2)	$17\,Ctl \Rightarrow Atg + 3\,Brc$	Last chrysotile
	(3)	$Atg + 20\,Brc \Rightarrow 34\,Fo + 51\,H_2O$	First forsterite (low-T limit of olivine)
	(4)	$Atg \Rightarrow 18\,Fo + 4\,Tlc + 27\,H_2O$	Last antigorite (high-T limit of serpentinites)
	(5)	$9\,Tlc + 4\,Fo \Rightarrow 5\,Ath + 4\,H_2O$	Tlc-out (at lower P)
	(6)	$2\,Tlc + 2\,Fo \Rightarrow 5\,En + 2\,H_2O$	Tlc-out (at higher P)
	(7)	$2\,Ath + 2\,Fo \Rightarrow 9\,En + 2\,H_2O$	Ath-out
CMSH	(8)	$Atg + 8\,Di \Rightarrow 18\,Fo + 4\,Tr + 27\,H_2O$	Upper limit of Di in serpentinites
	(9)	$2\,Tr + 2\,Fo \Rightarrow 5\,En + 4\,Di + 2\,H_2O$	Tremolite-out, lherzolite assemblage
MASH	(10)	$Chl \Rightarrow Fo + En + Spl + 4\,H_2O$	Chlorite-out
	(11)	$Py + Fo \Rightarrow 2\,En + Spl$	Garnet-spinel boundary
CMASH	(12)	$En + Di + Spl \Rightarrow 2\,Fo + An$	Spinel-plagioclase boundary
	(13)	$4\,Spl + 2\,Tr \Rightarrow 6\,Fo + En + 4\,An + 2\,H_2O$	Tr + Spl-out
	(14)	$4\,Chl + 2\,Tr \Rightarrow 10\,Fo + 5\,En + 4\,An + 18\,H_2O$	Tr + Chl-out

the triangle Fo–En–Di are usually termed lherzolites (undepleted mantle peridotite). Rocks falling on the Fo-En join are harzburgites (depleted mantle peridotites) and rocks falling close to the Fo corner in Fig. 5.1 are referred to as dunites. Pyroxenites are rocks with compositions along the Di-En join (including the Di and En corners).

Most of the accessible mantle material (tectonic fragments in the crust, xenoliths in basalts and other mantle-derived volcanic rocks) and of the ultramafic cumulate material is olivine normative. In this chapter, we will not consider metamorphism of pyroxenites and their hydrated equivalents. In addition, mantle lherzolites are often rich in Ol + Opx, whereas Cpx (Di) occurs in modal amounts below 20–30%. The composition of the majority of mantle rocks is restricted to the heavily shaded range (labeled "M") in Fig. 5.1. The present chapter will deal only with such rock compositions.

All important minerals occurring in ultramafic rocks are shown in Fig. 5.1. The minerals chrysotile and lizardite are structural varieties of serpentine of very similar composition, whereas antigorite, a third form of serpentine, is slightly less magnesian than Ctl or Lz. The minerals of the serpentine group are related by Eq. (2) in Table 5.1. Ctl and Lz represent the low-temperature serpentine minerals and typically occur at metamorphic grades below the middle greenschist facies. Antigorite is the typical greenschist, blueschist, eclogite and lower amphibolite facies serpentine mineral. All three carbonate minerals in Fig. 5.1 occur in ultramafic rocks. It is apparent from the figure that a large number of potential mineral assemblages in this system are excluded on rock compositional grounds. For example, the assemblage En + Tr + Tlc is outside the composition range of "normal" ultramafic rocks; it will not occur in such rocks and all mineral reactions leading to this assemblage are irrelevant for meta-lherzolites and have not to be considered on the subsequent phase diagrams. Other examples of assemblages outside the lherzolite composition (with reference to Fig. 5.1) are Brc + Fo and Di + En + Tlc. We also will not consider assemblages of the type Dol + Fo + Brc for the same reason.

5.3
Metamorphism in the MSH System

5.3.1
Chemographic Relations in the MSH System

Figure 5.2 shows the chemography of the three-component $MgO–SiO_2–H_2O$ (MSH) system. The baseline is anhydrous. Mantle rocks are normally restricted to the composition range between Fo and En (= harzburgites). Hydrated versions of harzburgites occupy the shaded area in Fig. 5.2. The H_2O in the system may be stored in solid hydrates such as amphiboles (Oam, anthophyllite), sheet silicates (talc, antigorite), hydroxide (brucite), or as a free fluid phase, depending on P–T conditions and hydrologic situation. In an equilibrium situation, many three-phase assemblages are possible that represent hydrated ultramafic rocks, some of which contain three solid phases but no fluid (fluid absent situation); some others contain two solids (minerals) and an aqueous fluid phase (fluid-present condition). For instance, the three minerals Fo + Tlc + Atg may represent the composition of an H_2O-bearing ultramafic rock. The same composition can also be made up from Fo + Tlc + H_2O. In the first case, there is no free fluid present; in the second case, H_2O is present as a distinct fluid phase. In the following, we only deal with fluid-present metamorphism.

Mantle harzburgites may, during weathering or other near-surface hydration processes, reach a state of maximum hydration. Three maximum hydrated assemblages are possible in the MSH system. The assemblages are shown in Fig. 5.2: (1) brucite + chrysotile, (2) chrysotile + talc, and (3) talc + quartz. The latter assemblage is outside the meta-harzburgite composition. However, a "super" hydrated assemblage, chrysotile + quartz (tie line chrysotile-quartz in Fig. 5.2), occasionally occurs in near-surface modifications of harzburgites. This assemblage is probably metastable under all geological conditions relative to chrysotile + talc (the assemblage brucite + quartz is still more hydrous than Ctl + Qtz, but it is definitely metastable under metamorphic P–T conditions). All three hydrated assemblages coexist with free water at subcritical conditions.

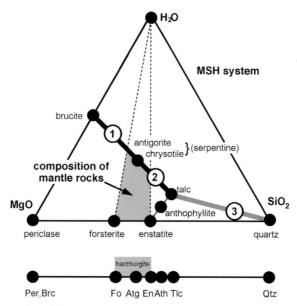

Fig. 5.2. Chemography of the system MSH showing mineral compositions relevant to rocks derived from harzburgite (depleted subcontinental upper mantle; *shaded area*). Maximum hydrated ultramafics contain the assemblages labeled 1, 2, and 3. The projection of the MSH system from H_2O onto the $MgO-SiO_2$ binary is shown in the lower half of the figure together with the composition range of harzburgites

Meta-harzburgites are often discussed by using projections onto the binary $MgO-SiO_2$ from H_2O. The resulting figures can be used for fluid-present conditions. Figure 5.2 also shows such a binary projection of the chemical compositions of the relevant phases in the MSH system for H_2O-excess situations. The harzburgite restriction is also shown on the binary chemography. It shows that assemblages such as En + Ath, Tlc + Ath, and Fo + Brc fall outside the composition range of mantle rocks and they will, therefore, be ignored here.

5.3.2
Progressive Metamorphism of Maximum Hydrated Harzburgite

Understanding progressive metamorphism of ultramafic rocks is somewhat more difficult than metamorphism of sedimentary rocks. Most sedimentary rocks are normally in a maximum hydrated state when formed and progressive metamorphism systematically reduces the water content of sediments as the P–T conditions increase progressively. Ultramafic rocks, in contrast, are often nonhydrous mantle rocks and a first process converts them to hydrous low-grade serpentinites. Serpentinization is a retrograde process and normally occurs at the ocean floor (see Sect. 5.7). As a result, anhydrous or partially hydrated mantle assemblages are first converted into the maximum hydrated equivalents (1), (2) and (3) before the onset of progressive metamorphism (Fig. 5.2). This, in turn, means that prograde metamorphism of serpentinites always affects rocks that have been metamorphosed beforehand.

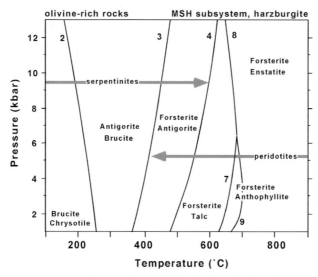

Fig. 5.3. Pressure–temperature diagram for equilibria in the MSH (harzburgite) system [reactions (1)–(9); Table 5.1]. The figure is valid if pure H_2O is present as a fluid phase. *4* Upper boundary of serpentinite; *3* lower limit of peridotite

Let us now examine prograde metamorphism of the two possible maximum hydrated versions of harzburgite. The rocks shall be in structural and chemical equilibrium in a very low-grade terrain. Two assemblages are possible: Brc + Ctl and Ctl + Tlc (or metastable Ctl + Qtz), respectively. The equilibrium distribution of assemblages is shown in Fig. 5.3 and the relevant reactions are listed in Table 5.1.

The assemblage chrysotile + talc is replaced by antigorite (+ talc) at the onset of prograde metamorphism [reaction (1), Table 5.1]. Antigorite forms already at temperatures below 200 °C. At 250–300 °C the maximum temperature for chrysotile is reached and the fibrous serpentine is replaced by antigorite [reaction (2)]. Chrysotile represents a very low-temperature weathering or alteration product of ultramafic rocks and the stable serpentine mineral with the beginning of the greenschist facies is antigorite. Chrysotile and antigorite coexist (stably or metastably) over a wide temperature range of about 100 °C in the subgreenschist facies. Serpentinites in low-grade Alpine ophiolite complexes may contain both, chrysotile and antigorite.

Antigorite serpentinites occur over a wide temperature range (Fig. 5.3). The lower limit is given by reaction (2) and the antigorite breakdown reaction (4) marks the upper boundary. In regional-scale orogenic metamorphism (Ky-type), Atg-schist may occur over a T-interval of >300 °C. Above about 250 °C, MSH rocks may contain two different assemblages depending on bulk composition: Brc + Atg and Atg + Tlc. The bulk chemistry control on the assemblages can be seen in Fig. 5.2. Harzburgite with a molar Fo/En ratio >0.7 will contain Brc + Atg, harzburgite with Fo/En <0.7 consists of Atg + Tlc, and finally harzburgites with Fo/En = 0.7 will be present as monomineralic antigorite serpentinites.

At about 400 °C, the assemblage Brc + Atg is replaced by Fo + Atg [reaction (3), Table 5.1]. Brucite-antigorite schists are diagnostic rocks for the lower to middle greenschist facies. Note, however, that brucite-bearing antigorite schists normally contain little brucite and you need to be careful at the microscope in order not to overlook the mineral. It is recommended that brucite be examined in a reference section before working with unknown samples. The low modal content of brucite in Brc-Atg schists is a consequence of the chemographic relationships shown in Fig. 5.2.

The product assemblage Fo + Atg is diagnostic for upper greenschist to lower amphibolite facies conditions. The upper T-limit of the assemblage Fo + Atg in orogenic metamorphism is near 570 °C. Antigorite serpentinite, therefore, occurs well within the middle amphibolite facies, but also under blueschist and low-T eclogite conditions (Fig. 5.3). In Alpine ophiolites, for example, that underwent subduction zone metamorphism, serpentinites occur together with eclogites and blueschists. These are isofacial occurrences and all three rock types formed at the same P–T conditions (e.g., 20 kbar/ 600 °C, e.g., in the Zermatt ophiolite of the Central Alps).

Reaction (3) also represents the low-temperature limit of forsterite (olivine) in the presence of an aqueous fluid phase. Olivine represents a stable mineral in ultramafic rocks at conditions as low as middle greenschist facies (~400 °C) even if water is present. Pure dunite will not develop any new mineral assemblage and dunite will not be serpentinized above the reaction curve (3). This is indicated in Fig. 5.3 by the peridotite arrow.

It is important to remember that reaction (3) only affects rocks with normative Fo/En >0.7. The widespread assemblage Atg + Tlc is unaffected by reaction (3); such rocks are, in contrast to Brc + Atg rocks, inappropriate for locating the 400 °C boundary in terrains with abundant ultramafic rocks.

The upper thermal stability for antigorite is given by reaction (4) (Table 5.1). The high-temperature limit for serpentinites is at about 570 °C in regional orogenic metamorphism, ~550 °C in low-P high-T metamorphism, and near 510 °C in a contact aureole at 2 kbar. Generally, the last serpentinites disappear approximately with the beginning of middle amphibolite facies conditions in orogenic belts. However, because of the positive slope of equilibrium 3 in Fig. 5.3, serpentinites are stable at fairly high temperatures if pressure is high, for instance, in subduction and crustal thickening settings. In high-pressure low-temperature ophiolite complexes produced by subduction of oceanic lithosphere, antigorite schist (serpentinites) often occurs in isofacial associations with 600–650 °C eclogite. In recent years, new experimental studies have defined the upper Atg-limit at super-high pressures (see references to this chapter). Furthermore, field evidence supported by new experimental data indicates that the periodicity of the antigorite structure and, consequently, its composition systematically varies with increasing grade (see reference list).

Two reactions theoretically define the high-T limit of Atg + Tlc and Atg, respectively, at pressures greater than 14 kbar. Both reactions produce enstatite from antigorite. The assemblage Atg + En has not been reported from rocks as a stable equilibrium assemblage.

The assemblage Fo + Tlc occupies a wide temperature interval (about 100–150 °C) along P–T paths typical of regional orogenic metamorphism. Tlc + Fo is the characteristic assemblage at middle amphibolite facies conditions; it will be replaced by either En + Fo [reaction (6)] or by Fo + Ath [reaction (5)], depending on the P–T path taken by prograde metamorphism. The upper T-limit of the Tlc + Fo assemblage is near 670 °C and rather independent of pressure.

The invariant point at the intersection of reactions (5), (6), and (7) (see Fig. 5.3) defines the maximum pressure for anthophyllite in meta-harzburgites. The precise pressure position of the invariant point depends, among other factors, on the amount of iron-magnesium exchange in the involved minerals. The effect is very small on the temperature position of the equilibria (5), (6), and (7). However, because of the low-angle intersection of equilibria (5) and (7), the invariant point may displaced by several kbars along the slightly displaced equilibrium (6).

However, anthophyllite + forsterite schists and felses occur in contact aureoles and low-P-type metamorphism. The assemblage is often related to partial hydration (retrogression) of En + Fo rocks during cooling and uplift in regional orogenic metamorphism. Prograde sequences in typical collision belts (such as the Scandinavian Caledonides) are characterized by enstatite forming from Fo + Tlc; anthophyllite is always a product of retrogression. Reactions (5) and (6) mark the approximate beginning of upper amphibolite facies conditions.

In contact aureoles, anthophyllite decomposes in the presence of forsterite according to reaction (7) at temperatures near 700 °C. The harzburgite assemblage enstatite + forsterite has its low temperature limit (~670 °C) in the presence of an aqueous fluid. The limit is defined by reactions (6) and (7), respectively. Towards higher temperatures, the assemblage forsterite + enstatite remains stable at all geologically possible temperatures in the crust.

5.4
Metamorphism in the CMASH System

5.4.1
Progressive Metamorphism of Hydrated Al-Bearing Lherzolites

CMASH equilibria appropriate for ultramafic rocks are presented in Fig. 5.4. **Calcium,** if present, is stored in ultramafic rocks in three minerals; calcic clinoamphibole (Cam), calcic clinopyroxene (Cpx) and, under rather extreme conditions, plagioclase (An). The pyroxene diopside is, perhaps somewhat surprising, the stable calcic phase at low temperatures (at extremely low temperatures, tremolite probably replaces diopside). Diopside is widespread in serpentinites and the white pyroxene occurs as a stable Ca-mineral in Atg + Brc, Atg + Tlc and Atg + Fo schists. The pure diopside found in many serpentinites as part of the equilibrium assemblage at greenschist and even sub-greenschist facies conditions should not be confused with relic augitic Cpx that is also often present in the same rocks. The latter Cpx, often displaying diallag lamellae, represents a relic mineral from the mantle assemblage.

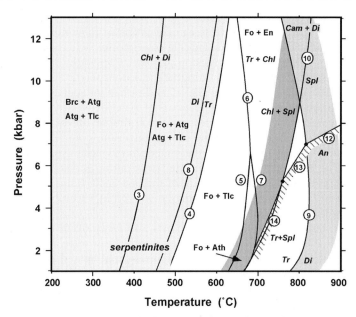

Fig. 5.4. Equilibrium P–T position of all reactions from Table 5.1 for P_{H_2O} = total pressure. *Shaded field above curve 9* divariant field for simultaneous presence of calcic amphibole (Cam) and Cpx. *Shaded field below curve 10* divariant field for coexistence of chlorite and spinel in ultramafic rocks. *Field with ruled pattern* lower limit of plagioclase in olivine-bearing rocks

Diopside is consumed by reaction (8) at about 20–30 °C below the upper limit of antigorite (reaction 4). This has the consequence that rocks containing the product assemblage of reaction (8), namely Atg + Fo + Tr, are diagnostic for a narrow range of temperature. The temperature for this interesting assemblage is near 540 °C in regional orogenic metamorphism and 500 °C in the 2-kbar contact aureole.

Tremolite is the only calcic mineral in isofacial ultramafic rocks in the entire amphibolite facies. The calcic amphibole is removed from En + Fo rocks at the onset of granulite facies conditions. Reaction (9) replaces tremolite with diopside in lherzolitic rock compositions. The product assemblage Fo + En + Di corresponds to the minerals of mantle lherzolite; Ol + Opx + Cpx. The onset of Cpx production in the pure CMSH system is at about 800 °C. However, pure CMSH ultramafics are very rare. The presence of Al (Chl, Spl), Na (fluid), Ti (Ilm) and other elements in natural lherzolites has the consequence that calcic amphibole (Cam) becomes increasingly aluminous, sodic and titaniferous (etc.) at high temperatures. The product Cpx of reaction (9) will also not be pure diopside. However, the partitioning of the "extra components" between the two minerals (Cpx and Cam) results in a substantial increase in the amphibole field towards higher temperatures. The precise position of equilibrium (9) is, therefore, composition-dependent. Reaction (9) corresponds to the amphibolite/granulite facies transition in ultramafic rocks. Curve 9 in Fig. 5.4 is valid for pure end-member Tr and Di (Table 5.1). The upper limit of the shaded area to the right of curve 9 corresponds to the breakdown of amphibole with a moderate departure from tremolite end-member compositions (~30% tremolite+70% pargasite). The shaded area above curve 9 stands for the amphibolite/granulite facies transition zone in H_2O-saturated regimes. Fo+En+Tr+Di (or Ol+Opx+Cam+Cpx) is the characteristic assemblage in this zone.

Aluminum, if present, is stored in low-temperature hydrated ultramafic rocks almost exclusively in chlorite. Mg-chlorite is a very stable mineral in quartz-free rocks such as ultramafic rocks. It may be present at temperatures exceeding 800 °C at 10 kbar, corresponding to the base of the crust! It decomposes to Fo+En and spinel according to reaction (10) in orogenic metamorphism. The equilibrium conditions for reaction (10) in the MASH system are shown in Fig. 5.4. However, the presence of "extra" components has a dramatic effect on the thermal stability of chlorite. In the case of chlorite breakdown, iron strongly partitions into the spinel phase (hercynite, magnetite component). This significantly alters the position of equilibrium (10) in Fig. 5.4. The breakdown of end-member Mg-chlorite is described by equation (10). The onset of the spinel-producing reaction in meta-lherzolites is, however, about 100–150 °C lower than in the pure MASH system. In Fig. 5.4 the transition zone of the chlorite-spinel conversion is indicated by a shaded band to the left (low-T side) of the end-member reaction (10). Within the gray band both chlorite and prograde spinel is present in addition to olivine and Opx. The first appearance of Mg-rich spinel formed by the prograde decomposition of chlorite is marked by the low-T end of the gray band. This prograde spinel often can be observed as spinel-hercynite overgrowth on low-grade magnetite, the latter originally produced by the serpentinization process. Note, by reference to Fig. 5.4, Chl+Spl may occur at low pressures together with Fo+En+Tr, with Fo+En+Cam+Cpx in the pressure range 8–12 kbar, and with the assemblage Fo+En+Cpx at relatively high pressures and high temperatures.

In regional orogenic metamorphism, chlorite begins to decompose near 700 °C, and amphibole is not removed from the ultramafics before a temperature of about 850 °C is reached. Pure tremolite decomposes at a temperature that is about 50 °C lower than the maximum temperature for pure Mg-clinochlore. In "real" rocks, chlorite begins to decompose 150 °C below the last amphibole disappears. "Extra" components in ultramafic rocks have opposite effects on the tremolite- and chlorite-out reactions, respectively. They tend to increase the upper limit for calcic clinoamphibole but decrease the upper limit for chlorite.

In the low-pressure high-temperature region of Fig. 5.4 three additional reactions are shown that delineate the field for plagioclase in ultramafic rocks. The reactions are ruled at the high-T side in such a way defining the plagioclase field. Textural evidence for all three reactions has been found in rocks. Commonly, evidence for reactions (12), (13) and (14) in olivine- (and Opx-) bearing rocks is given by the widespread occurrence of Spl+Cpx and Spl+Cam±Chl symplectites and rinds around decomposing plagioclase. We are not aware of reports of prograde plagioclase formation by the three reactions. Plagioclase + olivine formed in the first place by crystallization from a melt. The pair Pl + Ol is also widespread in basalts and gabbros (Ol-basalt, Ol-gabbro). The same three reactions, i.e., (12), (13), and (14), limit the mineral pair in these mafic rocks.

5.4.2
Effects of Rapid Decompression and Uplift Prior to Cooling

In many orogenic belts metamorphic rocks experience a period of rapid decompression and uplift after equilibration at maximum temperature of the reaction history. Possible effects related to rapid decompression must always be considered. Suppose an isofacial ultramafic rock body has equilibrated (Fig. 5.4) at 700 °C and 8 kbar. It contains the assemblage Fo + En + Tr + Chl. The rock may subsequently follow an isothermal decompression path. In that case the chlorite breakdown reaction (10) may begin to produce some Mg-Fe spinel at pressures below about 6 kbar. The decompression and cooling path may also cross the stability field of the assemblage anthophyllite + forsterite. Production of the assemblage, however, requires access of H_2O during cooling. Late Ath + Fo rocks often develop along fractures and vein systems that provide access for late aqueous fluids during the decompression and cooling history of a regional metamorphic terrain.

5.5
Isograds in Ultramafic Rocks

The successive sequence of diagnostic mineral assemblages in progressively metamorphosed ultramafic rocks is shown in Fig. 5.5. The metamorphic field gradient is that of collision-type orogenic metamorphism. The assemblage boundaries (zone boundaries) shown on the figure are all very well suited for isograd mapping and isograd definitions. They are equivalent to reaction isograds because all boundaries are related to well-defined reactions (Table 5.1) in relatively simple chemical systems. As discussed above, there is some overlap of the chlorite and spinel fields, as well as of the tremolite and diopside fields, respectively. In low-pressure metamorphism, the reaction isograds at lower grades are displaced to somewhat lower temperatures and a Ath + Fo field may appear near 700 °C. In such settings, amphibole decomposes at higher temperatures compared with Fig. 5.5.

The mineral assemblages and the positions of reaction isograds in ultramafic rocks heated by a magmatic intrusion at 1 kbar can be seen along the bottom axis of Fig. 5.4. Reactions (3), (4) and (8) occur in the andalusite field. The latter mineral may occur in meta-pelites, meta-sandstones and metamorphic quartzo-feldspathic rocks associated with ultramafic rocks. The first occurrence of sillimanite in felsic rocks is in the middle of the Tlc + Fo field in ultramafic rocks. Chlorite will decompose to spinel assemblages at temperatures in the vicinity of 650 °C, a temperature often attained at the immediate contact to granitoid intrusions. Figure 5.4 suggests that the assem-

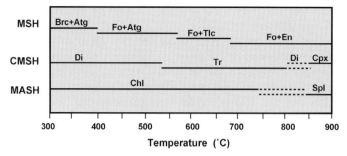

Fig. 5.5. Compilation of mineral assemblages and position of isograds in ultramafic rocks in orogenic metamorphism (Barrowian metamorphism)

blages Ath + Fo and En + Fo should, if present at all, be restricted to a very narrow zone at the immediate contact to shallow-level granitoid intrusions.

5.6
Mineral Assemblages in the Uppermost Mantle

The top of the upper mantle is situated at a depth of 30–40 km beneath stable continental crust. In tectonically active areas (subduction, collision, rifting), the boundary to the upper mantle may be found anywhere between the surface and 70 km depth or even deeper. The precise location of the crust–mantle boundary depends on the state and geological history of the continent. The average depth in stable continental areas corresponds to about 10 kbar pressure. Since the subcontinental mantle predominantly consists of harzburgite with subordinate lherzolite, a number of interesting conclusions regarding mantle petrology can be drawn from Fig. 5.4.

The 10-kbar isobar of the continent-mantle boundary on Fig. 5.4 implies that spinel lherzolite is not stable in the presence of H_2O at temperatures below ~850 °C. Such unusually high temperatures occur in active rift zones or in collision belts undergoing thermal relaxation or in other anomalous regions only. Harzburgite in the MSH system (Opx-Ol) is not stable below ~670 °C if an H_2O-rich fluid is present at 35 km depth. The temperature at the crust-mantle boundary in Central Europe, for instance, is typically between 600 and 700 °C. Any water reaching the mantle beneath the crust at T <670 °C will be consumed by hydration reaction (6); reactions (9) and (10) consume H_2O at even higher temperatures but affect only lherzolites and/or Al-phases in the rocks. The assemblage Talc + Fo + Tr + Chl represents the stable mantle assemblage under fluid-present conditions (H_2O fluid) beneath most of the young continental areas. Under the Precambrian shields the upper mantle temperature is usually in the range 350–450 °C (blueschist facies conditions). The upper mantle rocks will, therefore, be transformed into stable serpentinites (Brc + Atg; Atg + Tlc; Atg + Fo) if water crosses the crust-mantle boundary along deep faults and shear zones. Very prominent seismic reflectors observed in the mantle beneath the Celtic Sea and Great Britain

(Warner and McGeary 1987) are probably caused by partially hydrated mantle rocks (serpentinites).

Beneath most continental areas there are two types of mantle rocks present: (1) weakly hydrated or completely anhydrous peridotites may occur under fluid-absent conditions or in areas with absolutely abnormally high geothermal gradients; and (2) partially or fully hydrated ultramafic rocks (serpentinites and talc-schists) under fluid-present conditions.

5.7
Serpentinization of Peridotites

Serpentinization of mantle rocks occurs normally in three different environments: (1) in the mantle itself (see above), (2) in oceanic ophiolite complexes where serpentinization may be related to oceanic metamorphism, and (3) in the crust during collision belt formation. Some aspects of low-temperature alteration of ultramafics in ophiolite and other associations will be discussed below.

Generally, if an ultramafic rock of the MSH system (e.g., mantle harzburgite) is subjected to conditions to the left of curve 4 in Figs. 5.3 and 5.4, it will be partially or fully serpentinized in the presence of an aqueous fluid. The reversed reaction (4) (Table 5.1) describes the equilibrium serpentinization process. Metastable serpentinization (or stable at P >14 kbar) occurs through the reversed reactions:

$$Atg + 14\ Tlc \Rightarrow 45\ En + 55\ H_2O \qquad\qquad (15)$$

$$Atg \Rightarrow 14\ Fo + 10\ En + 31\ H_2O \qquad\qquad (16)$$

These two reactions describe the metastable serpentinization of orthopyroxene. It is uncertain whether Atg + En forms a stable mineral pair under any condition. To our knowledge, stable coexistence of antigorite and orthopyroxone has not been reported from any field occurrence.

The presence of iron in ultramafic rocks complicates the description of the general process somewhat. Both chrysotile and antigorite are extremely iron-pure minerals unable to accommodate the iron present in olivine and orthopyroxene. Although reactions (3), (4), (15) and (16) describe the serpentinization of the Mg end members of olivine + orthopyroxene, they do not describe the decomposition of the fayalite and ferrosilite components in olivine and orthopyroxene, respectively. The following reaction scheme illustrates the serpentinization of Fe-bearing olivine (e.g., typical $Fo_{90}Fa_{10}$ olivine).

$$3\ Fe_2SiO_4 + O_2 \Rightarrow\quad 2\ Fe_3O_4 + 3\ SiO_{2aq}$$
$$\text{fayalite} \qquad\qquad\ \text{magnetite} \qquad\qquad\qquad (17)$$

$$3\ Mg_2SiO_4 + SiO_{2aq} + 2\ H_2O \Rightarrow\quad 2\ Mg_3Si_2O_5(OH)_4$$
$$\text{forsterite} \qquad\qquad\qquad\quad\ \text{chrysotile} \qquad\qquad (18)$$

Reaction (17) oxidizes the fayalite component to magnetite and releases SiO_2 that, in turn, is consumed by reaction (18) to form serpentine from the forsterite component. Because olivine in mantle rocks always contains much more Fo than Fa, all SiO_2 released by reaction (17) will be consumed by reaction (18). Equilibrium of reaction (17) depends on the oxygen pressure in the fluid. Fayalite oxidation to magnetite can also be written with H_2O as an oxidizing substance and with H_2 gas as a reaction product: 5 vol% magnetite typically occurs in many serpentinites. Magnetite often nucleates on primary chromites (Cr-spinel). The resulting assemblage of the total process is chrysotile (serpentine) + magnetite; unreacted olivine (with resorption structures) may remain present in the rock. The reaction of 1 mol fayalite produces enough silica to serpentinize 3 mol of forsterite. Hence, a $Fo_{75}Fa_{25}$ olivine can be turned into serpentine and magnetite without leaving residual unreacted olivine. Most olivine is more magnesian than that, however, and a small amount of brucite may appear in the end product or the extra Mg component combines with additional components in the serpentinization fluid to form various secondary minerals such as dolomite, sepiolite or saponite, for example.

In the progression of metamorphism of such a Ctl + Mag rock the iron remains fixed in magnetite and does not dissolve in the product phases of the reactions in Table 5.1. As a result, olivine and orthopyroxene in rocks produced by prograde metamorphism of serpentinites are usually very Mg-rich in contrast to primary mantle peridotites that may have lower X_{Mg} in Ol and Opx.

Low-temperature alteration of high-temperature clinopyroxene (Al, Ti-rich Cpx) results in the decomposition of the pyroxene and the formation of chlorite and a pure diopside (the latter is stable in very low-temperature environments, as discussed above).

5.8
Reactions in Ultramafic Rocks at High Temperatures

At temperatures above about 700 °C, ultramafic rocks contain some very characteristic mineral assemblages. Some of these are presented in Fig. 5.4. Additional assemblages are shown in Fig. 5.6 together with previously presented ones.

Ultramafic rocks with characteristic anhydrous assemblages occupy three P–T fields in Fig. 5.6. The stability fields for the assemblages are bounded towards higher temperatures by melt-involving reactions (solidus). Towards lower temperatures the three anhydrous assemblages are limited by the hydration reactions (9), (10) and (13), respectively. The stoichiometric equations of the garnet-involving equivalents of reactions (9) and (11) that limit the Grt-Cam peridotite field are not listed.

The low-pressure assemblage is Cpx + Pl (+ Opx + Ol) and rocks containing it are referred to as plagioclase lherzolites. Rocks containing olivine + plagioclase are diagnostic for low pressure of formation (<6 kbar). Because the assemblage also requires very high temperatures (>800 °C), olivine + plagioclase is most often found in shallow-level ultramafic or gabbroic intrusions (olivine gabbros, troctolites). The assemblage is igneous and formed by crystalli-

Fig. 5.6. Distribution of mineral assemblages in lherzolitic rocks: lherzolite = olivine + hypersthene + augite + Al-phase (Pl, Spl, or Grt), peridotite = olivine + hypersthene + Ca-amphibole + Al-phase (Chl, Spl, or Grt)

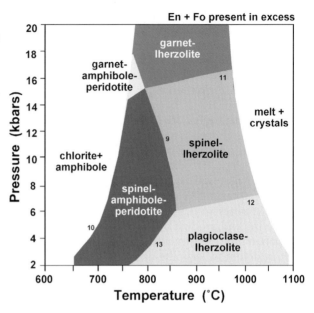

zation from a silicate melt. As mentioned earlier, no prograde metamorphic example of plagioclase formation in olivine-rich rocks has, to our knowledge, been reported to date. It could be expected to form in a low-pressure granulite facies terrain containing ultramafic rocks.

Spinel-lherzolites contain Spl + Cpx in addition to Opx + Ol. Spinel-lherzolites are related to plagioclase-lherzolites by reaction (12). Spinel-lherzolites occur widespread in granulite facies metamorphism and in hot regions of the uppermost mantle.

Garnet replaces spinel as the Al-bearing phase in anhydrous ultramafic rocks at higher pressures. Reaction (11), written in Table 5.1 for pure MASH components, limits the stability field for garnet + olivine in Al-bearing harzburgites towards lower pressure. The presence of calcium in lherzolites has the consequence that garnet which is a reactant mineral of reaction (11) contains Ca in the form of a grossular component, thus expanding the stability field for Ca–Mg–garnet + olivine. In addition, iron is strongly fractionated into spinel, the melt phase and garnet, which has rather significant effects on the high-temperature equilibria in ultramafic rocks. All field boundaries of Fig. 5.6 are calculated (or estimated) for typical iron contents found in Alpine-type peridotites (X_{Mg} of rock ~0.9). For more iron-rich rock compositions, garnet-amphibole- and spinel-amphibole-peridotites may occur at lower temperatures than indicated in Fig. 5.6. Garnet-amphibole-peridotites occasionally occur in association with high-temperature eclogites (see Chap. 8).

The high-grade assemblages of ultramafic rocks can also be shown on a series of chemographies (Fig. 5.7). The CMASH system has been projected from forsterite and H_2O onto the mole fraction triangle MAS. The black dot near the enstatite corner on all triangles shows the bulk rock composition of

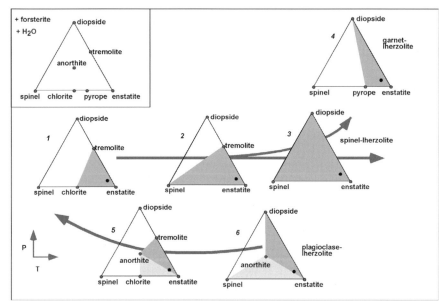

Fig. 5.7. High-grade mineral assemblages in lherzolites projected from forsterite and H_2O. The arrows indicate paths of successive evolution of Chl-Cam-peridotite to Pl-, Spl-, and Grt-Lherzolite, respectively. *Black circle* Composition of typical mantle lherzolite

a typical Ca-Al-bearing peridotite. Triangle 1 represents the situation at about 700 °C and 6 kbar. The characteristic assemblage is Opx + Cam + Chl (+Ol). At lower temperature, Opx (En apex) may be replaced by Oam and Tlc. Chemography 2 at higher temperature shows the consequences of chlorite breakdown and the new assemblage is Spl + Cam + Opx (+Ol). At still higher temperature, Cam decomposes and triangle 3 shows the typical spinel-lherzolite assemblage Spl + Cpx + Opx (+Ol) (e.g., at 900 °C and 8 kbar). At high pressures, the typical assemblage Grt + Cpx + Opx (+Ol) is making up garnet-lherzolite, for example, at 900 °C and 20 kbar (chemography 4). Garnet-lherzolite (garnet-peridotite) is the very characteristic mantle sample. Garnet-lherzolite bodies (lenses of some 100 m size) originating from >50 km or often even >100 km depth occur sporadically in mountain belts. Garnet-peridotites from the Western Gneiss region in Norway and from the Alps have attracted much interest and research. Grt-lherzolite nodules can occasionally be found as mantle xenoliths in basalts. Access to the assemblage 4 may be gained in prograde metamorphism from either chemography 3 by reaction (11) or from triangle 2 by the reaction Cam + Spl ⇒ Grt + Cpx.

Plagioclase lherzolite (chemography 6) with Pl + Cpx + Opx (+Ol) has only been found altered to Pl + Cam + Opx (+Ol) and Pl + Chl + Opx (+Ol) of chemography 5. Both assemblages are ultimately retrogressed to Chl + Cam + Opx (Oam, Tlc) (+Ol). On the other hand, it has been observed that Pl-Ol rocks (6) may develop assemblages of triangle (3) and ultimately (4) when pressure increased progressively, e.g., in connection with a subduction process.

5.9
Thermometry and Geobarometry in Ultramafic Rocks

Geothermobarometers applicable to ultramafic rocks include: intercrystalline FM exchange between any pair of anhydrous FM phases in ultramafic rocks (e.g., Ol-Spl, Ol-Grt, Opx-Cpx), intracrystalline FM exchange in Cpx, temperature dependence of miscibility gap between En-Di, Mg-Tschermak content in Opx in various defining assemblages (e.g., Opx-Grt), Ca-Tschermak content in Cpx in a number of assemblages (e.g., Cpx-Ol-Spl).

Most of the geothermobarometers listed can obviously only be used for estimating equilibration P and T for ultramafic rocks metamorphosed in the eclogite and granulite facies (also pyroxene-hornfelses). Reliable P–T estimates can be calculated for rocks that equilibrated within the shaded areas of Fig. 5.6. In greenschist and amphibolite facies ultramafic rocks, thermobarometers are lacking. P–T estimates from such rock types depend entirely on petrogenetic grids and reactions involving volatile species. This requires some assumptions or estimates on the activities of volatile species (H_2O). In progressive metamorphism it is often justified to assume H_2O-saturated conditions (as in Figs. 5.3, 5.4, and 5.6). In terrains with a polymetamorphic (polyphase) evolution, it may be difficult to estimate the activity of H_2O during equilibration of the ultramafic rocks. In many ultramafic rocks the presence of carbonate minerals (magnesite, dolomite) confirms that the fluid cannot be treated as a pure H_2O fluid.

The problem appears severe; however, it is probably not. Let us consider as a specific example: a Fo-En-fels contains scattered flakes of talc apparently replacing enstatite but no carbonates. The rock also contains occasional talc veins. Is it possible to estimate the temperature at which talc formation occurred? The presence of vein structures can be used as evidence that the talc formation is related to the infiltration of an aqueous (H_2O-rich) fluid. By reference to Fig. 5.3, direct talc formation from enstatite by reaction (6) requires pressures in excess of 6.5 kbar and, provided that metamorphism occurred along a Barrowian field gradient, temperatures below 670 °C. The minimum temperature is given by the position of reaction (4) in Fig. 5.3 because the alteration did not produce antigorite.

5.10
Carbonate-Bearing Ultramafic Rocks

Carbonate minerals (magnesite, calcite and dolomite) are found in many ultramafic rocks. In the Scandinavian Caledonides, for example, almost all Alpine-type ultramafic rocks contain carbonate minerals in an amazing variety of associations with silicates and oxides. The carbonate minerals are in some instances probably of mantle origin. However, most ultramafic rocks in the crust are carbonated in the same way as they are hydrated. Addition of external CO_2 to ultramafic rocks leads to saturation of the rocks with carbonate phases. Maximum hydrated and carbonated low-temperature varieties of ultramafic rocks such as talc + magnesite + dolomite rocks or antigorite + talc

+dolomite rocks encompass the cation composition range of typical mantle rocks (see chemographic relationships shown in Fig. 5.1).

Phase relationships in carbonate-bearing ultramafic rocks are relatively complex and can provide excellent information about P–T regimes in a particular metamorphic terrain. Two specific geologic settings will be used to demonstrate the usefulness of analyzing phase relationships in carbonate-bearing ultramafic rocks. It is obvious that the two examples cannot cover all possible geological situations. We strongly recommend modifying our presented phase diagrams (Figs. 5.8 and 5.9) in order to use them in other environments.

5.10.1
Metamorphism of Ophicarbonate Rocks

Serpentinites containing carbonate minerals in appreciable amounts are referred to as ophicarbonates (ophicalcite, ophimagnesite, etc.). Ophicarbonates form from carbonate-free serpentinites by reaction with crustal CO_2. Carbon dioxide evolves from progressing decarbonation reactions in carbonate-bearing metasediments. Serpentinites are very efficient CO_2 buffers. Small amounts of CO_2 in the fluid are sufficient to convert serpentine assemblages into carbonate-bearing assemblages.

As an example, Fig. 5.8 shows an isobaric T–X section at 3 kbar with phase equilibria of ophicarbonate rocks. Stoichiometries of antigorite-involving reactions are given in Table 5.2. The two dehydration reactions, (4) and (8), have been discussed previously, and are listed in Table 5.1. The diagram can be used to analyze assemblages in contact metamorphic aureoles. The positions of the equilibria are displaced to higher temperature with increasing pressure and, at lower pressure, other topologic configurations may become stable (it is therefore necessary to recalculate the figure for higher or lower pressure terrains).

The figure shows that antigorite-bearing rocks do not tolerate much CO_2 in the fluid phase. Reaction (19) limits the stability field for antigorite in the presence of CO_2-bearing fluids. Under the conditions of Fig. 5.8, 13 mol% CO_2 in the fluid is the maximum CO_2 content of a fluid coexisting with antigorite (point e). Therefore, serpentinites emplaced in the crust may behave as very efficient CO_2 absorbers. The ophicarbonate field is dark shaded and carbonate-free serpentinites occur in the lightly shaded field. Antigorite breaks down to Tlc+Fo at 520 °C (3 kbar, Fig. 5.8), thus point a limits the field of serpentinite (compare also with Fig. 5.4). Similarly, the dehydration reaction (8) that limits the assemblage Atg+Di (Table 5.1, Fig. 5.4) intersects the pure water axis at about 490 °C (point b on Fig. 5.8). However, if the fluid contains as little as 3 mol% CO_2, reaction (8) becomes metastable relative to Atg+Cal involving assemblages at point f. The temperature maximum for antigorite+calcite (ophicalcite) is given by point c, antigorite+dolomite (ophidolomite) reaches its maximum T and X_{CO_2} at point d and, lastly, antigorite+magnesite (ophimagnesite) does not occur at T and X_{CO_2} greater than

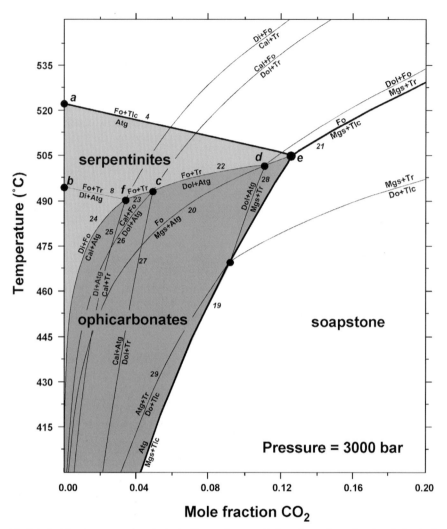

Fig. 5.8. Ophicarbonate rocks: *dark shading* carbonate (Cal, Dol, Mgs)+antigorite; *light shading* antigorite (carbonate-free serpentinites). *a* max. T of serpentinite; *b* max. T of Cpx-bearing serpentinite; *c* max. T of ophicalcite; *d* max. T of ophidolomite; *e* max. T of ophimagnesite

point *e*. The assemblage antigorite + calcite is diagnostic for extremely H_2O-rich fluids.

Six assemblages define a sequence of five reaction isograds in the field. The six assemblages with increasing grade are (Fig. 5.8): (1) Atg + Fo + Di + Cal (upper limit = *f*), (2) Atg + Fo + Tr + Cal (upper limit = *c*), (3) Atg + Fo + Di (upper limit = *b*), (4) Atg + Fo + Tr + Dol (upper limit = *d*), (5) Atg + Fo + Tr + Mgs (upper limit = *e*), and (6) Atg + Fo + Tr (upper limit = *a*). The po-

Fig. 5.9. Magnesite-bearing ultramafic rocks. *Shaded fields* Sagvandite = En+Mgs, soapstone = Tlc+Mgs, ophicarbonate = Atg+Mgs. Note: very different P–X topologies at a DT of only 25 °C. Harzburgite is not stable at mantle pressures in the presence of a fluid phase at T <650 °C. At T = 675 °C, harzburgite may coexist with an extremely H_2O-rich fluid at mantle depth (it does not tolerate CO_2)

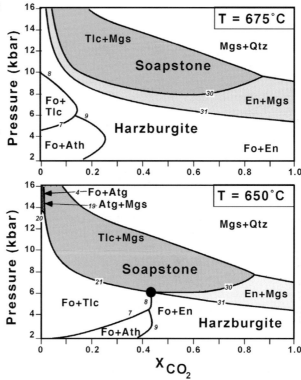

Table 5.2. Selected carbonate-involving reactions in serpentinite, soapstone and sagvandite

(19)	45 Mgs + 17 Tlc + 45 H_2O ⇒	45 CO_2 + 2 Atg
(20)	Atg + 20 Mgs ⇒	31 H_2O + 20 CO_2 + 34 Fo
(21)	Tlc + 5 Mgs ⇒	4 Fo + 5 CO_2 + H_2O
(22)	13 Atg + 40 Do ⇒	383 H_2O + 80 CO_2 + 20 Tr + 282 Fo
(23)	11 Atg + 40 Cc ⇒	321 H_2O + 40 CO_2 + 20 Tr + 214 Fo
(24)	3 Atg + 20 Cc ⇒	93 H_2O + 20 CO_2 + 62 Fo + 20 Di
(25)	45 Cc + 31 Tr ⇒	45 CO_2 + 107 Di + Atg
(26)	Atg + 20 Do ⇒	31 H_2O + 20 CO_2 + 34 Fo + 20 Cc
(27)	107 Do + 17 Tr + 107 H_2O ⇒	73 CO_2 + 141 Cc + 4 Atg
(28)	141 Mgs + 17 Tr + 107 H_2O ⇒	73 CO_2 + 34 Do + 4 Atg
(29)	30 Do + 47 Tlc + 30 H_2O ⇒	60 CO_2 + 15 Tr + 2 Atg
(30)	Tlc + Mgs ⇒	2 En + CO_2 + H_2O
(31)	2 Mgs + En ⇒	2 Fo + 2 CO_2

sitions of a similar set of isograds have been successfully mapped by Trommsdorff and Evans (1977) in a contact aureole of the Central Alps and the temperature resolution of the isograds was within a few °C. A petrologic tool with such a high-temperature resolution is very unusual and we strongly recommend studying and mapping ophicarbonate assemblages when investigating the thermal fine structure of a metamorphic terrain.

5.10.2
Soapstone and Sagvandite

Soapstone is an ultramafic rock containing predominantly talc + magnesite (or dolomite). Sagvandite is an ultramafic rock consisting of abundant enstatite + magnesite. Both rock types usually develop from ordinary carbonate-free ultramafic rocks by interaction with an externally derived CO_2–H_2O fluid phase. Fragments of ultramafic mantle rocks emplaced in the crust are normally accompanied by partly carbonated versions of these rocks.

Relevant reactions in meta-harzburgite involving a binary CO_2–H_2O fluid are shown on two isothermal P–X_{CO_2}-diagrams at 650 and 675 °C, respectively (Fig. 5.9). Magnesite is the only carbonate mineral considered. The two figures show the distribution of diagnostic assemblages for harzburgitic (Fo + En) total rock compositions at middle amphibolite facies temperatures as a function of pressure (depth). The familiar magnesite-absent reactions (4), (7), (8), and (9) have been discussed above (e.g., Fig. 5.4). Reactions (20), (21) and (31) separate magnesite-bearing from carbonate-free assemblages. The three major types of carbonate-bearing ultramafic rocks are Atg + Mgs (ophimagnesite), Tlc + Mgs (soapstone) and En + Mgs (sagvandite; shaded fields in Fig. 5.9). The ophicarbonate rocks are linked to the soapstones by reaction (21), the soapstones are separated from the sagvandites by reaction (30).

At low pressures carbonate saturation is not reached even in pure CO_2 fluids. At 2 kbar fluid pressure and at 650 °C, for instance, the stable assemblage in pure H_2O fluids is Fo + Tlc (cf. Fig. 5.3). In the fluid composition range 0.05–0.38, Fo + Ath replaces Fo + Tlc. Fo + En represents the stable assemblage in CO_2-rich fluids (X_{CO_2} up to 1.0). At higher pressures, carbonate-free rocks coexist with H_2O-rich fluids and magnesite forms at the expense of forsterite if the fluid contains sufficient CO_2. With increasing pressure (depth), forsterite breakdown in CO_2-bearing fluids produces first enstatite (rxn 31), then talc (rxn 21), and finally antigorite (rxn 20). In most carbonate-bearing ultramafitites, the composition of the fluid phase is controlled by reactions (20), (21), and (31). Magnesite + quartz assemblages are extremely rare which suggests that, in most ultramafic rock bodies, forsterite is not completely used up by the three reactions above.

At 650 °C, an isothermal invariant point exists at 6 kbar and $X_{CO_2} = 0.45$. The invariant assemblage Fo + En + Tlc + Mgs occurs widespread in alpine-type ultramafic rocks from many orogenic belts. Different processes may lead to this assemblage: During an episode of increasing pressure, harzburgite may, in the presence of a CO_2–H_2O fluid, undergo partial carbonation by reaction (31) that produces En + Mgs from Fo. The progressing reaction buffers the fluid composition to the invariant point and, finally, the invariant reaction produces Tlc. In contrast, soapstone and a CO_2–H_2O fluid may reach reaction (30) along a decompression path by which enstatite is formed. The rock may eventually reach the invariant reaction and new forsterite may form in addition.

At a temperature that is only 25 °C higher than that of Fig. 5.9 a, the invariant assemblage is at very high pressure outside the P–X window of Fig. 5.9 b (at the intersection of rxn 30 + 31). The dramatic difference between the two figures can best be appreciated by focusing on the field for Fo + En (= harzburgite field). At 650 °C, harzburgite is not stable at mantle depth in the presence of a fluid phase. This seems strange; however, it means that, in the mantle, harzburgite would efficiently bind all CO_2–H_2O fluids and convert to soapstone or ophicarbonate depending on depth. It also means that no free fluid phase can be present in the mantle at this temperature (because there is a lot of forsterite present). At 675 °C, in contrast, harzburgite at mantle depth can coexist with a fluid phase, but the composition of the fluid is restricted to the H_2O-rich side. CO_2-rich fluids would be consumed by reaction (31) to form sagvandite. No CO_2-rich fluids can survive in the mantle under these conditions. If a typical crust-mantle interface is at about 10 kbar, the 650–675 °C temperature range is an important switch for fluid-absent to fluid-present conditions in the uppermost mantle beneath continents. Temperatures of 650–675 °C at the crust-mantle boundary are characteristic for younger continental areas; beneath Precambrian shields it is much colder and in areas that experience active extension and rifting it can be much warmer (and even CO_2-rich fluids may cross the interface).

The discussion of mineral reactions and phase relations in soapstone and sagvandite above has been restricted to the MS-HC system. Ca-bearing rocks give rise to a number of additional reactions and diagnostic assemblages. Particularly important are reactions involving Dol, Tr, and Di in addition to the phases already considered. The relationships, reactions and assemblages presented above for the MS–HC system are, of course, also valid in a more complex system (e. g., CMAS–HC). Phase equilibria of carbonate-bearing ultramafic rock of spinel-lherzolitic bulk compositions are very useful for deciphering fine details of the metamorphic evolution of a field area.

References

Banfield JF, Bailey SW, Barker WW, Smith RC (1995) Complex polytypism: relationships between serpentine structural characteristics and deformation. Am Mineral 80:1116–1131
Barnes I, O'Neil JR (1969) The relationship between fluids in some fresh alpine-type ultramafics and possible modern serpentinization, western United States. Geol Soc Am Bull 80:1947–1960
Berman RG, Engi M, Greenwood HJ, Brown TH (1986) Derivation of internally-consistent thermodynamic data by the technique of mathematical programming: a review with application to the system MgO–SiO$_2$–H$_2$O. J Petrol 27:1331–1364
Brey G, Brice WR, Ellis DJ, Green DH, Haris KL, Ryabchikov ID (1983) Pyroxene-carbonate reactions in the upper mantle. Earth Planet Sci Lett 62:63–74
Bucher K (1991) Mantle fragments in the Scandinavian Caledonides. Tectonophysics 190:173–192
Burkhard DJM, O'Neil JR (1988) Contrasting serpentinization processes in the eastern Central Alps. Contrib Mineral Petrol 99:498–506
Carswell DA (1981) Clarification of the petrology and occurrence of garnet lherzolites and eclogite in the vicinity of Rödhaugen, Almklovdalen, West Norway. Norsk Geol Tidsskrift 61:249–260
Carswell DA, Gibb FG (1980) The equilibrium conditions and petrogenesis of European crustal garnet lherzolites. Lithos 13:19–29

Carswell DA, Gibb FG (1987) Evaluation of mineral thermometers and barometers appli-
    cable to garnet lherzolite assemblages. Contrib Mineral Petrol 95:499–511
Carswell DA, Curtis CD, Kanaris-Sotiriou R (1974) Vein metasomatism in peridotite at Kals-
    karet near Tafjord, South Norway. J Petrol 15:383–402
Chidester AH, Cady WM (1972) Origin and emplacement of alpine-type ultramafic rocks.
    Nature Phys Sci 240:27–31
Coleman RG (1971) Plate tectonic emplacement of upper mantle peridotites along continen-
    tal edges. J Geophys Res 76:1212–1222
Coleman RG, Keith TE (1971) A chemical study of serpentinization – Burro Mountain, Cali-
    fornia. J Petrol 12:311–328
Dawson JB (1981) The nature of the upper mantle. Mineral Mag 44:1–18
Dymek RF, Boak JL, Brothers SC (1988) Titanic chondrodite- and titanian clinohumite-bear-
    ing metadunite from the 3800 Ma Isua supracrustal belt, West Greenland: chemistry,
    petrology, and origin. Am Mineral 73:547–558
Eckstrand OR (1975) The Dumont serpentinite: a model for control of nickeliferous opaque
    mineral assemblages by alteration reactions in ultramafic rocks. Econ Geol 70:183–201
Evans BW (1977) Metamorphism of Alpine peridotite and serpentine. Annu Rev Earth Plan-
    et Sci 5:397–447
Evans BW, Trommsdorff V (1970) Regional metamorphism of ultramafic rocks in the Cen-
    tral Alps: parageneses in the system CaO–MgO–SiO$_2$–H$_2$O. Schweiz Mineral Petrograph
    Mitt 50:481–492
Evans BW, Trommsdorff V (1974) Stability of enstatite+talc, and CO$_2$-metasomatism of me-
    taperidotite, Val d'Efra, Lepontine Alps. Am J Sci 274:274–296
Evans BW, Trommsdorff V (1978) Petrogenesis of garnet lherzolite, Cima di Gagnone, Le-
    pontine Alps. Earth Planet Sci Lett 40:333–348
Evans BW, Johannes W, Oterdoom H, Trommsdorff V (1976) Stability of chrysolite and anti-
    gorite in the serpentinite multisystem. Schweiz Mineral Petrogr Mitt 56:79–93
Ferry JM (1995) Fluid flow during contact metamorphism of ophicarbonate rocks in the
    Bergell Aureole, Val Malenco, Italian Alps. J Petrol 36:1039–1053
Field SW, Haggerty SE (1994) Symplectites in upper mantle peridotites: development and
    implications for the growth of subsolidus garnet, pyroxene and spinel. Contrib Mineral
    Petrol 118:138–156
Frost BR (1975) Contact metamorphism of serpentinite, chlorite blackwall and rodingite at
    Paddy-go-easy pass, Central Cascades, Washington. J Petrol 16:272–313
Frost BR (1985) On the stability of sulfides, oxides and native metals in serpentinite. J
    Petrol 26:31–63
Hermann J, Müntener O, Scambelluri M (2000) The importance of serpentinite mylonites
    for subduction and exhumation of oceanic crust. Tectonophysics 327:225–238
Katzir Y, Avigad D, Matthews A, Garfunkel Z, Evans BW (1999) Origin and metamorphism
    of ultrabasic rocks associated with a subducted continental margin, Naxos (Cyclades,
    Greece). J Metamorph Geol 17:301–318
Lieberman JE, Rice JM (1986) Petrology of marble and peridotite in the Seiad ultramafic
    complex, northern California, USA. J Metamorph Geol 4:179–199
Medaris LG (1984) A geothermobarometric investigation of garnet peridotites in the Wes-
    tern Gneiss region of Norway. Contrib Mineral Petrol 87:72–86
Mellini M, Trommsdorff V, Compagnoni R (1987) Antigorite polysomatism: behaviour dur-
    ing progressive metamorphism. Contrib Mineral Petrol 97:147–155
Miller JA, Cartwright I (2000) Distinguishing between seafloor alteration and fluid flow dur-
    ing subduction using stable isotope geochemistry: examples from Tethyan ophiolites in
    the Western Alps. J Metamorph Geol 18:467–482
Naldrett AJ, Cabri LJ (1976) Ultramafic and related mafic rocks: their classification and gen-
    esis with special references to the concentrations of nickel sulfides and platinum-group
    elements. Econ Geol 71:1131–1158
Nishiyama T (1990) CO$_2$-metasomatism of a metabasite block in a serpentine melange from
    the Nishisonogi metamorphic rocks, southwest Japan. Contrib Mineral Petrol 104:35–46
O'Hanley DS, Dyar MD (1993) The composition of lizardite 1T and the formation of magne-
    tite in serpentinites. Am Mineral 78:391–404
O'Hara MJ, Mercy ELP (1963) Petrology and petrogenesis of some garnetiferous peridotites.
    Trans R Soc Edinb 65:251–314
Scambelluri M, Hoogerduijn Strating EH, Piccardo GB, Vissers RLM, Rampone E (1991) Al-
    pine olivine- and titanian clinohumite-bearing assemblages in the Erro-Tobbio peridotite
    (Voltri Massif, NW Italy). J Metamorph Geol 9:79–91

Scambelluri M, Piccardo GB, Philippot P, Robbiano A, Negretti L (1997) High salinity fluid inclusions formed from recycled seawater in deeply subducted alpine serpentinite. Earth Planet Sci Lett 148:485–499

Schmädicke E (2000) Phase relations in peridotitic and pyroxenitic rocks in the model system CMASH and NCMASH. J Petrol 41:69–86

Schmädicke E, Evans BW (1995) Phase relations in CMASH and NCMASH, with applications to ultramafic rocks. Boch Geol Geotech Arb 44:204–208

Schreyer W, Ohnmacht W, Mannchen J (1972) Carbonate-orthopyroxenites (sagvandites) from Troms, northern Norway. Lithos 5:345–363

Soto JI (1993) PTMAFIC: software for thermobarometry and activity calculations with mafic and ultramafic assemblages. Am Mineral 78:840–844

Stalder R, Ulmer P (2001) Phase relations of a serpentine composition between 5 and 14 GPa: significance of clinohumite and phase E as water carriers into the transition zone. Contrib Mineral Petrol 140:670–679

Trommsdorff V (1983) Metamorphose magnesiumreicher Gesteine: Kritischer Vergleich von Natur, Experiment and thermodynamischer Datenbasis. Fortschr Mineral 61:283–308

Trommsdorff V, Connolly JAD (1990) Constraints on phase diagram topology for the system CaO–MgO–SiO$_2$–CO$_2$–H$_2$O. Contrib Mineral Petrol 104:1–7

Trommsdorff V, Evans BW (1972) Progressive metamorphism of antigorite schist in the Bergell tonalite aureole (Italy). Am J Sci 272:487–509

Trommsdorff V, Evans BW (1977) Antigorite–ophicarbonates: contact metamorphism in Valmalenco, Italy. Contrib Mineral Petrol 62:301–312

Trommsdorff V, Evans BW (1980) Titanian hydroxyl–clinohumite: formation and breakdown in antigorite rocks (Malenco, Italy). Contrib Mineral Petrol 72:229–242

Ulmer P, Trommsdorff V (1995) Serpentine stability to mantle depths and subduction related magmatism. Boch Geol Geotech Arb 44:248–249

Viti C, Mellini M (1998) Mesh textures and bastites in the Elba retrograde serpentinites. Eur J Mineral 10:1341–1359

Warner M, McGeary S (1987) Seismic reflection coefficients from mantle fault zones. Geophys J R Astron Soc 89:223–230

Wunder B, Schreyer W (1997) Antigorite: high pressure stability in the system MgO–SiO$_2$–H$_2$O (MSH). Lithos 41:213–227

Wyllie PJ, Huang W-L, Otto J, Byrnes AP (1983) Carbonation of peridotites and decarbonation of siliceous dolomites represented in the system CaO–MgO–SiO$_2$–CO$_2$ to 30 kbar. Tectonophysics 100:359–388

Zhang RY, Liou JG, Cong BL (1995) Talc-, magnestite- and Ti-clinohumite-bearing ultrahigh-pressure meta-mafic and ultramafic complex in the Dabie Mountains, China. J Petrol 36:1011–1037

they are identical for the two composition groups 1 and 2. Specifically, reactions (1), (2) and (3) all involve Dol + Qtz as reactants and, if the reactions run to completion, the rock will contain the product assemblage plus unused reactants that can be either dolomite *or* quartz.

The equilibrium conditions of the reactions listed in Table 6.1 depend on P, T and, in the presence of a binary CO_2–H_2O fluid phase, on the composition of this fluid. The fluid composition can be expressed as X_{CO_2} (see Chap. 3). At a given pressure and temperature (e.g., along an orogenic geotherm), the assemblages of the reactions in Table 6.1 coexist with a fluid of a fixed given composition.

6.2
Orogenic Metamorphism of Dolomites

The P–T–X relationships of dolomite-rich marbles in orogenic belts are shown in Fig. 6.2. The section has been constructed for a field gradient typical of collision belts (Ky-type geotherm). The figure is valid for assemblages that contain dolomite + calcite throughout; the sedimentary assemblage is, therefore, represented by the quartz field. Dol + Qtz will react to form talc, tremolite or diopside according to reactions (1), (2), or (3) depending on the composition of the fluid phase initially present in the pore space or introduced into the marble during metamorphism. Note that both the talc- and tremolite-forming reactions consume H_2O. The widespread occurrence of coarse and abundant tremolite in marbles is compelling evidence of the availability and presence of H_2O in prograde metamorphism.

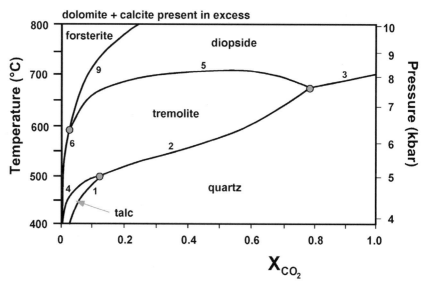

Fig. 6.2. P–T–X phase relationships in dolomite marbles containing excess dolomite and calcite. The vertical axis represents an orogenic geotherm characteristic of regional metamorphism in collision belts

Talc will form in H_2O-rich fluids at T <500 °C. Talc-bearing marbles are restricted to T <500 °C at P <5 kbar. However, talc in marbles of orogenic belts is usually associated with late hydrothermal activity or it is of retrograde origin. Talc is the only metamorphic mineral in dolomites at lower metamorphic grades. The steep equilibrium position of reaction (1) in P–T–X space (Fig. 6.2) has the consequence that only little modal talc is produced by this reaction under closed-system conditions. Tremolite is, therefore, usually the first recognizable metamorphic mineral in marbles. Talc is removed from the marbles by reaction (4).

Tremolite first occurs at the invariant point involving Qtz + Tlc + Tr (+ Dol + Cal) at about 500 °C and 5 kbar. The first occurrence of tremolite represents an excellent mappable isograd in the field and coincides approximately with the beginning of the amphibolite facies (= tremolite-in isograd). The temperature of the isograd is rather independent of the precise position of the geotherm and close to 500 °C. Reaction (2) will continue to produce tremolite with increasing temperature and the rock will effectively control the composition of the fluid phase along the boundary of the quartz and tremolite fields of Fig. 6.2 (the rock will contain the assemblage Dol + Cal + Qtz + Tr). Finally, the rock will run out of the reactant quartz and will enter the divariant tremolite field. The assemblage Dol + Cal + Tr is **the** characteristic assemblage in lower to middle amphibolite facies dolomite marbles. At a typical mid-amphibolite facies temperature (say 600 °C, 6.5 kbar; Fig. 6.2) dolomite marbles may contain three assemblages: Dol–Cal–Tr (common), Dol–Cal–Qtz–Tr [reaction (2)], or Dol–Cal–Qtz (rare).

Diopside forms in dolomite marbles from reactions (3) and (5), and from the reaction at the invariant point Qtz + Tr + Di (+ Dol + Cal). The position of this invariant point defines the lowest temperature of possible occurrence of diopside in dolomite marbles (closed system). In other words, the first occurrence of diopside represents a sharp mappable boundary in the field and coincides with a temperature of about 670 °C along the P–T section of Fig. 6.2 (= diopside-in isograd).

If a rock does not reach the invariant point along a prograde path because reaction (2) consumed all quartz at lower temperature, then diopside will form from reaction (5). This reaction replaces the divariant assemblage Dol + Cal + Tr by Dol + Cal + Di. Reaction (5) separates tremolite from diopside in Fig. 6.2 (and also in the field). The maximum temperature of the assemblage Dol + Cal + Tr is given by the T_{max} of reaction (5). Along a collision-type P–T path, no tremolite-bearing dolomite marbles can be stable at T >720 °C (= tremolite-out isograd). In the P–T interval between the diopside-in and the tremolite-out isograd (ΔT is about 40 °C), tremolite- and diopside-bearing marbles may occur side by side and the four minerals of reaction (5) may be found in rocks.

Diopside may also form from reaction (3) if the rock is sufficiently quartz-rich to leave the invariant point along the quartz-diopside boundary (= equilibrium of reaction 3) in the course of prograde metamorphism or if the initial fluid contained more than about 80 mol% CO_2. Reaction (3) will also produce diopside in rocks devoid of a fluid phase. Reaction (3) will create a pure CO_2 fluid in such rocks and impose the maximum limit for the

Dol + Qtz assemblage (~700 °C at 8 kbar). No mineralogical changes will occur in fluid-free siliceous dolomitic limestone until P–T conditions of the upper amphibolite facies are reached.

Forsterite will not form in regional metamorphism of siliceous dolomites along the P–T path of Fig. 6.2. In principle, diopside will be removed from dolomitic marbles by reaction (9). The reaction replaces the Dol + Cal + Di assemblage of the diopside field by the Dol + Cal + Fo assemblage of the forsterite field. Reaction (9) represents the boundary between the two fields in Fig. 6.2. At temperatures <800 °C, forsterite can only be produced by interaction of the marble with an externally derived H_2O-rich fluid phase (e.g., along fractures or shear zones, veins). Closed-system marbles will be free of forsterite if metamorphism occurred at P–T conditions of Fig. 6.2. At pressures in the range <5 to 8 kbar (sillimanite-type metamorphism), forsterite may form in dolomite at granulite facies conditions [T >800 °C (8 kbar) or T >700 °C (5 kbar)].

In summary, in regional metamorphic terrains of orogenic belts, dolomite-rich marbles (Fig. 6.1) may contain small amounts of talc in addition to Dol + Cal + Qtz in the upper greenschist facies. Tremolite-bearing marbles are characteristic of lower to middle amphibolite facies. Diopside-bearing marbles are stable from the middle amphibolite facies onwards. Forsterite marbles are diagnostic for granulite facies conditions or for infiltration of the marble by H_2O-rich fluids.

6.3
Orogenic Metamorphism of Limestones

Orogenic metamorphism of quartz-rich marble or calc-silicate marbles (composition range 2 in Fig. 6.1) is modeled by Fig. 6.3. The figure is valid for rocks with excess calcite + quartz. Because dolomite is the limiting mineral in reactions (1), (2), and (3) for these compositions, the field for the sedimentary assemblage (Dol + Cal + Qtz) is labeled "dolomite" in Fig. 6.3. The production of Tlc, Tr, and Di by reactions (1), (2), and (3) is identical to the dolomite-rich rocks above. The upper limit for talc is, however, given by reaction (14) in such rocks.

The **tremolite**-producing reaction (2) will consume all dolomite and the resulting assemblage, after completion of reaction (2), is Cal + Qtz + Tr. Here, tremolite will be removed from the rocks at higher grades by reaction (15). It occurs at significantly lower temperatures than reaction (5) that removes tremolite in dolomite-excess rocks. The tremolite field for these rocks is smaller in size than for dolomite-rich marbles.

Diopside is usually produced by reaction (15). For the same reason as above, diopside may appear in calc-silicate marbles at lower temperature than in dolomite marbles (~650 °C). The highest-grade assemblage in calc-silicate marbles is Cal + Qtz + Di (diopside field of Fig. 6.3). In contrast to dolomite-rich rocks, there is very little overlap between the tremolite and the diopside field in calc-silicate marbles.

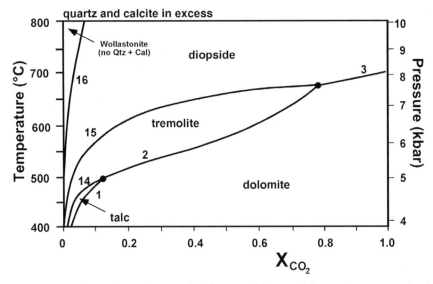

Fig. 6.3. P–T–X phase relationships in calc-silicate marbles containing excess quartz and calcite. The vertical axis represents an orogenic geotherm characteristic of regional metamorphism in collision belts

A small complication may occur if reaction (15) consumes all quartz, leaving a rock with the assemblage Cal + Tr + Di (composition field 1c in Fig. 6.1). Tremolite is then removed from the rocks at higher grade by reaction (5). This reaction produces new metamorphic dolomite, the rock again becomes saturated with dolomite and the phase relations shown in Fig. 6.2 apply. This conversion from quartz-saturated to dolomite-saturated conditions as metamorphism progresses affects only rocks of the restricted composition range 1c (Fig. 6.1).

Wollastonite does not form in regional metamorphic rocks under closed-system conditions. Even under granulite facies conditions the assemblage Cal + Qtz remains stable. Figure 6.3 shows that wollastonite may only form from reaction (16) by interaction of the marble with an H_2O-rich fluid. Such externally derived fluids may invade the marbles along fractures or shear zones.

In summary, in regional metamorphic terrains, quartz-rich marbles (Fig. 6.1) may contain small amounts of talc in addition to Dol + Cal + Qtz in the upper greenschist facies. Tremolite-bearing marbles are characteristic of lower to middle amphibolite facies. Diopside-bearing marbles are stable from middle amphibolite facies upward. Wollastonite marbles are diagnostic for interaction with externally derived fluids or for certain P–T paths after maximum metamorphic conditions (see below).

6.4
Contact Metamorphism of Dolomites

Dolomitic limestone sediments may also be metamorphosed by adding heat from an igneous heat source at shallow levels in the crust. Commonly, the intrusive body is of granitic, granodioritic or quartz-dioritic composition and the temperature at the contact to the country rocks is in the order of 600–700 °C. Stability relationships of siliceous dolomites at 2 kbar are shown in Fig. 6.4. The T–X section is valid for rocks with dolomite + calcite present throughout (like the companion Fig. 6.2). It models progressive metamorphism of dolomites in a typical contact aureole of a granitic pluton at a depth of about 7 km (2 kbar). The size (thickness) of the distinct mineralogical zones in the aureole depends on the heat given off by the pluton that, in turn, depends on the size of the intrusion, the composition of the magma and its temperature. Typically, contact effects of shallow-level intrusions can be recognized in the country rocks to distances ranging from a few hundred meters to 1 or 2 km.

Reactions (1), (2) and (3), Table 6.1, define the upper thermal limit for the sedimentary Dol + Cal + Qtz assemblage (as in metamorphism along orogenic geotherms). However, the invariant point Tlc + Tr + Qtz (+ Dol + Cal) is at $X_{CO_2} = 0.6$ and reaction (1) has a low dT/dX slope in Fig. 6.4. These features potentially result in the development of a recognizable zone with **talc**-bearing dolomites in the outer (cooler) parts of the aureole. Such a talc zone is diagnostic for temperatures below 450 °C.

A zone with **tremolite**-bearing dolomites may develop from reactions (2), (4), or the reaction at the invariant point. The low-temperature limit for tremolite coincides with the upper limit for talc (~450 °C). All tremolite will subsequently be removed from dolomite marbles by reactions (5) and (6). The maximum temperature for tremolite in dolomite marble is given by the position of the invariant point Tr + Di + Fo (+ Dol + Cal) at about 600 °C. Reaction (6), which is metastable in Fig. 6.2, produces forsterite in tremolite-bearing marbles in low-pressure aureoles.

Forsterite appears in marbles in modally recognizable amounts at about 570–600 °C, depending on the quartz content of the original dolomite. The assemblage Fo + Dol + Cal is characteristic of dolomites close to the contact to the intrusion. Reaction (6), if run to completion, results in the assemblage Dol + Cal + Fo in dolomite-rich rocks of composition 1a in Fig. 6.1. The rock remains dolomite-saturated. However, reaction (6) may also consume all dolomite present in the rock and the resulting assemblage is Cal + Fo + Tr. This applies to rocks of the composition field 1b in Fig. 6.1. The assemblage cannot be represented in Fig. 6.4 because it lacks dolomite. In this assemblage, tremolite will be removed from the marble by reaction (7). The equilibrium curve for this reaction emerges from invariant point Tr + Fo + Di (+ Dol + Cal) in Fig. 6.4 and follows the tremolite-forsterite boundary inside the forsterite field at slightly higher temperatures (maximum ΔT ~15 °C). This means that the maximum temperature for tremolite in rocks of composition 1b roughly coincides with that of composition 1a. However, the high-temperature assem-

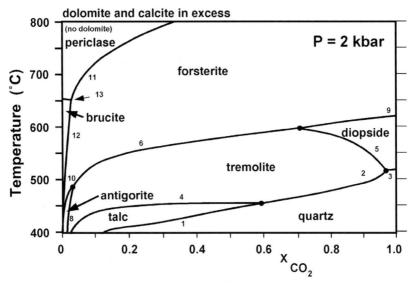

dolomite and calcite in excess

Fig. 6.4. T–X phase relationships in dolomite marbles containing excess dolomite and calcite at a constant pressure of 2 kbar (contact metamorphic aureoles)

blage for these compositions is Cal + Di + Fo instead of Dol + Cal + Fo. The stability field for the Cal + Di + Fo assemblage is about coincident with the field labeled "forsterite" in Fig. 6.4.

Diopside, if present in dolomite-excess rocks at all, is restricted to a narrow T interval near 600 °C under fluid-present conditions. It will form by reaction (5) or at the invariant point Qtz + Tr + Di (+ Dol + Cal). Diopside may also occur in situations where H_2O had no access to the marbles and the Dol + Cal + Qtz assemblage is replaced by Di + Dol + Cal according to reaction (3) near 520 °C. Diopside will be removed from the marbles by reaction (9) in the temperature interval 600–620 °C. Rocks of the composition range 1a will, after completion of the reaction, contain Dol + Cal + Fo; rocks of the composition range 1b, the assemblage Cal + Di + Fo (see also above).

The assemblage Dol + Cal + Fo represents the high-temperature assemblage in most contact aureoles. In many contact aureoles periclase- or brucite-bearing marbles occur close to the contact to the intrusives or in marble inclusions in the intrusives themselves (xenoliths, roof pendants). **Periclase** forms usually from decomposition of dolomite according to reaction (11). **Brucite** may form by direct decomposition of dolomites by reaction (12) or, more commonly, by retrograde hydration of previously formed periclase by the reversed reaction (13). It is apparent from Fig. 6.4 that both periclase and brucite can only form by interaction of forsterite-marble with an externally derived H_2O-rich fluid phase. Under closed-system conditions, dolomite remains stable in "normal" contact aureoles (around granitoid intrusions).

Periclase and brucite are most often found in pure dolomites (SiO_2-free rocks) because such rocks do not have the capacity to create and control their own (relatively CO_2-rich) fluid phase. Pure dolomites are, therefore, most exposed to periclase formation by H_2O flushing from magmatic fluids released by the intrusives close to the contact.

A stability field for **antigorite** is presented in Fig. 6.4 which also cannot be accessed by progressively metamorphosed siliceous dolomites under closed-system conditions. However, antigorite may form in dolomites at peak conditions by interaction of the marble with an external fluid. Reaction (8) replaces Tr + Dol + Cal by Atg + Dol + Cal. Antigorite may be destroyed in more H_2O-rich fluids by reaction (10). Antigorite also forms by the reversed reaction (10) during retrograde replacement of the Fo + Dol + Cal assemblage.

In summary, in low-pressure contact aureoles, siliceous dolomites may develop the following mineral zonation (from low to high T, listing only the silicates): (1) dolomite-rich compositions (1a): Qtz, Tlc, Tr, (Di), Fo; (2) dolomite-poor compositions (1b): Qtz, Tlc, Tr, Tr + Fo, Fo + Di or Qtz, Tlc, Tr, Di, Fo + Di. Periclase may form in pure dolomite close to the igneous heat source. Periclase, brucite and antigorite also often form by interaction of the marble with an externally controlled H_2O-rich fluid. A model of contact metamorphism of dolomite-rich marbles is also presented in Chapter 3.8.3.

6.5
Contact Metamorphism of Limestones

Rocks of composition field 2 in Fig. 6.1 undergo progressive transformations which can be described by the phase relationships shown in Fig. 6.5. The figure is valid for rocks containing quartz and calcite at all temperatures, except where indicated. The three low-T reactions (1), (2) and (3) limit the sedimen-

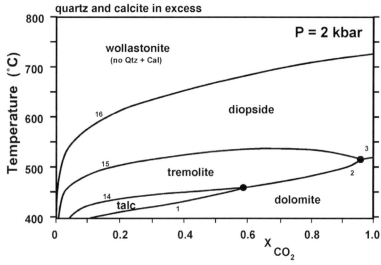

Fig. 6.5. T–X phase relationships in calc-silicate marbles containing excess quartz and calcite at a constant pressure of 2 kbar (contact metamorphic aureoles)

tary Qtz + Dol + Cal assemblage as in the previous cases. **Talc** is subsequently removed by reaction (14).

Tremolite has a distinctly lower maximum temperature of occurrence compared with dolomite-rich rocks. Tremolite is replaced by diopside by reaction (15), which has its temperature maximum at 540 °C. Tremolite occurs in a relatively narrow temperature interval of about 80 °C. For the sake of completeness, if reaction (15) consumes all quartz, the resulting assemblage will be Cal + Tr + Di. For this to happen, rock compositions in the narrow composition range 1c of Fig. 6.1 are required. For a discussion of the fate of this assemblage as metamorphism progresses, see previous section [reactions (7), (5), and (9)].

It follows from the above description that **diopside** first appears in calc-silicate marbles further away from the intrusive contact than in dolomite marbles (if it occurs at all in the latter). The assemblage Cal + Di + Qtz is typical for calc-silicate marbles of the CMS-HC system (diopside field of Fig. 6.5) near the contact to the intrusives.

Wollastonite may form in contact metamorphic calc-silicate marbles even under closed-system conditions. In siliceous limestones with the sedimentary assemblage Qtz + Cal and an H_2O-rich pore fluid, for example, wollastonite will form in small but petrographically detectable amounts at temperatures around 600 °C. In rocks that contained initially dolomite and consequently underwent a series of CO_2-producing reactions, the fluid may have the composition of the T maximum of reaction (15) when entering the diopside field under closed-system conditions. Here, a temperature of about 700 °C is required for the onset of the wollastonite-producing reaction (16). Such high temperatures are rarely reached even at the immediate contact to the intrusives. Thus, wollastonite forms by interaction of the Cal + Qtz + Di assemblage with an externally controlled H_2O-rich fluid phase.

If the wollastonite reaction (16) runs to completion the composition field 2 of Fig. 6.1 is subdivided into two subfields. Rocks of field 2a contain Cal + Wo + Di; reaction (16) has used up all the quartz. Rocks of field 2b contain Qtz + Wo + Di; reaction (16) has consumed all the calcite. The rock is devoid of calcite and, consequently, cannot be called a marble any longer. At a typical temperature of about 620 °C at the immediate contact to granitoid intrusions, three assemblages characterize calc-silicate marbles: Cal + Qtz + Di, Cal + Di + Wo and Qtz + Di + Wo.

In summary, calc-silicate marbles of the composition field 2 (Fig. 6.5) may contain talc in the outer aureole, tremolite in a middle zone and diopside in a relatively wide inner zone around the intrusive body. Wollastonite is often present in marbles close to the contact and usually forms from interaction of the marble with H_2O-rich (magmatic) fluids.

6.6
Isograds and Zone Boundaries in Marbles

Isograds and zone boundaries in mixed volatile systems such as marbles can be related to unique points on P–T–X sections. For example, the invariant

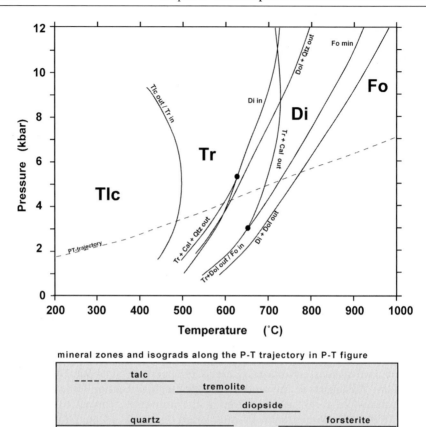

mineral zones and isograds along the P-T trajectory in P-T figure

	talc	
	tremolite	
	diopside	
quartz		forsterite

Fig. 6.6. Isograds and mineral zone boundaries in siliceous dolomites and limestones projected into the P-T plane. The P–T trajectory (*stipled line*) signifies a metamorphic field gradient typically found in regional metamorphic areas in collision belts

point involving $Qtz + Tr + Di + Dol + Cal$ marks the first occurrence of diopside in marbles in Figs. 6.2, 6.4 (triple point of Tr, Di, Qtz fields), 6.3, and 6.5 (triple point of Tr, Di, Dol fields), and it can be expected that the first appearance of diopside in marble shows up as an isograd on a map. Likewise, the last occurrence of tremolite in dolomite-rich marbles can be mapped as an isograd that is related to the P–T maximum of reaction (5) in Fig. 6.2. All singular points that limit a unique assemblage on a P–T–X section can be squeezed into a P–T graph (Fig. 6.6). For instance, the maximum P–T conditions of the assemblage $Tr + Dol + Cal$ given by the maximum of reaction (5) in Fig. 6.2 (last Tr in Dol-marble) is at about 700 °C and 8.3 kbar. If a flat P–T path (geotherm) had been chosen to construct Fig. 6.2, the T and P for the last Tr would have been lower too. A steeper P–T path results in a higher T at a higher P. All P–T conditions at which the last tremolite disappears from Dol-marbles are consequently given by a curve in Fig. 6.6. Note that also X_{CO_2} varies along the curves in Fig. 6.6. Other upper or lower limits of assemblages in marbles are shown in Fig. 6.6 as well. A prograde path of metamorphism in a given terrain will intersect the boundaries displayed in Fig. 6.6 at specific P–T points. These intersections are excellent and mappable isograds in marbles.

The **diopside-in isograd** shown in Fig. 6.6 represents coexistence of Dol + Cal + Tr + Qtz + Di. The assemblage coexists with increasingly H_2O-rich fluids ($X_{CO_2} = 0.96$ at 2 kbar, $X_{CO_2} = 0.76$ at 8 kbar) as pressure increases along the curve. This is a general relationship and holds for all polybaric traces of isobaric invariant assemblages. It can be seen from Fig. 6.6 that a diopside-in isograd mapped in a contact aureole at 2 kbar marks the 500 °C isotherm, in a low-P regional metamorphic area it coincides with the 600 °C isotherm, and in a Ky-type regional metamorphic field the diopside-in isograd is coincident with the 670 °C isotherm (as previously shown).

The assemblage Dol + Cal + Qtz + Tlc + Tr defines the **talc-out isograd** and **tremolite-in isograd**. Its temperature is about 500 °C and rather independent of the precise position of the P–T trajectory of a particular orogenic metamorphic region.

The upper limit for Tr + Cal + Qtz (**Tr + Cal + Qtz-out isograd**) coincides with the diopside-in isograd at pressures above about 5 kbar. Below this pressure (marked with a black dot on the figure), the isograd occurs at slightly higher T than the diopside-in isograd.

The **Dol + Qtz-out isograd** is defined by the T maximum of reaction (3) at a given pressure (pure CO_2 fluid). It marks the maximum conditions for the sedimentary Dol + Cal + Qtz assemblage.

The **Tr + Cal-out isograd** represents the T maximum of reaction (5) at a given pressure (>3.5 kbar). Along the curve, X_{CO_2} is constant at 0.5. Below this pressure (filled circle in Fig. 6.6) the upper limit of Tr + Cal is given by the polybaric trace of the isobaric invariant assemblage Dol + Cal + Tr + Di + Fo. Along this curve X_{CO_2} increases with decreasing pressure.

The same curve also marks approximately the **Fo + Cal-in isograd**. At higher pressures, the **Fo-min** boundary is taken at the approximate minimum temperature where petrographically detectable amounts of forsterite have been produced by reaction (9) ($X_{CO_2} = 0.6$).

The **Di + Dol-out isograd** is defined by the P–T position of equilibrium (9) in pure CO_2 fluids.

On Fig. 6.6, a P–T trajectory is shown that represents a possible metamorphic path in a low-P metamorphic region (And-Sil type path). The prograde sequence of metamorphic silicate minerals and the isograd temperature in marbles that contain dolomite and calcite in excess (dolomite marbles) along the trajectory are shown in the box below the P–T figure.

6.7
Metamorphic Reactions Along Isothermal Decompression Paths

As outlined in Chapter 3, the metamorphic evolution of rocks in an orogenic process follows distinct P–T time paths. After the maximum temperature has been reached during metamorphism associated with collision belt formation,

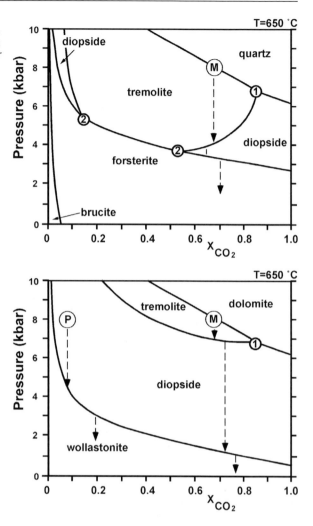

Fig. 6.7. P–X phase relationships for siliceous dolomites and limestones at a constant temperature of 650 °C. *Upper figure* shows dolomite-rich marbles (Cal+Dol excess), *lower section* represents silicate-rich marbles (calc-silicate marbles; Cal+Qtz excess). *M* rock with Dol+Cal+Qtz+Tr at 8 kbar and 650 °C; *P* rock with Cal+Qtz at 8 kbar and 650 °C

a marked period of rapid uplift and decompression often follows and a nearly isothermal decompression stage has been documented from many orogenic belts. Below we will examine transformations that may occur in middle amphibolite facies marbles during uplift and decompression.

This geological situation is best modeled by isothermal P–X sections. Figure 6.7 shows two model sections at a constant temperature of 650 °C, one for dolomite-rich marbles and one for calc-silicate marbles. The figure is valid for all carbonate rocks that were initially composed of Dol+Cal+Qtz. Before decompression begins, marbles may contain a number of different assemblages at 650 °C and, for instance, 8 kbar depending on the composition of the rock and the fluid initially present in the pore space (or introduced during metamorphism). Possible assemblages include: Dol+Cal+Qtz, Dol+Cal+Tr, Cal+Qtz+Tr, and Cal+Tr+Di (Fig. 6.7). Consider, for example, a rock composed of Dol + Cal + Qtz + Tr [reaction (2)]. The position of this rock is labeled "M" in Fig. 6.7. This rock will travel along a complex path through Fig. 6.7 upon decompression to a final pressure of say 2 kbar before the onset of cooling. The precise path depends on the modal abundance of Dol, Cal and Qtz in the initial rock.

Diopside will appear below 7 kbar in calc-silicate marbles and below 4 kbar in dolomite limestone. Tremolite will be removed at about 4 kbar. Ultimately, forsterite may appear in the marbles either at invariant point (2) or from reaction (9) at pressures below 4 kbar.

In Cal + Qtz marbles wollastonite may be produced by reaction (16) in detectable amounts at pressures below about 4 kbar (e.g., along a path taken by the rock P); the effect may further be enhanced by introducing H_2O-rich fluids during decompression. Wollastonite observed in granulite facies marbles can often be related to rapid uplift along a path of nearly isothermal decompression together with interaction of the marbles with late H_2O-rich fluids.

In summary, rapid uplift along isothermal decompression paths may drastically modify marble assemblages that originally formed during prograde orogenic metamorphism. Assemblages produced by reactions during isothermal decompression may be difficult to distinguish from assemblages produced at "peak" conditions. The best way to detect the presence of decompression assemblages in marbles is by an integral evaluation of the metamorphic evolution of the entire region (e.g., comparison with thermobarometric data from metapelites). Decompression assemblages may form in all type of rocks. Generally, they form if a rock follows a path from maximum P–T that crosses devolatilization reactions in prograde direction. The least hydrated state (or in the case of marbles: the state with the least volatile content) is typically reached at the low-P end of an isothermal decompression period before significant cooling sets in (especially if the isothermal decompression portion of the path follows maximum T). It is this point along the P–T path that may be best preserved in metamorphic rocks.

6.8
Marbles Beyond the CMS–HC System

In the following sections the consequences of additional components in marbles will be discussed.

6.8.1
Fluorine

As mentioned earlier, marbles are often extremely poor in iron. Consequently, hydrous Mg-silicates such as Tlc and Tr are usually very close to their end-member composition. However, metamorphic fluids commonly have a small but essential partial pressure of hydrofluoric acid (HF). HF is strongly partitioned into solid hydrate phases. In other words, a small amount of HF in the fluid coexists with rather fluorine-rich minerals. In addition, X_{Fe} and X_F in hydrate silicates (e.g., talc and tremolite) are negatively correlated, which means that, at a given P_{HF}, replacement of hydroxyl groups by fluorine rapidly decreases with increasing X_{Fe}. The consequence of the $F(OH)_{-1}$ exchange in talc and tremolite is a general increase in their respective stability fields on the phase diagrams presented above. The net effect on the talc-tremolite boundary is an increase in the talc field.

In forsterite-grade marbles a small HF pressure may lead to the formation of minerals of the humite group. The two types of humite group minerals occasionally found in forsterite grade marbles are clinohumite (low P_{HF}) and chondrodite (high P_{HF}). Both humites are not stable in pure CO_2-H_2O fluids. The stability fields for the two minerals roughly coincide with the forsterite fields on the P–T–X diagrams above. However, their occurrence is restricted to the H_2O-rich side of the phase diagrams. The precise location of the phase boundaries involving Chu or Chn depends on the prevailing HF pressure during metamorphism. Typical assemblages in dolomite marbles are: Dol + Cal + Fo + Chu (clinohumite marbles) or Dol + Cal + Chu + Chn (chondrodite marbles, humite marbles). Chn and Fo do not occur together. Chu and Chn may also be stabilized by the presence of titanium, but both minerals always contain some fluorine. It is strongly advised that minerals are analyzed for fluorine when studying marble samples with the electron microprobe.

6.8.2
Aluminum

Dolomitic limestones usually contain various amounts of Mg-chlorite (clino-chlore). The mineral is stable at all metamorphic grades except in the granulite facies and in low-pressure contact aureoles where it may decompose according to the two spinel-producing reactions:

$$2\,Dol + Chl \Rightarrow 2\,Cal + 3\,Fo + Spl + 2\,CO_2 + 4\,H_2O \tag{17}$$

or

$$3\,Chl + 2\,Cal \Rightarrow 2\,Di + 5\,Fo + 3\,Spl + 2\,CO_2 + 12\,H_2O \tag{18}$$

At 2 kbar, reaction (17) has its T maximum at 610 °C. It marks the chlorite-out isograd in rocks containing Dol + Cal + Fo + Chl. Reaction (18) has its maximum temperature at 640 °C at 2 kbar. The temperature corresponds to the chlorite-out isograd in rocks containing Cal + Di + Fo + Chl.

The maximum temperature for the two reactions increases with increasing pressure, which means that chlorite remains stable in regionally metamorphosed rocks up to granulite facies conditions.

In contact aureoles, the mineral clintonite (Cli), a trioctahedral calcium brittle mica, occasionally occurs in dolomite marbles of the forsterite zone. Typical assemblages are (e.g.): Cal + Fo + Di + Spl + Cli and Cal + Fo + Chu + Spl + Cli.

The presence of aluminum in calc-silicate marbles may lead to the appearance of a number of additional minerals such as margarite, zoisite, grossular, scapolite, vesuvianite and anorthite. The phase relationships of such rocks will be discussed in Chapter 8 on metamorphism of marls.

6.8.3
Potassium

Potassium is usually stored in sedimentary dolomites in the form of autigenic K-feldspar. It is one of the first minerals of the sedimentary assemblage that is removed from marbles undergoing progressive metamorphism. The relevant reaction consumes K-feldspar and replaces it by phlogopite (biotite).

$$3\ Dol + Kfs + H_2O \Rightarrow Phl + 3\ Cal + 3\ CO_2 \tag{19}$$

In natural systems, the equilibrium of this reaction runs approximately parallel to and a few degrees below reactions (1) and (2) on the (P)–T–X phase diagrams above. After completion of reaction (19), dolomite marble contains the assemblage Dol + Cal + Qtz + Phl. The assemblage undergoes subsequently reactions (1) or (2), as discussed above. Reaction (19) increases X_{CO_2} in the fluid and may run to completion at X_{CO_2} greater than that of the invariant point Tlc + Tr + Qtz + Dol + Cal. Thus, rocks undergoing reaction (19) are often guided around the talc field and talc will not form in phlogopite-bearing marbles. Such rocks contain the following assemblages (with increasing grade): Dol + Cal + Qtz + Kfs, Dol + Cal + Qtz + Phl, Dol + Cal + Tr + Phl (phlogopite marbles). Phlogopite remains stable in rocks containing excess Dol and Cal to very high grades. For example, the assemblage Dol + Cal + Fo + Chu (fluorine) + Spl (aluminum) + Phl (potassium) often occurs in contact metamorphic marbles near the contact to the intrusive body. Note that also phlogopite is significantly affected by the F(OH)$_{-1}$ exchange.

The phase relationships involving muscovite, phlogopite and K-feldspar in calc-silicate rocks will be discussed in Chapter 8 on metamorphism of marls.

6.8.4
Sodium

Sodium may be present as NaCl in the metamorphic fluid (albite in low-grade dolomites is rare). Its presence complicates the phase relationships in Al-bearing dolomite marbles because it continuously modifies the composition of tremolite with increasing temperature. The compositional changes of calcic amphibole are closely described by the tremolite-pargasite exchange reaction (# Mg 2Si = Na 3Al; # denotes a vacancy). The exchange reaction produces pargasitic amphiboles from tremolite at high temperature. Consequently, the upper limit of amphibole in marbles is displaced to higher temperatures compared to the pure CMS–HC system.

6.9
Thermobarometry in Marbles

There are only two useful fluid-independent equilibria available for geological P–T estimates in marble.

6.9.1
Calcite-Aragonite Phase Transition

$CaCO_3$ undergoes a phase transition from trigonal calcite to orthorhombic aragonite at high pressures. At 400 °C, the transition pressure is close to 10 kbar, at 800 °C aragonite is stable at pressures greater than 15 kbar. However, aragonite is easily converted to calcite during decompression of carbonate-bearing high-pressure rocks. Occasionally, the mineral is preserved in rocks that were metamorphosed in a subduction zone and rapidly returned to the surface without being substantially heated. The presence of aragonite in metamorphic rocks indicates minimum pressures in excess of 10 kbar. See also comments on the Arg=Cal transition in Chapter 4.

6.9.2
Calcite-Dolomite Miscibility Gap

In rocks containing excess dolomite, the solubility of $MgCO_3$ in calcite is a function of the temperature and to a lesser degree also of the pressure. The temperature dependence of the $MgCa_{-1}$ exchange in calcite (and because of the asymmetry of the miscibility gap with some restrictions also the $CaMg_{-1}$ exchange in dolomite) represents a widely used geological thermometer. Various calibrations exist in the geological literature (see Chap. 4).

References

Anovitz LM, Essene EJ (1987) Phase equilibria in the system $CaCO_3$–$MgCO_3$–$FeCO_3$. J Petrol 28:389–414

Bickle MJ, Powell R (1977) Calcite-dolomite geothermometry for iron-bearing carbonates. The Glockner area of the Tauern window, Austria. Contrib Mineral Petrol 59:281–292

Bowen NL (1940) Progressive metamorphism of siliceous limestones and dolomite. J Geol 48:225–274

Bucher K (1981) Petrology of chlorite-spinel marbles from NW Spitsbergen (Svalbard). Lithos 14:203–214

Bucher K (1982a) Mechanism of mineral reactions inferred from textures of impure dolomitic marbles from East Greenland. J Petrol 23:325–343

Bucher K (1982b) On the mechanism of contact aureole formation in dolomitic country rock by the Adamello intrusion (N Italy). Am Mineral 67:1101–1117

Cartwright I, Buick IS (1995) Formation of wollastonite-bearing marbles during late regional metamorphic channelled fluid in the Upper Calcsilicate Unit of the Reynolds Range Group, central Australia. J Metamorph Geol 13:397–417

Castelli D (1991) Eclogitic metamorphism in carbonate rocks: the example of impure marbles from the Sesia-Lanzo zone, Italian Western Alps. J Metamorph Geol 9:61–78

Connolly JAD, Trommsdorff V (1991) Petrogenetic grids for metacarbonate rocks; pressure-temperature phase-diagram projection for mixed-volatile systems. Contrib Mineral Petrol 108:93–105

Cook SJ, Bowman JR (2000) Mineralogical evidence for fluid-rock interaction accompanying prograde contact metamorphism of siliceous dolomites: Alta Stock Aureole, Utah, USA. J Petrol 41:739–758

Dunn SR, Valley JW (1992) Calcite-graphite isotope thermometry: a test for polymetamorphism in marble, Tudor gabbro aureole, Ontario, Canada. J Metamorph Geol 10:487–502

Eggert RG, Kerrick DM (1981) Metamorphic equilibria in the siliceous dolomite system: 6 kbar experimental data and geologic implications. Geochim Cosmochim Acta 45:1039–1049

Ehlers K, Hoinkes G (1987) Titanian chondrodite and clinohumite in marbles from the Ötztal Crystalline Basement. Mineral Petrol 36:13–25

Eveney B, Shermann D (1962) The application of chemical staining to the study of diagenesis in limestones. Proc Geol Soc Lond 1599:10 pp

Ferry JM (1976) P, T, f_{CO_2} and f_{H_2O} during metamorphism of calcareous sediments in the Waterville-Vassalboro area, south-central Maine. Contrib Mineral Petrol 57:119–143

Ferry JM, Rumble D (1997) Formation and destruction of periclase by fluid flow in two contact aureoles. Contrib Mineral Petrol 128:313–334

Flowers GC, Helgeson HC (1983) Equilibrium and mass transfer during progressive metamorphism of siliceous dolomites. Am J Sci 283:230–286

Gieré R (1987) Titanian clinohumite and geikielite in marbles from the Bergell contact aureole. Contrib Mineral Petrol 96:496–502

Glassley WE, William E (1975) High grade regional metamorphism of some carbonate bodies: significance for the orthopyroxene isograde. Am J Sci 275:1133–1163

Goldsmith JR, Newton RC (1969) P–T–X relations in the system $CaCO_3$–$MgCO_3$ at high temperatures and pressures. Am J Sci 267A:160–190

Greenwood HJ (1962) Metamorphic reactions involving two volatile components. Annual Report Director Geophysical Laboratory Carnegie Inst Wash Year Book, pp 82–85

Holness MB (1997) Fluid flow path and mechanisms of fluid infiltration in carbonates during contact metamorphism: the Beinn an Dubhaich aureole, Skye. J Metamorph Geol 15:59–70

Holness MB, Bickle MJ, Graham CM (1991) On the kinetics of textural equilibration in forsterite marbles. Contrib Mineral Petrol 108:356–367

Jamtveit B, Dahlgren S, Austrheim H (1997) High grade contact metamorphism of calcareous rocks from the Oslo Rift, southern Norway. Am Mineral 82:1241–1254

Letargo CMR, Lamb WM, Park J-S (1995) Comparison of calcite+dolomite thermometry and carbonate+silicate equilibria: constraints on the conditions of metamorphism of the Llano uplift, central Texas, U.S.A. Am Mineral 80:131–143

Liu L-G, Lin C-C (1995) High-pressure phase transformations of carbonates in the system CaO–MgO–SiO_2–CO_2. Earth Planet Sci Lett 134:297–305

Molina JF, Poli S (2000) Carbonate stability and fluid composition in subducted oceanic crust: an experimental study on H_2O–CO_2-bearing basalts. Earth Planet Sci Lett 176:295–310

Piazolo S, Markl G (1999) Humite- and scapolite-bearing assemblages in marbles and calcsilicates of Dronning Maud Land, Antarctica: new data for Gondwana reconstructions. J Metamorph Geol 17:91–107

Powell R, Condliffe DM, Condliffe E (1984) Calcite-dolomite geothermometry in the system $CaCO_3$–$MgCO_3$–$FeCO_3$: an experimental study. J Metamorph Geol 2:33–41

Rice JM (1977 a) Contact metamorphism of impure dolomitic limestone in the Boulder Aureole, Montana. Contrib Mineral Petrol 59:237–259

Rice JM (1977 b) Progressive metamorphism of impure dolomitic limestone in the Marysville aureole, Montana. Am J Sci 277:1–24

Skippen G (1971) Experimental data for reactions in siliceous marbles. J Geol 70:451–481

Skippen G (1974) An experimental model for low pressure metamorphism of siliceous dolomitic marbles. Am J Sci 274:487–509

Skippen G, Trommsdorf V (1975) Invariant phase relations among minerals on T–X_{fluid} sections. Am J Sci 275:561–572

Skippen G, Trommsdorff V (1986) The influence of NaCl and KCl on phase relations in metamorphosed carbonate rocks. Am J Sci 286:81–104

Slaughter J, Kerrick DM, Wall VJ (1975) Experimental and thermodynamic study of equilibria in the system CaO–MgO–SiO_2–H_2O–CO_2. Am J Sci 275:143–162

Tilley CE (1951) A note on the progressive metamorphism of siliceous limestones and dolomites. Geol Mag 88:175–178

Tracy RJ, Hewitt DA, Schiffries CM (1983) Petrologic and stable-isotopic studies of fluid-rock interactions south-central Connecticut. I. The role of infiltrations in producing reaction assemblages in impure marbles. Am J Sci 283A:589–616

Trommsdorff V (1966) Progressive Metamorphose kieseliger Karbonatgesteine in den Zentralalpen zwischen Bernina und Simplon. Schweiz Mineral Petrogr Mitt 46:431–460

Trommsdorff V (1972) Change in T–X during metamorphism of siliceous dolomitic rocks of the central Alps. Schweiz Mineral Petrogr Mitt 52:567–571

Valley JW, Peterson EU, Essene EJ, Bowman JR (1982) Fluorphlogopite and fluortremolite in Adirondack marbles and calculated C-O-H-F fluid compositions. Am Mineral 67:545–557

CHAPTER 7

Metamorphism of Pelitic Rocks (Metapelites)

7.1
Metapelitic Rocks

Metapelites are probably the most distinguished family of metamorphic rocks. Typical metapelites include well-known metamorphic rocks such as, for example: chlorite-kyanite-schists, staurolite-garnet micaschists, chloritoid-garnet micaschists, kyanite-staurolite schists, biotite-garnet-cordierite gneisses, sillimanite-biotite gneisses and orthopyroxene-garnet granulites. Many distinct metamorphic minerals are found in metapelitic rocks (e.g., staurolite, chloritoid, kyanite, andalusite, sillimanite and cordierite).

In many metamorphic terrains, characteristic minerals in metapelites show a regular spatial distribution that can be readily related to the intensity of metamorphism. The distribution pattern of metamorphic minerals in metapelites reflects the general metamorphic style and structure of almost all metamorphic terrains in any orogenic belt. Mineral zones and phase relationships in metapelites often permit a fine-scale analysis of the intensity and nature of metamorphism in a given area. A number of well-established and calibrated geological thermometers and barometers can be applied to mineral assemblages found in metapelitic rocks.

7.2
Pelitic Sediments

7.2.1
General

Metapelites are metamorphic rocks derived from clay-rich sediments including unconsolidated sediments such as mud and clay, and consolidated sediments, e.g., shales, claystones and mudstones. After incipient metamorphism, all these sedimentary rocks are collectively termed argillite. Weak metamorphism transforms these rocks into slates. In general, pelitic sediments are characterized by very small grain size (often < 2 µm) and by a mineralogy that is dominated by clay minerals.

The term pelitic rock is used in metamorphic geology in a general way to designate fine-grained clay-rich sediments. It includes sediments that fall under the general terms mudstone and shale used by sedimentary geologists. Clay-rich pelites often develop a sequence of characteristic minerals and mineral assemblages during prograde metamorphism. Mudstones with a high proportion of silt (siltstones) and less abundant clay transform into metamorphic equivalents with less interesting and diagnostic mineral assemblages. They are often referred to as semi-pelites in the metamorphic literature (e.g., semi-pelitic gneisses). Shales represent >80% of all sedimentary rocks (Table 2.2).

7.2.2
Chemical Composition

Compositions of two types of pelites are listed in Table 2.3. The characteristic compositional features of pelitic rocks are best represented by the analysis of typical pelagic clay. Aluminum is very high, total iron is up to 10 wt%, there is a fair amount of magnesium; however, CaO is extremely low. The water content of pelites is high (5 mol of H_2O/kg rock in the case of the pelagic clay of Table 2.3). From the viewpoint of a metamorphic geologist, this is a positive aspect because it can be expected that H_2O released during metamorphism helps to maintain the rocks in chemical equilibrium. Prograde metamorphism of pelitic sediments starts with rocks at a maximum hydrated state. This is often not the case in other rock compositions (e.g., ultramafic and mafic rocks). Shales deposited on platforms often deviate from pure clay compositions as a result of higher silt fractions (higher SiO_2, more quartz and feldspar) or the presence of carbonate minerals (higher CaO and CO_2). All transitions between pure clay compositions and arkoses or marls exist. This chapter deals with carbonate-free, CaO-pure claystones and shales (such as pelagic clay in Table 2.3).

7.2.3
Mineralogy

The mineralogy of clays and shales is dominated, as expected, by clay minerals. Clay minerals are aluminous sheet silicates of variable compositions (important representatives are: montmorillonite, smectite, kaolinite). Muscovite and paragonite of detrital or autigenic origin are the next important group of minerals in shales. Abundant chlorite carries much of the Mg and Fe found in shales. All sheet silicates together usually represent more than 50 vol% of the rock. Quartz occurs in various modal proportions (typically 10–30 vol%). As a consequence, most of the metamorphic equivalents of shales contain free quartz despite the fact that many prograde mineral reactions in pelitic rocks consume quartz. Most shales also contain detrital and/or autigenic feldspars and sporadically also a substantial amount of zeolites. An additional complication arises from high modal proportions of Fe-oxides–hydroxides (hematite–limonite–goethite) in some shales and sulfides and organic carbonaceous matter in others (oil shale). This has the consequence that redox reactions and sulfur-involving reactions can be important in metapelitic rocks.

7.3
Pre-metamorphic Changes in Pelitic Sediments

During compaction and diagenesis, primary clays and shales undergo significant chemical and mechanical changes. The very large (e.g., >50%) porosity of clays is continuously reduced during burial and compaction. Typical shales may still contain several % pore space filled with formation water when metamorphism starts (at approximately 200 °C and about 6 km depth = 1.6 kbar). The original clay minerals, such as smectite, are replaced by illite (a precursor mineral of the white K-micas), and chlorite. The lattice ordering of the sheet silicates, particularly of illite, progressively increases with increasing temperature and pressure. Illite "crystallinity" can in fact be used as a measure of the degree of diagenetic and very low-grade metamorphic recrystallization. Carbonaceous matter undergoes a series of reactions that ultimately destroy the organic compounds. The organic carbon compounds are replaced finally by graphite or are completely transferred to the vapor phase as CO_2 or CH_4 under oxidizing and reducing conditions, respectively. The optical reflectivity of the carbonaceous matter is also used as a sensitive indicator of the thermal regime during diagenesis and incipient metamorphism (Chap. 4.10.2.2). Compaction and recrystallization during burial and diagenesis also produce a distinct fissility and cleavage (slaty cleavage) resulting from a pressure-induced parallel orientation of the sheet silicate minerals.

Consequently, shales at the beginning of metamorphism have been transformed to slates and phyllites. The mineralogy most typically includes: kaolinit, illite (muscovite), chlorite, quartz and feldspar (K-feldspar and albite), hematite or sulfide and graphite. The chemical composition of the sediments remains more or less preserved during pre-metamorphic processes with the exception of H_2O loss.

Below, we shall explore what happens to such slates and phyllites if they experience an increase in temperature and pressure during a tectono-metamorphic event such as a typical orogenic cycle.

7.4
Intermediate-Pressure Metamorphism of Pelitic Rocks

Metapelites formed by this type of metamorphism include the classic metamorphic rocks mentioned above (Sect. 7.1). The distinct zonal pattern of mineral assemblages in metapelitic rocks that is produced by intermediate-pressure orogenic metamorphism has been found and extensively studied in a great number of metamorphic terrains from many orogenic belts (fold belts, mountain belts, mobile belts on Precambrian shields) ranging in age from the Precambrian to the Tertiary. This type of metamorphism is also referred to as "Barrovian" metamorphism after its first description from the Scottish Highlands by G. Barrow. It corresponds roughly to a prograde metamorphic path along a kyanite-type field gradient.

7.4.1
Chemical Composition and Chemographies

Reactions and assemblages in metapelites can be discussed by means of se-lected simple chemical subsystems. By reference to the composition of pelagic clays (Table 2.3), the components occur in the following sequence of abun-dance: SiO_2, Al_2O_3, $FeO-Fe_2O_3$, H_2O, MgO, K_2O, Na_2O, TiO_2, and CaO. This complex ten-component system can be broken up into manageable subsystems. Metapelites are often approximated by a simple six-component system includ-ing the six most abundant components above and representing total iron as FeO. This system can be graphically represented in AFM diagrams as outlined in Chapter 2. Na_2O is mainly stored in feldspar (albite) and paragonite. Reac-tions involving Na-bearing minerals will be discussed separately below. CaO is typically low in pure clays. The component is stored in metapelitic rocks in gar-net (grossular component) and in plagioclase (anorthite component). Compli-cations arising from CaO and other minor components in metapelites will be discussed in Section 7.9. Metamorphism of calcareous clay (marls) such as the platform shale listed in Table 2.3 is outlined in Chapter 8.

7.4.2
Mineral Assemblages at the Beginning of Metamorphism

The AFM chemography of shales at the onset of metamorphism is repre-sented in Fig. 7.1. The requirements given by the AFM projection are met by many shales. As outlined above, most shales contain modally abundant quartz and illite (white mica, muscovite, sericite) and excess H_2O. Because pelitic sediments are at a maximum hydrated state (at least when metamor-phosed for the first time!), they will experience a series of dehydration reac-tions when heat is added to the rocks during prograde metamorphism. Thus, during much of the prograde metamorphic evolution of metapelites, the rocks may remain water-saturated.

The composition of typical shales falls into the dark-shaded area of Fig. 7.1. The recalculated composition of the pelagic clay (Table 2.3) has the fol-lowing AFM coordinates: $A=0.41$; $F=0.62$. It is represented by point P in Fig. 7.1. The composition P can be regarded as reference "normal" pelite composition. Such a shale contains kaolinite/pyrophyllite + chlorite in addi-tion to quartz, illite and water at the beginning of metamorphism. The pre-requisite of the AFM diagram requires that quartz and K-white-mica (musco-vite) is present at all times in the course of prograde metamorphism (at high-grade Ms may be replaced by Kfs, see below). This means that "true" metapelites always must contain muscovite (or Kfs). The metapelite P shown on Fig. 7.1 has the TS component in chlorite buffered by the assemblage, whereas the FM component is constrained by the bulk composition. This fol-lows from the fact that the point P is in the univariant field kaolin-ite + chlorite. Shale compositions falling into the divariant chlorite field con-tain chlorite as the only AFM phase. The composition of the chlorite is given by the bulk composition and can vary within the shaded field representing

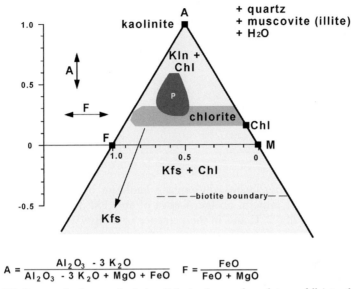

$$A = \frac{Al_2O_3 - 3\,K_2O}{Al_2O_3 - 3\,K_2O + MgO + FeO} \qquad F = \frac{FeO}{FeO + MgO}$$

Fig. 7.1. AFM diagram for low-grade shales. Pelagic clays and mudstones fall into the dark shaded area and the average pelagic clay (Table 2.3) projects to point P. AFM coordinates of minerals and rocks can be calculated from the equations given at the bottom of the figure (molar basis)

the range of compositional variation observed in low-temperature chlorites (chlorite solid solution). Shales with A-values below about 0.17 (clinochlore composition) contain the univariant assemblage chlorite + feldspar. Such rocks will not develop the characteristic and diagnostic mineral assemblages that one usually associates with metapelites unless the bulk composition is much more magnesian than for the compositions of the shaded area. If the A-coordinate is below –0.5, diagnostic assemblages will not develop in such rocks irrespective of their F-coordinate. This compositional restriction may be designated the **biotite boundary.** Biotite-gneisses and schists, and garnet–biotite gneisses are typical metamorphic products of low-Al shales/siltstones. They are often referred to as semi-pelitic gneisses and shales.

7.4.3
The ASH System

The two most abundant components of pelites (SiO_2 and Al_2O_3), together with H_2O, make up a series of typical minerals in metapelites: kaolinite, pyrophyllite, the three aluminosilicate polymorphs (kyanite, sillimanite, andalusite) and quartz (and its polymorphs). The ASH system is a subsystem of the KFMASH system portrayed by AFM diagrams. It describes all phase relations and reactions in the A-apex of an AFM diagram. Reaction equilibria in the ASH system are shown in Fig. 7.2 and listed in Table 7.1. Only reactions affecting bulk compositions with excess quartz and H_2O (as for AFM diagrams) are shown.

Table 7.1. Reactions in the ASH system

Sillimanite, kyanite, andalusite	Al_2SiO_5
Pyrophyllite	$Al_2Si_4O_{10}(OH)_2$
Kaolinite	$Al_2Si_2O_5(OH)_4$
Quartz	SiO_2

(1) And = Ky
(2) And = Sil
(3) Sil = Ky
(4) Kln + 2 Qtz = Prl + H_2O
(5) Prl = Ky + 3 Qtz + H_2O
(6) Prl = And + 3 Qtz + H_2O

Prograde metamorphism will replace kaolinite by pyrophyllite at about 300 °C. The stable mineral on the A-apex of an AFM diagram becomes pyrophyllite. At about 400 °C and in the presence of a pure H_2O fluid, pyrophyllite reaches its upper thermal stability and decomposes to aluminosilicate + quartz (kyanite at P > 2.7 kbar, andalusite at P < 2.7 kbar). At higher temperatures, the stable divariant assemblage in the ASH system is quartz + H_2O + aluminosilicate. The nature of the stable aluminosilicate polymorph depends on pressure and temperature (along a Ky-type path: Qtz + Ky + H_2O). The stable mineral on the A-apex of an AFM diagram is kyanite, sillimanite or andalusite, depending on the P–T conditions selected for the AFM diagram. Andalusite is not stable at pressures greater than 4 kbar, sillimanite is not stable below about 500 °C. Aluminosilicates are not stable below about 350–400 °C in the presence of water and quartz [reactions (5) and (6), Table 7.1]. In low-P orogenic metamorphism, pyrophyllite decomposes to andalusite + quartz at about 380 °C. Andalusite is replaced by sillimanite at about 530 °C.

Note, however, that in rocks containing abundant organic material (a situation that is not uncommon in pelites), the fluid phase may contain much CH_4 at low temperature. Under this typical condition, the equilibrium of any dehydration reaction is displaced to lower temperatures compared with the pure H_2O fluid case shown in Fig. 7.2. Therefore, pyrophyllite usually forms at much lower temperature than shown in Fig. 7.2. Pyrophyllite is common at temperatures below 200 °C, whereas kaolinite is extremely rare in metapelites (>200 °C).

The polymorphic transitions [reactions (1), (2) and (3)] require a complete reconstruction of the aluminosilicate structure. Because these "solid-solid" reactions involve only small changes in free energy, metastable persistence of one polymorph in the field of another more stable polymorph is very common. For example, prograde metamorphism along a Ky-type path to conditions near 600 °C and 6.5 kbar produces a rock with the assemblage kyanite + quartz (+ H_2O). From these P–T conditions that rock may follow a typical decompression and cooling path (shown in Fig. 7.2 as dashed line). Excess water conditions cannot be taken for granted along this path (retrograde metamorphism). However, reactions (1), (2) and (3) are water-absent reactions. At about 6 kbar, coarse-grained kyanite that has originally formed

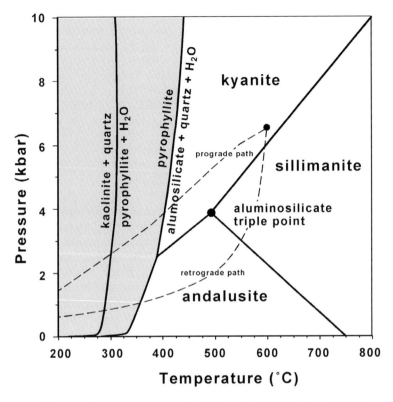

Fig. 7.2. Equilibria in the ASH system calculated from the RB88 database. Typical prograde and retrograde paths of metamorphism are shown as *dashed lines*. Maximum P–T conditions are at 6.5 kbar and 600 °C. The chosen retrograde path has a distinct section of initial isothermal decompression

along the prograde path becomes less stable than sillimanite. Under equilibrium conditions all kyanite should now be replaced by sillimanite. This is normally not the case in real rocks. Fine sillimanite needles may grow on kyanite or precipitate from aluminosilicate-saturated fluids in fractures and nodules. Most of the originally formed kyanite remains in the rock. At about 550 °C and 3 kbar along the decompression and cooling path, both kyanite and sillimanite become less stable than andalusite. However, in most rocks, no spontaneous formation of andalusite from kyanite or sillimanite occurs in the matrix of the rock. Andalusite, like sillimanite, may form in nodules, veins and fissures by direct precipitation from a fluid phase. Thus, all three aluminosilicates (kyanite, sillimanite and andalusite) can often be observed at a given locality or even on the hand-specimen or thin-section scale. Obviously, this does not necessarily mean that the rocks have been metamorphosed at or near the aluminosilicate triple point. The three aluminosilicates found at one locality are normally not isochronous and formed each at distinct sections along a metamorphic P–T–t path.

The observation of chronic metastable persistence of one polymorph mineral in the stability field of another polymorph mineral suggests that water, although not a species in the stoichiometric reaction equation, needs to be present in the rock as a solvent and catalyst. If water is present along the cooling and decompression path, metastable kyanite (for example) can be dissolved and stable sillimanite may precipitate. This is further supported by the frequent observation of sillimanite or andalusite in veins, fractures and nodules in otherwise massive kyanite-bearing rocks; i.e., in structures that are clearly late in the metamorphic history.

7.4.4
Metamorphism in the FASH System

Addition of FeO to the ASH system introduced above permits discussion of phase relationships among a number of further minerals: Fe-chlorite (daphnite, amesite), chloritoid, staurolite, almandine garnet, Fe-cordierite, hercynite-spinel. Additional minerals in the FASH system include Fe-ortho-ambiboles (Oam) and Fe-orthopyroxene (Opx). Phase relationships among Oam- and Opx-involving assemblages will be presented in Sections 7.5 and 7.6.

Some important equilibria in the FASH system are shown in Fig. 7.3. The grid can be directly applied to Fe-rich pelites. Fe-chlorite has the lowest X_{Al} of all minerals of interest here. This means that the decomposition of chlorite at higher temperature produces an iron oxide in addition to a product silicate. However, Fe-oxides in metapelitic rocks are either hematite and/or magnetite that both contain trivalent iron. Consequently, terminal chlorite decomposition depends on the oxygen activity in the rocks. Two important chlorite-involving redox reactions are shown in Fig. 7.3 and they will be explained below. The stoichiometric equations of all reaction equilibria shown in Fig. 7.3 are listed in Tables 7.2 and 7.1.

Let us now consider prograde Ky-type metamorphism of a pelitic rock containing the assemblage chlorite + pyrophyllite in addition to excess quartz and H_2O (dashed line in Fig. 7.3).

The chlorite-pyrophyllite pair is replaced at about 280 °C by chloritoid by reaction (7); the equilibrium is outside the P–T window of Fig. 7.3. The appearance of chloritoid in metapelites coincides with the transition from the subgreenschist to the greenschist facies. In the lower greenschist facies (~350 °C), Fe-rich metapelites contain the assemblages chlorite-chloritoid or chloritoid-pyrophyllite depending on the X_{Al} of the bulk rock. In "normal" metapelites chloritoid will usually appear at somewhat higher temperatures (see below).

At about 370 °C, chlorite reacts with hematite; the latter is often present in low-grade slates and phyllites, and produces magnetite and more chloritoid (reaction 8). The oxygen activity is controlled by the hematite–magnetite pair through the reaction: $6\ Fe_2O_3 = 4\ Fe_3O_4 + O_2$. This is known as the hematite–magnetite oxygen buffer (Chap. 3.6.2.4). It defines relatively high oxygen activities at a given temperature. Since chlorite is modally more abundant

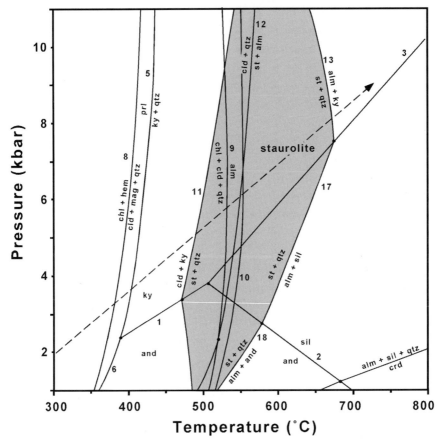

Fig. 7.3. Equilibria in the FASH system. Reaction numbers refer to Tables 7.1 and 7.2. *Shaded field* Coexistence of staurolite+quartz (HP98 database). *Dashed line* Metamorphic path discussed in text

than Fe-oxides in low-grade phyllites, all hematite will normally be consumed by reaction (8). The assemblage chlorite–chloritoid–magnetite is now stable above 360 °C.

At 400 °C, pyrophyllite is replaced by kyanite and chloritoid-kyanite becomes the diagnostic assemblage in Al-rich rocks [reaction (5)].

At about 480 °C, chloritoid and kyanite react according to reaction (11) and form staurolite. The first occurrence of staurolite and almandine–garnet marks the transition from the greenschist facies to the amphibolite facies in pelitic rocks. The temperature at the beginning of the amphibolite facies is about 500 °C. Chloritoid-staurolite and staurolite–kyanite are the diagnostic assemblages in the lower amphibolite facies. Almandine–garnet is formed by chlorite–chloritoid decomposition [reaction (9)]. Two new diagnostic assemblages are chlorite–almandine and almandine–chloritoid. Garnet formation

Table 7.2. Reactions in the FASH system

Sillimanite, kyanite, andalusite	Al_2SiO_5
Quartz	SiO_2
Pyrophyllite	$Al_2Si_4O_{10}(OH)_2$
Kaolinite	$Al_2Si_2O_5(OH)_4$
Chlorite (daphnite)	$Fe_5Al_2Si_3O_{10}(OH)_8$
Chloritoid	$FeAl_2SiO_6(OH)_2$
Staurolite	$Fe_4Al_{18}Si_{7.5}O_{44}(OH)_4$
Almandine	$Fe_3Al_2Si_3O_{12}$
Hercynite	$FeAl_2O_4$
Magnetite	$FeFe_2O_4$
Hematite	Fe_2O_3

Reactions with excess quartz + H_2O
  (1) And = Ky
  (2) And = Sil
  (3) Sil = Ky
  (4) Kln + 2 Qtz = Prl + H_2O
  (5) Prl = Ky + 3 Qtz + H_2O
  (6) Prl = And + 3 Qtz + H_2O
  (7) Chl + 4 Prl = 5 Cld + 2 Qtz + 3 H_2O
  (8) Chl + 4 Hem = Cld + 4 Mag + 2 Qtz + 3 H_2O
  (9) Chl + Cld + 2 Qtz = 2 Alm + 5 H_2O
  (10) 3 Chl = 3 Alm + 2 Mag + 12 H_2O (+ QFM)
  (11) 8 Cld + 10 Ky = 2 St + 3 Qtz + 4 H_2O
  (12) 23 Cld + 7 Qtz = 2 St + 5 Alm + 19 H_2O
  (13) 75 St + 312 Qtz = 100 Alm + 575 Ky + 150 H_2O
  (14) 3 Cld + 2 Qtz = Alm + 2 Ky + 3 H_2O
  (15) 3 Cld + 2 Qtz = Alm + 2 And + 3 H_2O
  (16) 8 Cld + 10 And = 2 St + 3 Qtz + 4 H_2O
  (17) 75 St + 312 Qtz = 100 Alm + 575 Sil + 150 H_2O
  (18) 75 St + 312 Qtz = 100 Alm + 575 And + 150 H_2O

Decomposition of staurolite in qtz-free rocks
  (19) 2 St = Alm + 12 Sil + 5 Hc + 4 H_2O
  (20) 2 St = Alm + 12 And + 5 Hc + 4 H_2O

in the pure FASH system requires a temperature of about 520 °C. At a few °C higher, chloritoid disappears from Qtz-bearing rocks by reaction (12). This means that chloritoid, which was the most characteristic mineral in the greenschist facies, extends into the lower amphibolite facies. The disappearance of chloritoid-bearing assemblages marks the upper boundary of the lower amphibolite facies. The greater part of the amphibolite facies is characterized by the assemblages St + Ky and St + Alm.

Chlorite decomposes in quartz-saturated rocks at about 540 °C by reaction (10) under production of almandine–garnet and magnetite. The disappearance of chlorite coincides with that of chloritoid. The oxygen activity along reaction (10) in Fig. 7.3 is controlled by the QFM buffer. It is implicitly assumed that oxygen activities in metapelites at the beginning of the amphibolite facies are close to those defined by the QFM assemblage. However, this appears to be reasonable. The stable assemblages at this stage depending on the X_{Al} of the bulk rock are: Mag + Alm, Alm + Cld, Cld + St and St + Ky.

Finally, at 650 °C, staurolite decomposes in quartz-bearing rocks and the characteristic high-grade assemblage is almandine-garnet + kyanite (reaction

13). The P–T field with stable staurolite + quartz is shaded in Fig. 7.3. It follows from Fig. 7.3 that metamorphism along prograde paths at very low and very high pressures will remain in the staurolite + quartz field only in a small temperature interval. Chloritoid is replaced directly by almandine-garnet + andalusite at very low pressure [reaction (15)] and by almandine-garnet + kyanite at pressures greater than 14 kbar [reaction (14)]. Staurolite in quartz-free rocks survives to much higher temperatures than in quartz-bearing rocks. Reactions (19) and (20) describe staurolite decomposition in quartz-free rocks. The product assemblage is almandine-garnet + aluminosilicate + hercynite-spinel. In high-T contact metamorphism a number of Fe-cordierite- and hercynite-involving assemblages may appear in Fe-rich metapelites. Cordierite-spinel assemblages will be discussed separately in Section 7.5.

The sequence of diagnostic assemblages and their P–T limits during metamorphism along other prograde P–T paths can be taken from Fig. 7.3 directly. For example, the temperature interval with stable staurolite + quartz is much smaller along a low-pressure metamorphic P–T path than along the Ky-type path. In fact, extensive zones with staurolite-bearing metapelites are most characteristic for typical "Barrovian-style" (Ky-type) regional metamorphic terrains, in agreement with Fig. 7.3. In Al-rich rocks, the prograde sequence of two-mineral assemblages is: Chl + Prl, Cld + Prl, Cld + Ky, St + Ky and Alm + Ky. These diagnostic assemblages can also be used to define characteristic metamorphic zones in a "Barrovian-style" regional metamorphic terrain.

7.4.5
Mica-Involving Reactions

Micas are the most important group of minerals in metapelitic rocks. Typical rocks are, for this reason, micaschists and micagneisses. The principal micas in metapelites are muscovite, paragonite and biotite. The minerals muscovite and paragonite are commonly designated white micas (dioctahedral micas). Their distinction under the microscope is difficult (use X-ray patterns). All micas show at least compositional variations along the $FeMg_{-1}$ (FM) and $MgSiAl_{-2}$ (TS) exchange vectors. Sedimentary and diagenetic illite rapidly recrystallizes to K-white mica with various amounts of TS and FM components. Low-temperature (high-pressure) K-white mica contains a significant amount of TS component. Such micas are termed phengites. Increasing temperature removes much of the TS component and high-grade white micas are close to the muscovite end member composition. NaK_{-1} exchange is important in metapelites because muscovite and paragonite commonly coexist in many micaschists; it is also significant in ordinary rocks containing the widespread assemblage muscovite and albite.

Reaction (21) in Table 7.3 replaces the sedimentary assemblage K-feldspar + chlorite by biotite + muscovite. The reaction affects rocks with low X_{Al} that do not contain Prl (Fig. 7.1). Equilibrium conditions of this reaction in the pure KFASH system are shown in Fig. 7.4. The product biotite is Fe-rich biotite (often green biotite under the microscope). The first prograde biotite

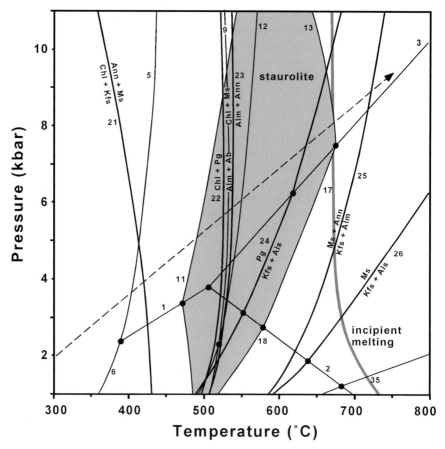

Fig. 7.4. Mica reactions. Reaction numbers refer to Tables 7.1–7.3. *Dashed line* Metamorphic path discussed in text

appears at about 420 °C. Since Kfs is modally subordinate to Chl in Al-poor metapelites, all Kfs is used up by reaction (21) and such rocks typically contain Ms + Chl + Bt + Pg + Qtz at T > 420 °C.

Reaction (22) replaces the paragonite + chlorite pair with garnet and Na-feldspar. The temperature at which this happens along the chosen model metamorphism (dashed path in Fig. 7.4) is about 520 °C. The reaction may use up either one of the reactants leaving two possible product assemblages: Chl + Alm + Ab (chlorite + garnet + feldspar) or Pg + Alm + Ab. Since metapelites proper contain excess Ms + Qtz, the two minerals are present in addition.

Reaction (23) limits the presence of chlorite in rocks containing excess muscovite, i.e., in normal metapelites. The assemblage is replaced by garnet + biotite at a temperature of about 540 °C.

Following the metamorphic path in Fig. 7.4, the sedimentary (diagenetic) Na-white mica, paragonite, breaks down in the presence of quartz at about

Table 7.3. Additional important reactions in the KNFASH system (in addition to reactions in Tables 7.1 and 7.2)

Annite (biotite)	$KFe_3AlSi_3O_{10}(OH)_2$
Muscovite	$KAl_3Si_3O_{10}(OH)_2$
Paragonite	$NaAl_3Si_3O_{10}(OH)_2$
K-feldspar	$KAlSi_3O_8$
Albite	$NaAlSi_3O_8$

All reactions with excess quartz + H_2O
(21) 3 Chl + 8 Kfs = 5 Ann + 3 Ms + 9 Qtz + 4 H_2O
(22) 2 Pg + 3 Chl + 6 Qtz = 5 Alm + 2 Ab + 14 H_2O
(23) 1 Ms + 3 Chl + 3 Qtz = 4 Alm + 1 Ann + 12 H_2O
(24) Pg + Qtz = Ab + Als + H_2O
(25) 1 Ms + 1 Ann + 3 Qtz = 1 Alm + 2 Kfs + 2 H_2O
(26) Ms + Qtz = Kfs + Als + H_2O

Discontinuous reactions in the KFMASH system
(27) Ctd = St + Grt + Chl
(28) Grt + Chl = St + Bi
(29) St + Chl = Bt + Als
(30) St = Grt + Bt + Als
(31) St + Bt = Grt + Als
(32) St + Chl = Als + Grt
(33) Grt + Chl = Bt + Als

630 °C [reaction (24)]. The product assemblage is albite + kyanite. It follows that paragonite is present in large portions of the stability field of staurolite + quartz. Staurolite-kyanite-paragonite schists are one of the classical metamorphic rocks of the mid-amphibolite facies zone of the Central Alps.

Biotite is removed from rocks containing much muscovite by reaction (25) at T > 700 °C in the pure KFASH system (Fig. 7.4). The Ms + Bt breakdown takes place at conditions where partial melting occurs at water-saturated conditions (Fig. 7.4).

Muscovite remains stable along the Ky-type path within the P–T window of Fig. 7.4. Reaction (26) represents the upper limit of muscovite in the presence of quartz. The product assemblage is K-feldspar + sillimanite (or andalusite, respectively). However, because reactions (22) and (24) produce albite (plagioclase) and reaction (25) produces K-feldspar, the assemblage at high temperatures is Kfs + Ab + Qtz. The assemblage will, in the presence of excess water, begin to melt at rather low temperatures (approximate position of the solidus is given in Fig. 7.4). Complications arising from partial melting in metapelites will be discussed separately in Section 7.6.

The sequence of stable two-phase assemblages in the pure KFASH system along the Ky geotherm is given in Fig. 7.5 together with the approximate upper temperature limits for the respective assemblages in the presence of excess quartz, muscovite and H_2O.

Fig. 7.5. Sequence of assemblages in the KFASH system projected from muscovite, quartz and H$_2$O onto the FeO–Al$_2$O$_3$ binary (Note: Kfs would not strictly project onto the binary). At 200 °C, metamorphism begins and the subsequent temperatures indicate the mineralogical changes along the dashed line in Figs. 7.3 and 7.4

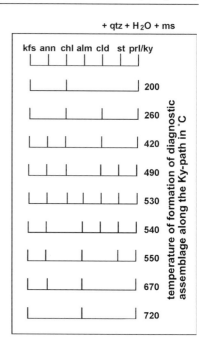

7.4.6
Metamorphism in the KFMASH System (AFM System)

The six-component system K$_2$O–FeO–MgO–Al$_2$O$_3$–SiO$_2$–H$_2$O is often used to discuss metamorphism in metapelites. AFM diagrams are popular graphic representations of phase relationships in this system (see Chap. 2 and above). In contrast to the discussion so far, the effects of the FM component in Fe-Mg-bearing minerals on mineral equilibria is now considered. The sequence of mineral assemblages in metapelitic rocks during prograde metamorphism along a medium-pressure geotherm ("Barrovian-type", Ky geotherm) is shown in Fig. 7.6 in a series of AFM projections. Prograde metamorphism will now be discussed, step-by-step, by reference to Fig. 7.6. As outlined above, all mineralogical changes occurring in the A-apex of the AFM figures are described in the ASH system. The FASH system, also discussed above, refers to the FA binary on the left-hand side of the AFM diagrams in Fig. 7.6. All reactions in the FASH system represent limiting reactions in the KFMASH system and give the basic framework of the following presentation. Therefore, many of the topological changes in the AFM diagrams can be understood by reference to Figs. 7.2, 7.3 and 7.4. Important note: the treatment refers strictly to the six-component KFMASH system. Metapelitic rocks, however, usually contain small but not negligible amounts of other components that may give rise to some "extra" complications (see Sect. 7.9).

1. At temperatures below about 260 °C, the stable assemblage in "normal" pelites is quartz–illite (muscovite)–chlorite–pyrophyllite (or kaolinite)–

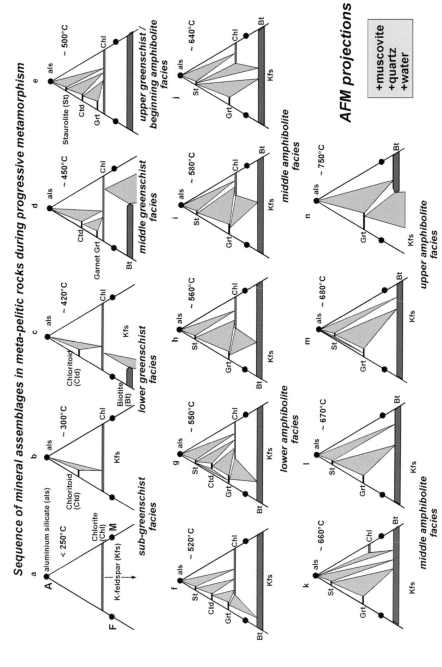

Fig. 7.6. Sequence of assemblages in the KFMASH system (AFM system) along a model path of metamorphism that corresponds to the dashed line on the previous figures

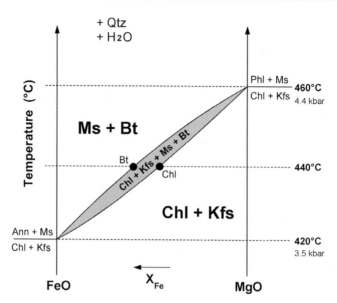

Fig. 7.7. P–T–X_{Fe} diagram representing the Kfs–Ms–Bt–Chl assemblage [reaction (21); Table 7.3]. The P–T gradient corresponds to the Ky-type path used before

paragonite. Al-poor semi-pelites may contain abundant K-feldspar instead of pyrophyllite.

2. The first truly metamorphic mineral formed is **chloritoid**. In "normal" pelites, chloritoid appears at about 300 °C.

3. Near 400 °C, the first **biotite** appears in Al-poor metapelites. Biotite forms at the expense of K-feldspar and chlorite. The reaction has equilibrium conditions of 420 °C at about 3.5 kbar along the Ky-type path in the pure KFASH system. The reaction, as most other reactions discussed here, is a continuous divariant reaction in the AFM system, however. The details of this reaction will be discussed as an example for all divariant equilibria in the AFM system below. The equilibrium conditions of such reactions depend on the Fe/Mg ratio of the rock. Therefore, the equilibrium conditions of reaction (21) can be discussed as a function of X_{Fe} (Fig. 7.7). The reaction equilibrium in the pure KFASH system is given on the left-hand side of Fig. 7.7 (420 °C). The reaction equilibrium in the pure KMASH system (pure Mg end member) can be found on the right-hand side of Fig. 7.7 (460 °C). In the temperature interval between the two limiting equilibria, all minerals (Chl + Kfs + Ms + Bt) may occur stably together. At, e.g., a temperature of 440 °C (Fig. 7.7), rocks with an appropriate A-coordinate may consist of biotite + muscovite, biotite + muscovite + K-feldspar + chlorite, or K-feldspar + chlorite, depending on the X_{Fe} of the rock. The composition of biotite and chlorite in the assemblage biotite + muscovite + K-feldspar + chlorite is given by the filled circles. It can be seen from Fig. 7.7 that the three-phase field biotite + K-feldspar +

chlorite moves across the entire AFM diagram from Fe-rich compositions to Mg-rich compositions within a temperature interval of only 40 °C. This can also be seen in Fig. 7.6 b–e, where at (b) Kfs-Chl is stable across the entire range of X_{Fe}, at (c) and (d) some intermediate positions of the three phase field are shown, and at (e) the entire range of X_{Fe} is covered by Chl-Bt (in the presence of excess muscovite + quartz).

4. The **first garnet** appears in metapelites at temperatures of around 450 °C. This temperature is in conflict with the information given for the pure FASH system. However, garnets in natural rocks preferentially incorporate manganese in the form of a spessartine component at low temperatures. Therefore, Mn–Fe–garnet appears at significantly lower temperatures than pure almandine. In addition, natural garnets in metapelites also always contain small amounts of Ca that further lowers the temperature of arrival of garnet in prograde metamorphism. This has been taken into account in Fig. 7.6.

5. At 500 °C, the K-feldspar–chlorite assemblage disappears also in Mg-rich compositions. The **first staurolite** grows. Its appearance marks the transition to the **amphibolite facies**. New diagnostic assemblages at this stage are: Ky-St-Chl and St-Cld-Chl.

6. Fe-rich chlorite begins to be replaced by **garnet + biotite** between 500 and 520 °C. The new assemblage garnet + biotite remains stable to very high grades, and the Fe-Mg partitioning between the two minerals is a frequently used geothermometer. Note, however, the succession of critical mineral assemblages shown in Figs 7.6 f–n probably has a better temperature resolution than the Grt-Bt thermometer. Because garnet is stabilized by the incorporation of Mn (that is preferentially fractionated into the first prograde garnet), the Grt + Bt assemblage may also occur in natural rocks at significantly lower temperatures. The first Grt + Bt pair evidently forms at temperatures as low as 470 °C.

7. Fe-chloritoid breaks down to **garnet + staurolite** and the diagnostic three-phase assemblage garnet + staurolite + chloritoid is present in a narrow temperature interval and in special bulk compositions (about 550 °C).

8. **Chloritoid disappears** from the rocks. The terminal reaction that removes chloritoid is the first discontinuous AFM reaction in the sequence [reaction (27), Table 7.3]. The equilibrium temperature of reaction (27) is only a few °C above equilibrium (12) in the pure FASH system (Fig. 7.3).

9. This is the characteristic AFM topology at the transition from lower- to mid-amphibolite facies conditions.

10. The AFM discontinuous reaction (28) has removed the garnet + chlorite tie line and replaced it with a two-phase field between staurolite + biotite. The **first staurolite + biotite** appears at temperatures slightly above 600 °C and marks the beginning of **middle amphibolite facies**. The discontinuous nature of reaction (28) makes it well suited for isograd mapping. In metamorphic terrains the chlorite + garnet zone is separated from the staurolite + biotite zone by a sharp isograd. The temperature at the boundary is about 600 °C and it is rather insensitive to pressure (along Sil or Ky paths).

11. The discontinuous AFM reaction (29) replaces the staurolite + chlorite tie line by the new assemblage **kyanite (sillimanite) + biotite**.

12. Mg-chlorite decomposes to biotite + kyanite (sillimanite). All chlorite disappears from quartz + muscovite saturated rocks at about 670 °C. Note, however, some additional complications turn up in extremely Mg-rich composition because talc and Mg-rich cordierite are additional stable minerals near the AM binary at various conditions. The phase relationships in extremely Mg-rich rocks will be discussed in Sections 7.5 and 7.7; they are irrelevant for "normal" pelite compositions (Fig. 7.1) in regional metamorphism.

13. Fe-rich staurolite breaks down and forms garnet + kyanite (sillimanite). The new diagnostic assemblage is kyanite + staurolite + garnet, which is restricted to a small range of bulk compositions, however.

14. Staurolite has disappeared from quartz-bearing rocks. The new diagnostic assemblage is **garnet + biotite + kyanite (sillimanite)**. The appearance of the assemblage by the discontinuous AFM reaction (30) also marks the beginning of the **upper amphibolite facies**. Note that this boundary also coincides with the production of the **first melt** in rocks of suitable composition at H_2O-saturated conditions. Note that metamorphism along a higher pressure path shows an inverted sequence of reactions limiting the assemblages Chl + Grt and Chl + St, respectively. Reaction (32) in Table 7.3 breaks down the St + Chl tie line at slightly lower temperature than reaction (33), which limits the occurrence of Grt + Chl. Fe-rich biotite decomposes to garnet + K-feldspar. From about 720 °C, K-feldspar + garnet + biotite constitutes a stable assemblage in semi-pelitic gneisses. Three assemblages cover the full range of Fe–Mg variations in Al-poor rocks: K-feldspar + garnet (in Fe-rich compositions), K-feldspar + garnet + biotite (in intermediate compositions), and K-feldspar + biotite (in Mg-rich compositions). Note, however, at the pressure of a Ky-type path, partial melting will form migmatite gneisses under these conditions.

The mineral assemblage diagram (Fig. 7.8) shows the distribution of metamorphic assemblages in a specific metapelite of typical Al-rich Ms-excess composition. The P–T trajectory of the previous figures is again shown as a dashed line. Under the conditions of the diagram, Chl + Cld is present in addition to Ms + Qtz up to 520 °C. Ky appears then in this rock and at 560 °C the first staurolite develops in the rock. At 570 °C, all chloritoid has disappeared and the diagnostic assemblage is St + Chl. In this Al-rich rock biotite does not form until the temperature has reached 610 °C. At 620 °C, all chlorite has gone and St + Bt is the typical assemblage in the middle amphibolite facies. Ky reappears in the rock at 670 °C and finally garnet forms from the breakdown of staurolite at about 680 °C. The assemblage Ky + Grt + Bt is representative of the upper amphibolite facies.

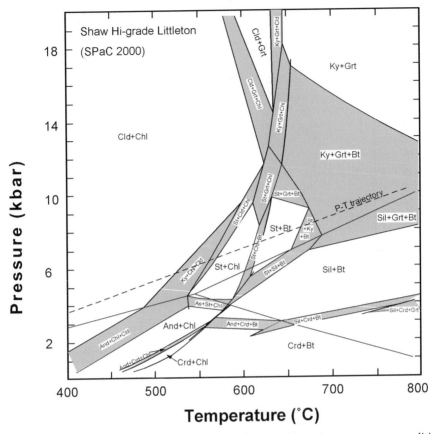

Fig. 7.8. Mineral assemblage diagram representing the assemblages in an average metapelitic rock computed on the basis of a petrogenetic grid by Spear et al. (2000) and downloaded from the website maintained by Prof. Spear (for address, see Chap. 4). The P–T gradient corresponds to the Ky-type path used before

7.5
Low-Pressure Metamorphism of Pelites

In low-pressure metamorphic terrains (<5 kbar at 700 °C), the sequence of stable mineral assemblages in metapelitic rocks is quite different from those found in orogenic "Barrovian" type terrains. The heat sources in low-pressure terrains normally are igneous intrusions that transfer heat from deeper to shallower levels of the crust. Diapiric plutons may rise up to 3 km (~1 kbar) below the surface before the magma loses dissolved H_2O and cools below the solidus. Thermal contact aureoles develop around such intrusive bodies. The pressure is often in the range 1–4 kbar (3–12 km depth). The maximum temperature to which sediments can be heated depends on a number of factors, including the composition of magma and the size of the intrusion. However, maximum contact temperatures at shallow-level "wet" granitic

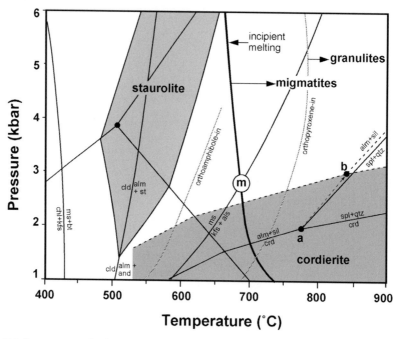

Fig. 7.9. Low-pressure high-temperature reactions in metapelites and metamorphism in the KFASH system (reactions are listed in Tables 7.1–7.4). For St reactions, see Fig. 7.3

and granodioritic plutons rarely exceed 650 °C. On the other hand, at somewhat deeper levels in the upper crust, large-scale accumulation of magma may result in temperatures up to 750 °C or so. Intrusion of large amounts of hot and dry magma from the mantle may result in extensive high-temperature contact aureoles with temperatures in the metamorphic envelope reaching 900–1000 °C. High-temperature contact aureoles are typically found around large mafic intrusions (gabbros, troctolites), intrusions of charnockite and mangerite batholiths, and anorthosite complexes that occur in Precambrian continental crust.

The consequences of isobaric heating of pelitic sediments of "normal" composition in a low-pressure contact aureole (e.g., at 1–2 kbar) can be derived from the mineral equilibria shown in Fig. 7.9.

7.5.1
KFASH System

Let us first consider equilibria in the pure KFASH system in order to derive a first overview. At low temperatures, the sequence of assemblages is the same as in orogenic metamorphism. Important here is reaction (21). It divides the cooler parts of a contact aureole into an outer zone with chlorite + K-feldspar and a higher-grade zone with biotite + muscovite. The temperature of the

Table 7.4. Cordierite–spinel–orthopyroxene–orthoamphibole–reactions (KFASH system)

Cordierite	$Fe_2Al_3[AlSi_5]O_{18} \cdot n\ H_2O$
Spinel (hercynite)	$FeAl_2O_4$
Orthopyroxene (ferrosilite)	$FeSiO_3$
Orthoamphibole (ferro-anthophyllite)	$Fe_7Si_8O_{22}(OH)_2$
Sapphirine (ferro-sapphirine)	$Fe_2Al_4SiO_{10}$
Osumilite (ferro-osumilite)	$KFe_2Al_3[Al_2Si_{10}]O_{30}$

All reactions with excess quartz + H_2O
(35) $2\ Alm + 4\ Als + 5\ Qtz + n\ H_2O = 3\ Crd$
(36) $2\ Spl + 5\ Qtz = Crd$
(37) $Alm + 2\ Als = 3\ Spl + 5\ Qtz$
(38) $Oam = 7\ Opx + Qtz + H_2O$
(39) $14\ Chl + 57\ Qtz = 8\ Oam + 7\ Crd + 48\ H_2O$
(40) $Ann + 3\ Qtz = 3\ Opx + Kfs + H_2O$

Some important cordierite reactions involving sheet silicates (in order of increasing grade)
Continuous reactions
(41) $2\ Chl + 8\ And/Sil + 11\ Qtz = 5\ Crd + 8\ H_2O$
(42) $Chl + Ms + 2\ Qtz = Crd + Bt + 4\ H_2O$
(43) $3\ Crd + 2\ Ms = 8\ Sil + 2\ Bt + 7\ Qtz$
(44) $3\ Crd + 2\ Bt = 4\ Grt + 2\ Ms + 3\ Qtz$
(45) $3\ Crd + 4\ Bt + 3\ Qtz = 6\ Grt + 4\ Kfs + 4\ H_2O$
(46) $2\ Bt + 6\ Ms + 15\ Qtz = 3\ Crd + 8\ Kfs + 8\ H_2O$
(47) $6\ Sil + 2\ Bt + 9\ Qtz = 3\ Crd + 2\ Kfs + 2\ H_2O$
(48) $2\ Grt + 4\ Ms + 9\ Qtz = 3\ Crd + 4\ Kfs + 5\ H_2O$

Discontinuous reactions
(49) $Chl + Sil + Ms + Qtz = Crd + Bt + H_2O$
(50) $Bt + Sil + Qtz = Crd + Grt + Kfs + H_2O$
(51) $Bt + Grt + Qtz = Crd + Opx + Kfs + H_2O$
(52) $Bt + Grt + Qtz = Opx + Sil + Kfs + H_2O$
(53) $Bt + Sil + Qtz = Crd + Opx + Kfs + H_2O$
(54) $Opx + Sil + Qtz = Crd + Grt$
(55) $Crd + Grt + Sil = Spl + Qtz$
(56) $Crd + Sil + Kfs + Qtz = Osm + Opx$

zone boundary is at about 400 °C. Around very shallow-level intrusions, such as, for example, in the Oslo rift zone (P ~500 bar), this is, in fact, the only mineralogical change taking place in many metapelitic sediments. The maximum temperature that can be reached at contacts to H_2O-rich granitic magmas that stop a few km under the surface is below 500 °C (magma temperature < 800 °C, country rock temperature T = 200 °C, ΔT < 600 °C. Rule of thumb: maximum T in sediments = T + 1/2 ΔT $\Rightarrow$ < 500 °C). The final assemblage is biotite + chlorite + muscovite + quartz.

The reactions at higher temperature replace the following assemblages present from lower grade (Fig. 7.5): Kfs + Bt, Bt + Chl, Chl + Cld, Cld + And (all assemblages with excess muscovite and quartz). Note that andalusite is the characteristic aluminosilicate polymorph in low-pressure terrains. Its maximum pressure stability limit varies with composition (Fe^{3+}) and structural details (e. g., dislocation density) but is around 3.5–4.5 kbar.

Slightly above 500 °C, three reactions remove chlorite and chloritoid from the assemblages. All reactions produce garnet. Of these reactions only $Cld = Alm + And$ (reaction 15) is shown in Fig. 7.9. The other two reactions,

(9) and (22), are shown in Figs. 7.3 and 7.4. No staurolite is formed in contact metamorphism below about 2 kbar. At about 600 °C, muscovite breaks down in the presence of quartz and the diagnostic high-grade assemblage K-feldspar + andalusite is produced.

At about 650 °C (1.5 kbar), the assemblage garnet + andalusite becomes unstable and it is replaced by cordierite by reaction (35), Table 7.4. The first stable sillimanite appears at about 650–700 °C. In most contact aureoles such high temperatures are not realized even at the immediate contact to the intrusives. Andalusite + cordierite + K-feldspar + biotite represents the highest-grade assemblage close to the contact of granitoid shallow-level intrusions. Metastable sillimanite occasionally can be found close to the contact to shallow-level intrusives as a result of structural disorder and chemical impurities that stabilize sillimanite relative to andalusite. Cordierite + sillimanite + K-feldspar + quartz is the stable assemblage at high temperatures in the pure KFAS system.

At slightly higher pressures (>2 kbar), a stability field for the assemblage spinel (hercynite) + quartz appears at very high temperatures (above 770 °C). The pair Spl + Qtz is diagnostic for very high temperatures.

The sequence of successive mineral assemblages in low-pressure metamorphism of iron-rich metapelites can be summarized as follows. At low temperatures, the sequence is identical to the one found in orogenic metamorphism. Between about 500 and 550 °C, cordierite appears in various assemblages involving chlorite, muscovite and biotite. Staurolite is absent and also garnet may not be present. Andalusite is the characteristic aluminosilicate. Above about 600 °C, metapelites contain the diagnostic assemblage cordierite + K-feldspar + biotite ± andalusite (at higher T sillimanite).

7.5.2
KFMASH System

The iron-magnesium exchange in the Fe–Mg minerals has profound effects in high-temperature metamorphism of metapelitic rocks. The cordierite-involving reactions (35)–(37) are fluid-absent reactions and as such strongly dependent on the compositions of the participating minerals (geological thermometers and barometers). Cordierite may contain molecular fluid species such as H_2O and CO_2 in the channels of the structure. The number of moles of water in cordierite ("n" in Table 7.4) varies between 0 and 2. The amount depends on the total pressure, the H_2O pressure, the composition of cordierite and the temperature. The presence of water in cordierite stabilizes the mineral relative to the respective reactant assemblage. H_2O is relatively loosely bound and can be compared with molecular water stored in many zeolite minerals. However, the reactions involving cordierite can also be formulated for H_2O-free environments (n = 0). In addition, the iron-magnesium content of the bulk rock is strongly fractionated between cordierite and garnet and spinel. Cordierite in metapelites is much more magnesian than coexisting garnet or spinel.

As a consequence, the stability field of cordierite is greatly expanded relative to the pure FAS system. The shaded area in the lower right P–T space of Fig. 7.9 shows the approximate expansion for the cordierite field for Fe-rich bulk compositions (X_{Fe} ~0.8). At a temperature of 600 °C near the contact of a pluton, cordierite may be found in 2 kbar rocks (instead of 1 kbar in pure FAS system). In addition, the Grt + Als + Qtz = Cord reaction (35) intersects a number of mica- and chlorite-involving reactions and this brings about some important cordierite-sheet silicate reactions (and orthoamphibole reactions). These reactions are not shown in Fig. 7.9 because their equilibrium coordinates vary strongly with X_{Fe} of the rock. Some of these reactions are listed in Table 7.4, however. Reactions (41) and (42) are continuous cordierite-producing dehydration reactions that are terminated by the discontinuous reaction Chl + And = Crd + Bt [reaction (49)] which intersects the 2-kbar isobar at about 530 °C. Above this temperature, cordierite + biotite may be present in metapelites. Reactions (43) and (44) are water-absent reactions; the equilibria are characterized by very low dp/dT slopes. Equilibrium conditions are independent of temperature and the assemblages have potential as geobarometers. The assemblages cordierite + muscovite and cordierite + biotite occur on the low-pressure side of the equilibria.

At higher temperatures, a number of continuous dehydration reactions (45–48) produce K-feldspar (that may enter a melt phase together with quartz and H_2O). The reactions are linked by the discontinuous reaction (50) that ultimately produces the characteristic assemblages cordierite + garnet + K-feldspar + biotite and cordierite + garnet + K-feldspar + aluminosilicate (andalusite or sillimanite). The assemblages are diagnostic for high-grade cordierite gneisses. The assemblages occur widespread in low- to medium-pressure terrains and typical P–T conditions deduced for such gneisses are in the range of 700 ± 50 °C at pressures from 2 to 5 kbar. Cordierite-garnet-K-feldspar-biotite-gneisses are transitional between upper-amphibolite and granulite facies conditions. The assemblage often indicates equilibration under conditions of reduced water pressure. Very commonly, cordierite–gneisses do not contain K-feldspar, however. Biotite is then the only K-bearing mineral in the rock and biotite assemblages *cannot* be represented in AFM diagrams (Kfs projection); biotite becomes an "extra" phase and typical assemblages involve four (or even more) "AFM" minerals. The representative assemblage in such rocks is Crd + Grt + Sil + Bt. It forms at P–T conditions similar to the Kfs-bearing Crd–gneisses. They are most typically found in terrains that were metamorphosed at 3–5 kbar and 650–750 °C and low water pressure.

Some additional important boundaries are shown in Fig. 7.9. At about 600–650 °C, orthoamphibole starts to form in low-pressure metamorphism of "pelites". Orthoamphibole-bearing assemblages are produced from chlorite- and biotite-consuming reactions. Reaction (39) may serve as an example of an Oam-producing reaction. The reaction is metastable in the pure FASH system, but it is important in natural systems. It generates cordierite-**anthophyllite** rocks that occur widespread in Precambrian shield areas. Note that the term **orthoamphibole** is used here for all Fe–Mg-amphiboles and includes also **cummingtonite** and **gedrite**. More than one type of Fe–Mg-amphibole may occur in high-grade rocks. Gedrite often occurs in rocks all the way up to granulite facies conditions. The precise conditions of the first occurrence of Fe–Mg–amphiboles in metapelites depend on the rock and fluid composition. The orthoamphibole-in curve in Fig. 7.9 represents an approximate boundary above which metapelites (and related rocks) may contain various Fe–Mg–amphiboles in a number of associations formed by several reactions. Note that there is little overlap of the orthoamphibole and muscovite fields, respectively. In addition, biotite appears to be distinctly more stable than the alternative assemblage K-feldspar + orthoamphibole in metapelites. Kfs + Oam have not been reported from metapelitic rocks to our knowledge, even if the mineral pair is predicted to be stable by the thermodynamic datasets (Chap. 3.8.1). A typical stable natural orthoamphibole assemblage at low- to intermediate-pressures (3–5 kbar) is Bt + Oam + Crd±Grt.

7.5.3
Cordierite–Garnet–Spinel Equilibria

The three FAS reactions (35), (36) and (37) define a stable invariant point (point "a" in Fig. 7.9). The assemblage spinel + quartz is a diagnostic high-temperature assemblage and the temperature of about 770 °C given by the invariant point "a" represents the lowest possible temperature for this assemblage. In Fe-rich metapelites, the corresponding equilibrium conditions of the three reactions are approximately given by the dashed curves in Fig. 7.9. It can be seen that the invariant point has moved to position "b" and the spinel + quartz assemblage requires at least 840 °C. The dashed arrow in Fig. 7.8 represents the equilibrium conditions of the discontinuous reaction (51) as a function of decreasing X_{Fe}.

The equilibrium constant of reaction (35) can be written as (assuming pure Als, Qtz and H_2O): $\ln K_{PT} = 3 \ln a_{Crd} - 2 \ln a_{Alm}$. Coexisting garnet and cordierite in medium-pressure high-grade gneisses may, for example, contain 80 mol% almandine and 40 mol% "dry" Fe-cordierite, respectively. The corresponding activities using a simple ideal site mixing solution model are: $a_{Alm} = 0.51$ and $a_{Crd} = 0.16$ and, hence, the calculated ln K for this Grt–Crd pair is equal to −4.159. On the schematic P-T diagram (Fig. 7.10), the P-T equilibria (35), (36) and (37) are contoured for fixed values of ln K. The garnet–cordierite pair from above is constrained to the ln K = −4.16 contour of equilibrium (35). It can be seen that the field for stable cordierite (light shading) is **enlarged** relative to the pure FAS system.

It also follows from the reaction equation (Table 7.4) that the presence of H_2O, which can be accommodated in the cordierite structural channels, will lead to a further stabilization of cordierite.

In addition, because there are two ferro-magnesian minerals involved in the transfer reaction (35), an Fe–Mg exchange reaction can be written between Grt and Crd: $FeMg_{-1}$ (Crd) = $FeMg_{-1}$ (Grt). The exchange reaction must be at equilibrium simultaneously with the transfer reaction (35). The equilibrium conditions of the Fe–Mg exchange are virtually dependent on the temperature alone. Simultaneous solution of the two equilibria (using GTB

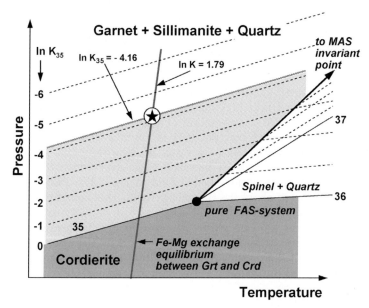

Fig. 7.10. Reactions among Grt-Crd-Spl (+Sil+Qtz). Stoichiometry of reactions listed in Table 7.4. Equilibrium constant of reaction (35) and the Grt–Crd FeMg exchange reaction are shown for a Grt with $Alm_{80}Prp_{20}$ and a Crd with $Crd_{60}FeCrd_{40}$. The equilibrium of reaction (35) is schematically contoured for constant $\ln K_{35}$. *Star* Simultaneous equilibrium of reaction (35) and FeMg exchange in a specified sample of Grt–Crd–Sil rock

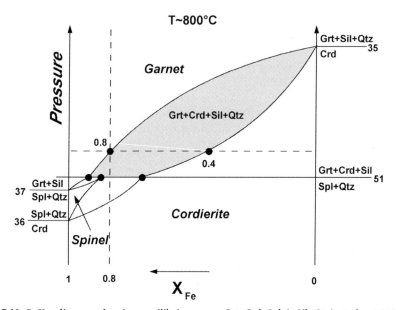

Fig. 7.11. P–X_{Fe} diagram showing equilibria among Grt–Crd–Spl (+Sil+Qtz) at about 800 °C. Reaction numbers and reaction stoichiometries listed in Table 7.4

software; see Chap. 4, and the calibration by Nichols et al. 1992) gives P = 3.5 kbar and T = 670 °C for the garnet–cordierite pair above coexisting with sillimanite and quartz. This is a very typical P–T condition for Grt–Crd rocks. Garnet–cordierite pairs in gneisses with excess sillimanite and quartz define then a unique equilibration point in P–T space (star in Fig. 7.10). Reactions among garnet–cordierite–spinel may also be represented, for example, by an isothermal pressure-composition diagram such as that shown in Fig. 7.11 (T ~800 °C). At a temperature of about 800 °C, the equilibrium pressure of the discontinuous reaction (51) is about 2 kbar. All three minerals tend to become more Mg-rich as a result of the continuous reactions (35), (36) and (37). There is a large pressure range over which the assemblage Grt + Crd + Sil + Qtz may occur (shaded field in Fig. 7.11). Pelitic rocks commonly have an X_{Fe} of about 0.8 (normal pelites). This composition is shown as a vertical dashed line in Fig. 7.11. Such a rock contains cordierite at low pressure and it will produce spinel by reaction (36) as pressure increases. Spinel will be removed from the rock by reaction (51) at a specific pressure. Above this pressure cordierite decomposes to garnet by reaction (35) until all cordierite is used up. In more Mg-rich rock compositions, the spinel fields are by-passed and cordierite is directly replaced by garnet. With decreasing X_{Fe}, cordierite decomposes at progressively higher pressures. Note: the pure MAS version of reaction (35) is metastable.

7.6
Very High-Temperature Metamorphism of Pelites–Metapelitic Granulites

7.6.1
Partial Melting and Migmatites

At about 650–700 °C, partial melting begins to be important in rocks containing feldspar and quartz under water-saturated conditions (Fig. 7.9). The curve of incipient melting marks the onset of partial melting in "granitic" systems and the beginning of migmatite formation. This means that at temperatures above this curve Kfs-component produced by dehydration reactions (Table 7.4) dissolves in a melt phase that may leave the site of production in some cases. The melt segregates and removes Kfs and Ab produced by mica-involving dehydration reactions. The rocks left behind may be devoid of alkali feldspar and consequently also alkalis and "granite" components. Particularly important is the decomposition of muscovite in the presence of quartz. The melt-absent reaction (25) that produces andalusite or sillimanite and Kfs at low pressures intersects the minimum melt curve at about 3 kbar (point "m" in Fig. 7.9). At higher pressures, muscovite decomposes to sillimanite and a Kfs-bearing melt. The situation is schematically shown in Fig. 7.12. The melt-absent decomposition of muscovite (+ Qtz) is an ordinary dehydration reaction and has a characteristic curve shape on a P–T diagram with a positive dp/dT slope at the pressures of interest here. At $a_{H_2O} = 1$, Ms + Qtz has its maximum stability. The equilibrium of reaction (25) can be contoured for decreasing a_{H_2O} similar to, e.g., increasing dilution of the aqueous fluid with CO_2 (Fig. 7.11). With decreasing a_{H_2O}, K-feldspar + sillimanite forms at progressively lower temperature. The formation of sillimanite by reaction (25) is occasionally referred to as "the second sillimanite isograd" in the geological literature (in contrast to sillimanite from the phase transitions Ky = Sil and And = Sil, respectively, that may or may not form Sil because of metastable survival of And or Ky!). The solidus for muscovite-granite at H_2O-saturated conditions has a negative slope in a P–T diagram at the pressures of interest. At the minimum melt curve shown in Figs. 7.9 and 7.12, quartz + feldspar begins to melt. Muscovite of the solidus assemblage dissolves in the melt and contributes Kfs-component and H_2O to the melt, leaving sillimanite as a solid residue. Compared with the melt-absent reaction, decreasing a_{H_2O} has an opposite effect on the Ms + Qtz breakdown curve and the minimum temperature of muscovite melting increases with decreasing a_{H_2O}. The consequences for the muscovite-out isograd in Qtz-bearing rocks are as follows: At pressures below the intersection of the water-saturated melt-present and melt-absent Ms-breakdown curves (point "m" in Fig. 7.12), muscovite disappears by reaction (25) and forms Kfs + Als + H_2O. At pressures above that intersection, muscovite disappears by partial melting. The intersection marks the highest possible temperature for stable Ms + Qtz. The dark-shaded area in Fig. 7.12 marks the muscovite field in Qtz-bearing rocks. Decreasing a_{H_2O} displaces the intersection point of the two Ms-out reactions towards higher pressures whilst the net temperature effect is minimal. Muscovite cannot be found in Qtz-bearing rocks outside the light-shaded area irrespective of fluid composition. The quantitative position of the intersection point "m" of the two reactions is shown in Fig. 7.9 (~2.8 kbar, 680 °C).

The melt produced by prograde metamorphism of mica schists is close to eutectic composition of the granite system and H_2O-saturated. The further fate of the produced melt phase depends primarily on the amount of melt produced in the partial melting process and also on accompanying deformation and other factors. Small amounts of melt may not migrate over long distances and will be found later as **leucocratic quartzo-feldspathic bands, pods, lenses** and **patches**. If much anatectic "granitic" melt is produced by local in situ partial melting of gneisses and schists, the melt may collect to larger masses and may cross cut the primary gneissic foliation or bedding, thus forming veins and irregular masses of **discordant granite** and the melt may eventually leave the site of production and collect in a larger magma chamber of crustal granitic melt that may rise as a pluton to shallow levels in the crust by buoyancy forces. The light Qtz–Fsp material in gneisses representing the anatectic melt phase typically shows randomly oriented minerals and fabrics typical of igneous rocks (**leucosome**). The **restite** material from which the "granitic" melt has been extracted has a high modal proportion of mafic minerals [mainly alumnosilicate (sillimanite), biotite and garnet] and has a dark gneissic appearance (**melanosome, restite**). The two components of gneisses that underwent partial melting and **anatexis**, the leucosome and the restite, may be present in countless proportions and ar-

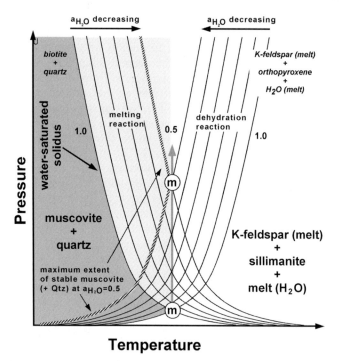

Temperature

Fig. 7.12. Effects of water-pressure variations on the melt-present and melt-absent musco-vite+quartz breakdown reaction (25) and the corresponding biotite+quartz reaction (40). Reaction stoichiometry listed in Tables 7.3 and 7.4. Point m marks the intersection of the vapor and melt producing Ms-breakdown reactions at equal a_{H_2O} (pure water present as phase). At the high pressure point m $a_{H_2O} = 1$ (e.g., a fluid phase present with H_2O and CO_2 in equal proportions). The line connecting the two points and projects to $a_{H_2O} = 0.5$ marks the maximum temperature limit of the assemblage Ms+Qtz

ranged in an infinite number of different light-dark patterns. These anatectic gneisses are commonly termed ***migmatites***. Migmatite is one of the prime rock materials found in continental crust.

7.6.2
Granulites

Orthoamphibole that may have formed previously decomposes to orthopy-roxene and quartz (reaction 38) at about 750–800 °C. The appearance of **ortho-pyroxene** in **quartz-bearing rocks** marks the transition from upper amphibolite facies to **granulite facies** conditions (see Fig. 7.9). However, the diagnostic granulite facies assemblage orthopyroxene + quartz may originate from a num-ber of different reactions. The most important of them is reaction (40), which eventually removes the last remaining sheet silicate, biotite, from metapelites. In general, biotite breaks down in the presence of quartz at about 800 °C to orthopyroxene + K-feldspar. Biotite replacement by orthopyroxene takes place

Fig. 7.13. a AFM projection (from K-feldspar) of typical granulite facies assemblages in metapelites and related rocks at about 6 kbar and 770 °C and $a_{H_2O} \sim 0.4$. **b** Schematic P–T diagram, showing five discontinuous high-grade (granulite-facies) reactions (listed in Table 7.4). The *shaded field* corresponds to the AFM topology of Fig. 7.13a; the generic metamorphic path leading to that field started from a typical upper amphibolite facies Grt+Sil+Bt assemblage. The position of the invariant point is at about 8 kbar and 800 °C (and $a_{H_2O} < 1$)

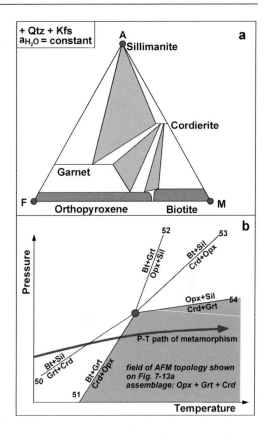

over a fairly wide T interval, depending on the composition of biotite and fluid. Biotite decomposition produces Kfs that, together with the H_2O released by reaction (40), commonly enters a melt phase. The melt-absent version of reaction (40) is metastable at pressures above a few hundred bars and in the presence of H_2O-rich fluids relative to corresponding biotite melting reaction. The relationships are about analogous to those of the muscovite breakdown shown in Fig. 7.12, which can also be used to portray biotite breakdown. Kfs may dissolve in a melt phase (H_2O-rich conditions) or it may remain in the rock (dry conditions). In the upper amphibolite facies, typical metapelitic gneisses with Bt + Grt + Sil + Kfs + Pl + Qtz may have lost a free fluid phase after the major dehydration reactions had gone to completion. Biotite remains as the last hydrate in the rocks and it survives in relatively low a_{H_2O} environments. However, if the rocks experience further heating, fluid-absent melting in high-grade gneisses becomes important and this process is responsible for much of the orthopyroxene found in granulite terrains.

Figure 7.13a shows a representative AFM chemography of phase relationships in high-grade metapelites. The chemography is meant to be representative of conditions of about 5 kbar and 750 °C and an H_2O activity smaller than 1. The assemblages contain excess K-feldspar and quartz. Figure 7.13a shows that under these conditions "normal" Fe-rich metapelites contain garnet + sillimanite or garnet + sillimanite + cordierite. More Mg-rich rocks contain orthopyroxene + cordierite + garnet, whereas biotite is restricted to fairly Mg-rich compositions. Biotite may occur in assemblages with cordierite + orthopyroxene. Note that the stable biotite + orthopyroxene assemblage shown in Fig. 7.13 a together with excess K-feldspar and quartz requires equilibrium of reaction (40). The Bt + Opx pair can in fact be used to estimate H_2O activity provided that an independent P–T estimate can be made for the assemblage from geothermobarometry. Figure 7.13 a depicts a stage in metamorphism where Fe-rich biotite has been replaced by orthopyroxene [reaction (40)], whereas Mg-rich biotite still persists. It is also immediately evident from Fig. 7.13 a that a great number of other possible assemblages can be expected involving the minerals garnet, sillimanite, cordierite, orthopyroxene and biotite in rocks containing excess K-feldspar + quartz. Using the topology given in Fig. 7.13 a, the five minerals are connected by five discontinuous reactions at a fixed water pressure. Reactions among the five most common minerals in high-grade metapelites are shown in Fig. 7.13 b, together with a possible P–T path of metamorphism that produced the assemblages shown in Fig. 7.13 a. At "low-temperature" (= higher amphibolite facies here), the typical assemblage is biotite + sillimanite. Garnet occurs as an additional mineral in "normal" metapelites. The high-grade K-feldspar + sillimanite assemblage is also present because Kfs is an excess phase above the Ms-out isograd. As discussed above, low- to intermediate-pressure metamorphism replaces biotite + sillimanite with garnet + cordierite [reaction (50)]. At higher grade, the garnet + biotite tie-line is replaced by the new orthopyroxene + cordierite assemblage shown in Fig. 7.13 b by reaction (51). If metamorphism follows a path that is on the high-pressure side of the invariant point of Fig. 7.13 b, then a new sequence of diagnostic assemblages can be observed. The most significant of the assemblages is orthopyroxene + sillimanite. It forms from reactions (52) and (54), respectively. Possible assemblages include Opx + Sil + Qtz + Grt + Kfs and Opx + Sil + Qtz + Bt + Kfs; both are diagnostic for high-pressure granulites. Opx + Sil + Qtz + Grt granulites typically form at pressures greater than 8 kbar and temperatures above 800 °C. The rocks are characteristic for metapelitic granulites in the lower continental crust. The pair Opx + Ky requires even higher pressures to form (P > 10 kbar). Note, however, that the precise position of the invariant point depends on a_{H_2O}. The diagnostic Opx + Sil assemblage is often (partly) replaced at later stages of metamorphism, e.g., during isothermal decompression, by cordierite-bearing assemblages such as sapphirine + cordierite symplectites.

Much of the biotite and also the assemblage biotite + sillimanite found in higher amphibolite facies rocks may form from K-feldspar + garnet and K-feldspar + orthopyroxene assemblages by retrograde rehydration of granulite facies rocks. H_2O-rich fluids often become available when the rocks cool through the "wet" granite solidus. H_2O dissolved in granitoid melts that formed during migmatization (see above) is released and may pervasively retrograde granulites to biotite + sillimanite + garnet gneisses or biotite + sillimanite + cordierite gneisses. Conventional geothermobarometry will yield 650–700 °C and 3–5 kbar for such rocks and the P–T conditions mimic the "wet" granite solidus that actually is responsible for the pervasive retrogression under these conditions.

At still higher temperatures, diagnostic assemblages are spinel + quartz (>850 °C) that may form by reaction (55), sapphirine + quartz (>900 °C), and osumilite + quartz (>950 °C). Pigeonite (ternary pyroxene) may also appear in high-grade rocks at temperatures above 850–900 °C. High-grade metamorphic rocks with assemblages that are diagnostic for crustal temperatures above 850–900 °C are rare and have markedly been reported from Precambrian terrains where massive heat transfer from the mantle to the middle continental crust took place by means of dry igneous intrusions (e.g., contact aureoles around anorthosite complexes).

Metamorphism of metapelitic rocks is strongly affected by partial melting reactions at very high temperatures. Increased formation of partial melts has severe consequences for the bulk composition of the rocks. The rocks become depleted in the "granite" component, i.e., in feldspar and quartz. Fe is also strongly fractionated into the melt phase and consequently the residuum is enriched in Mg and Al. Recurring dehydration reactions along a prograde P–T path also inevitably leave the rocks devoid of an aqueous fluid phase. Accordingly, extreme high-grade "metapelites" are often quartz-free Al–Mg-rich restites left behind by partial melting processes and they consist of various assemblages among distinct minerals including orthopyroxene, cordierite, sillimanite (kyanite), sapphirine, spinel, garnet, corundum and others. Water-deficient conditions are often reflected in disequilibrium microstructures including symplectites, replacement structures and coronites.

7.7
Metamorphism of Very Mg-Rich "Pelites"

Extremely Mg-rich primary sedimentary shale compositions are quite unusual (see Sect. 7.2). However, Mg-rich micaschists and other rocks with remarkably high Mg/Fe ratios can be found occasionally intercalated in other schists and gneisses in many metamorphic terrains and orogenic belts. The unusual bulk rock composition may have a number of feasible origins including evaporitic sediments, fluid/rock interaction (metasomatism) during metamorphism or restites after partial melting. Nevertheless, the assemblages of these rocks resemble those of "normal" metapelites and the rock compositions may show continuous transitions to more Fe-rich metapelites. A number of special features are, however, special to Mg-rich schists and some selected equilibria in the pure KMASH system are shown in Fig. 7.14. Note that the phase diagram is valid for rocks that contain quartz and for the condition $P_{total} = P_{H_2O}$. The latter condition has the consequence that some equilibria at high temperatures become metastable relative to melt-producing reactions (see also Sects. 7.5 and 7.6).

Mg–cordierite can be found over a wide range of P–T conditions and the mineral can be found in rocks that were metamorphosed at pressures in ex-

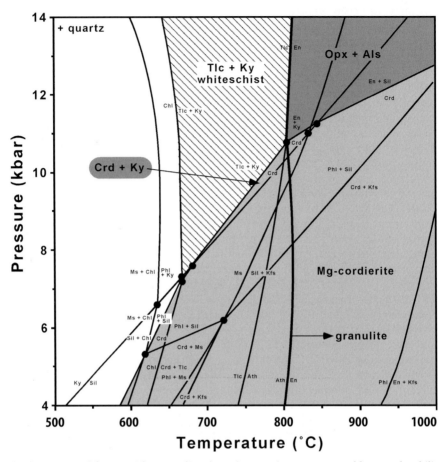

Fig. 7.14. P-T grid for Mg-rich metapelites (HP98). Some important assemblages and stability fields are highlighted. Granulite facies boundary = first Opx+Qtz. Tlc+Ky at P >7.5 kbar, Crd+Ky at P >8 kbar, Opx+Als at P >11 kbar

cess of 8 kbar. This is in sharp contrast to Fe–cordierite occurring in Fe-rich metapelites. The P–T range for Mg–cordierite in various assemblages is light shaded in Fig. 7.14. Note that there is some overlap of the Mg–cordierite and kyanite fields, respectively (horizontally ruled pattern in Fig. 7.14). The assemblage cordierite + kyanite has been observed in amphibolite facies schists from "Barrovian" metamorphic terrains. It is restricted to a small area in P–T space. Other observed natural assemblages are, in fact, diagnostic for even more restricted P–T ranges. For example, in the Tertiary amphibolite facies Ticino mountains of the Central Swiss Alps, the assemblage cordierite + kyanite + quartz + chlorite + phlogopite (Mg-biotite) has been reported by Wenk (1968). From Fig. 7.14 it is evident that the assemblage is tightly restricted to pressures between 7 and 8 kbar and a temperature close to 670 °C. Note that the geologic interpretation of such numbers (pairs of numbers) is a theme not covered in this volume. The diagnostic high-pressure granulite facies assemblage orthopyroxene + aluminosilicate (dark-shaded area in Fig. 7.14) is restricted to temperatures > 800 °C and pressures > 8 kbar. The diagnostic granulite facies assemblage orthopyroxene (enstatite) + quartz requires minimum temperatures of 800 °C. However, one must be aware that this is true only in the presence of a pure H_2O fluid. Also note that in extremely Mg-rich rocks *talc* may appear in a number of new assemblages. Talc is not present in Fe-rich metapelites and the mineral does not substitute much iron for magnesium. It therefore promptly disappears with increasing X_{Fe} of the bulk rock. The most interesting and diagnostic assemblage is talc + kyanite which is restricted to high pressures above 8 or 9 kbar at temperatures in the range of 700±50 °C. The so-called whiteschist (Schreyer 1977) assemblage Tlc + Ky appears in high-pressure low-temperature metamorphism associated with lithosphere subduction (see below) but also formed in many Precambrian granulite facies terrains by isobaric cooling of lower crustal granulites. In the lower right-hand corner of Fig. 7.14 the position of the metastable phlogopite + quartz breakdown curve is shown. The reaction takes place at about 100 °C above the corresponding annite + quartz breakdown. Both biotite-consuming Opx-forming reactions are metastable relative to the corresponding melt-producing reactions in rocks containing feldspar. As discussed above, the maximum temperature for biotite + quartz is about 850 °C.

7.8
High-Pressure Low-Temperature Metamorphism

High-pressure low-temperature metamorphism (HPLT) is most commonly associated with subduction of oceanic lithosphere along destructive plate margins. The few pelitic sediments that are found associated with HPLT ophiolite complexes and metapelites in continental HPLT complexes show very distinct metamorphic minerals and characteristic assemblages that are unique for HPLT metamorphism.

Some relationships in HPLT rocks are summarized in Fig. 7.15 and a selection of relevant continuous reactions are listed in Table 7.5. The calculated (RB88) phase relationships in the ASH system are shown in Fig. 7.15 as a reference frame. The constant 9 °C/km geotherm will represent the approximate boundary of geologically feasible P–T paths. This type of metamorphism is initially characterized by extremely steep dp/dT slopes that continuously decrease as HPLT metamorphism progresses (see also Chap. 3.5).

The mineral sudoite, a dioctahedral chlorite, appears to be typical in very low-grade though high-pressure rocks. With increasing pressure the distinctive mineral carpholite may replace sudoite. The H_2O-conserving reaction (56) describes this transition in terms of the end-member phase components given in Table 7.5. There is, as in the case of talc, the unusual situation that Fe–carpholite and Mg–carpholite apparently do not mix over the entire FM range, but rather form separate minerals. Carpholite also forms from chlorite + pyrophyllite by reaction (57). Nominally, reaction (57) is a dehydration

Fig. 7.15. Tentative P–T grid for low-temperature high-pressure metamorphism of metapelitic" rocks

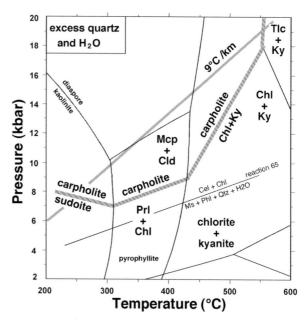

Table 7.5. Reactions at high pressure and low temperature

Chlorite (clinochlore, daphnite)	$(Mg,Fe)_5Al_2Si_3O_{10}(OH)_8$
Sudoite	$Mg_2Al_3[AlSi_3O_{10}](OH)_8$
Chloritoid	$(Mg,Fe)Al_2SiO_6(OH)_2$
Magnesiocarpholite	$MgAl_2[Si_2O_6](OH)_4$
Ferrocarpholite	$FeAl_2[Si_2O_6](OH)_4$
Celadonite (phengite)	$K(Mg,Fe)AlSi_4O_{10}(OH)_2$
Talc	$Mg_3Si_4O_{10}(OH)_2$
Pyrope	$Mg_3Al_2Si_3O_{12}$
Coesite	SiO_2

Reactions with excess quartz + H_2O
(56) $Sud + Qtz = 2\ Mcp$
(57) $5\ Mcp + 9\ Qtz = Chl + 4\ Prl + 2\ H_2O$
(58) $Fcp\ (Mcp) = Cld + Qtz + H_2O$
(59) $5\ Mcp = Chl + 4\ Ky + 3\ Qtz + 6\ H_2O$
(60) $3\ Chl + 14\ Qtz = 5\ Tlc + 3\ Ky + 7\ H_2O$
(61) $Chl + 3\ Qtz = 2\ Cld + Tlc + H_2O$
(62) $4\ Cel + Mcp = Chl + 4\ Kfs + 3\ Qtz + 2\ H_2O$
(63) $5\ Cel + Ms = Chl + 6\ Kfs + 2\ Qtz + 2\ H_2O$
(64) $3\ Cel = Phl + 2\ Kfs + 3\ Qtz + 2\ H_2O$
(65) $4\ Cel + Chl = 3\ Phl + Ms + 7\ Qtz + 4\ H_2O$

reaction that decomposes rather than forms carpholite. However, P–T paths in this kind of metamorphism are characterized by extremely steep dp/dT slopes and P–T paths may, in some situations, cross dehydration equilibria from the dehydrated to the hydrated side upon pressure increase even if temperature also slightly increases. The effect is also shown in Fig. 3.13, where the path "dehydration by decompression" is followed in the other direction in HPLT metamorphism ("hydration by compression"). For this reason, carpholite may even form by hydration of the chlorite + kyanite assemblage by reaction (59). Note, however, that the carpholite formation by these mechanisms requires that free H_2O fluid is available and carpholite-forming reactions will cease when water is used up.

Ferrocarpholite decomposes to chloritoid (reaction 58) whereas magnesiocarpholite is replaced by chlorite + kyanite (reaction 59) if the rocks are not carried to greater depth but rather follow a "normal" clockwise P–T path through the greenschist facies. If HPLT metamorphism continues, carpholite decomposition produces magnesiochloritoid (Mg version of reaction 58). Chloritoid may, in fact, become extremely Mg-rich in HPLT rocks, and chloritoid with X_{Mg} ~0.5 is not uncommon. It is a general observation that the typical Fe-rich AFM phases chloritoid, staurolite and garnet become increasingly magnesian at very high pressures. The carpholite boundaries are given tentatively in Fig. 7.15. Talc is also a characteristic mineral in HPLT metapelites and it often occurs together with K-white mica (the two minerals are difficult to distinguish under the microscope). Chloritoid + talc forms from the continuous chlorite-consuming reaction (61). The two minerals finally combine to form pyrope components in garnet at very high pressures. The diagnostic white schist assemblage Tlc + Ky results from the chlorite-breakdown reaction (60) or analogous chloritoid- and carpholite-involving reactions.

Micas in high-pressure metapelites also show some characteristic diagnostic features. Particularly notable is the absence or scarcity of biotite in HPLT metapelites. Biotite persists only in extremely Mg-rich high-pressure rocks in the form of phlogopite. The main reason for this is that potassium white mica experiences significant tschermak substitution with increasing pressure. Thus, Mg-Fe-free muscovite changes its composition systematically towards celadonite (Table 7.5) with increasing pressure. Celadonite-rich K-white micas are commonly termed phengite. Phengite is the characteristic mica in HPLT rocks. The celadonite component is produced by a number of continuous reactions such as (62)–(65), Table 7.5. The equilibria of all four continuous reactions have low dp/dT slopes in the P–T range of interest. The celadonite component is always on the high-pressure side of the equilibrium curve. As an example, the equilibrium of reaction (65) is shown for a certain fixed value of the equilibrium constant in Fig. 7.15. HPLT metamorphism that follows a P–T path similar to that in Fig. 7.15, therefore, consumes the phlogopite-(biotite) component and H_2O and increases the celadonite component in phengite. Phase relationships in high-pressure metapelites can, for example, be represented in chemographies such as shown in Fig. 7.16. The graphical representation permits the discussion of relations involving phengite and in particular shows the important tschermak variation (TS) in phengite. These relationships cannot be seen on AFM diagrams because they represent projections from the muscovite component. On the other hand, Fig. 7.16a is a projection along $MgFe_{-1}$ and consequently does not allow for representation of relationships that depend on the Fe-Mg variation. The three-phase assemblages shown in Fig. 7.16 will contain, in general, an additional phase because of the FM variation. Figure 7.16 shows isobaric isothermal figures that may be typical for conditions around 13 kbar

Fig. 7.16a,b. Isothermal isobaric composition phase diagrams depicting associations
in high-pressure meta-
morphism of metapelites

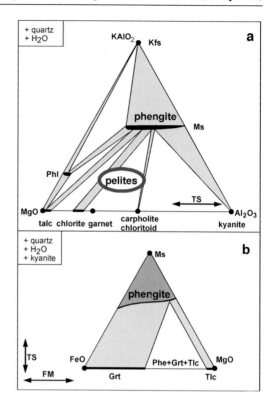

and 450 °C. "Normal" metapelites contain the assemblage chlorite + carpholite + chloritoid + phengite + quartz under these conditions. Phengite + kyanite
+ carpholite + chloritoid is present in Al-rich rocks and phengite + chlorite
+ talc + chloritoid in Al-poor schists at the same conditions. The projection
in Fig. 7.16b is from kyanite + quartz. It shows the HPLT assemblage pyrope-
rich garnet + phengite + talc + kyanite that is typical for very high pressures.
It is clear from Fig. 7.16a,b that changing pressure and temperature will result
in new assemblages and modified mineral compositions caused by many continuous and discontinuous reactions. Phase relationships in HPLT metapelites
offer an immense potential for analysis of metamorphic conditions and successive sequences of mineral assemblages that must be worked out for each HPLT
terrain in question. Ultra-high-pressure metamorphism of metapelites and
crustal schists and gneisses ultimately produces coesite from quartz and diamond from graphite.

7.9
Additional Components in Metapelites

Until now we have discussed metamorphism of pelites, step-by-step, by increasing the considered chemical complexity. The six-component AFM system typically accounts for more than 95 wt% of pure shales. However, the

compositional inventory of pelites lists a great number of additional chemical elements. These may be stored somewhere in "AFM" minerals that form complex solid solutions or they may give rise to separate minerals. Such "extra" phases in turn may give valuable information about the metamorphic evolution of a terrain.

With respect to a "normal" pelite, it is obvious that the widespread and abundant chemical elements calcium, sodium, manganese, titanium, and ferric iron will also be present to some degree in even the most mature pelitic sediments. These elements are found in any chemical analysis of a pelite and, hence, they must be found in some of the minerals that make up the rock. In the most common "AFM minerals" one finds many components that are not part of the AFM system.

Garnet normally incorporates Ca and Mn in significant amounts, but also Fe^{3+} and OH may be present. The spessartine component in prograde low-grade garnet is particularly important. Plagioclase is often present in metapelites and the mineral carries a great deal of the bulk rock's Na and Ca content. If no plagioclase is present in low- to medium-grade metapelites, the calcium content of the rock may be found entirely in garnet as grossular component or in a separate Ca-mineral, especially zoisite, clinozoisite or epidote. Biotite usually contains Ti, Mn, halogens and Fe^{3+}. Biotite can take up more than 5 wt% TiO_2 and it may be the only significant Ti-bearing mineral in the rock. If the bulk rock TiO_2 content exceeds Ti-saturation of biotite, some extra Ti-phase will be present (usually titanite or ilmenite). Staurolite in metapelitic rocks tends to grasp Mn and particularly Zn. Staurolite may contain several wt% zinc! Cordierite often accommodates Li and Be that replace Mg, Fe, and Al on regular crystallographic cation sites and Na, CO_2, and H_2O in the structural channels. These "extra" elements in cordierite are difficult or impossible to analyze with standard microprobe techniques. Muscovite, as discussed above, is often phengitic. Spinel may contain a large variety of "non-AFM elements" including Fe^{3+}, Zn, Mn, Cr, and V. Spinels often contain high concentrations of zinc (several wt%).

There are two major effects of extra components in "AFM minerals". The equilibrium conditions of mineral reactions can be displaced in P–T space relative to the pure AFM system and, in the most extreme case, the topology of phase relationships can be altered and inverted. In other words, mineral assemblages that are not stable in the pure AFM system can become conceivable as a result of strong fractionation of one "extra" element (e.g., zinc) into one "AFM mineral". Strong partitioning of manganese into garnet, for example, causes the first garnet to appear in prograde metamorphism at a temperature that is about 50–100 °C lower than in the pure AFM system. The other major effect of the presence of extra components is that it causes more minerals in an assemblage to appear than are usually expected in the pure AFM case. For example, in a typical general AFM situation, the minerals chlorite + biotite + staurolite can be found in a rock with muscovite + quartz in excess. In such a rock it will not be unusual to find also garnet as an additional mineral and as a true member of the assemblage owing to the garnet's manganese content. The coexistence of the four minerals does not mean, in this case, that the rock equilibrated at the conditions of the AFM discontinuous reaction (28) (Table 7.3). Garnet is in reality not a coplanar phase in terms of an AFM diagram but rather defines a phase volume in the AFM + Mn space.

Particularly difficult are high-grade assemblages involving spinel because the mineral often deviates significantly from pure Mg–Al–spinel–hercynite solutions and because it is usually involved in fluid-absent reactions which are especially sensitive to small deviations from the pure AFM system. Therefore, great care must be taken in the interpretation of spinel-bearing assemblages.

In conclusion, the observed common deviation of metapelites from the pure AFM composition may have significant consequences on phase relationships. The interpretation of phase relationships in metapelites, therefore, requires common petrologic sense and a great deal of professional experience. These difficulties, however, certainly do not invalidate the use of an AFM model for the prediction of phase relationships in metapelites.

References

Aranovich LY, Podlesskii KK (1983) The cordierite-garnet-sillimanite-quartz equilibrium: experiments and applications. In: Saxena SK (ed) Kinetics and equilibrium in mineral reactions. Springer, Berlin Heidelberg New York, pp 173–198

Àrkai P, Mata MP, Giorgetti G, Peacor DR, Tóth M (2000) Comparison of diagenetic and low-grade evolution of chlorite in associated metapelites and metabasites: an integrated TEM and XRD study. J Metamorph Geol 18:531–550

Àrkai P, Sassi FP, Sassi R (1995) Simultaneous measurements of chlorite and illite crystallinity: a more reliable tool for monitoring low- to very low grade metamorphism in metapelites. A case study from the Southern Alps (NE Italy). Eur J Mineral 7:115–128

Ashworth JR (1985) Migmatites. Blackie, Glasgow, 301 pp

Barrow G (1893) On an intrusion of muscovite biotite gneiss in the S.E. Highlands of Scotland and its accompanying metamorphism. Q J Geol Soc Lond 49:330–358

Barrow G (1912) On the geology of lower Deeside and the southern Highland border. Proc Geol Assoc 23:268–284

Bhattacharya A, Sen SK (1986) Granulite metamorphism, fluid buffering and dehydration melting in the Madras charnockites and metapelites. J Petrol 27:1119–1141

Black PM, Maurizot P, Ghent ED, Stout MZ (1993) Mg-Fe carpholites from aluminous schists in the Diaphot region and implications for preservation of high-pressure/low-temperature schists, northern new Caledonia. J Metamorph Geol 11:455–460

Blenkinsop TG (1988) Definition of low-grade metamorphic zones using illite crystallinity. J Metamorph Geol 6:623–636

Bohlen SR (1987) Pressure–temperature–time paths and a tectonic model for the evolution of granulites. J Geol 95:617–632

Bohlen SR (1991) On the formation of granulites. J Metamorph Geol 9:223–230

Bucher K, Droop GTR (1983) The metamorphic evolution of garnet–cordierite–sillimanite gneisses of the Gruf-Complex, eastern Pennine Alps. Contrib Mineral Petrol 84:215–227

Carmichael DM (1968) On the mechanism of prograde metamorphic reactions in quartz-bearing pelitic rocks. Contrib Mineral Petrol 20:244–267

Carrington DP (1995) The relative stability of garnet–cordierite and orthopyroxene–sillimanite–quartz assemblages in metapelitic granulites: experimental data. Eur J Mineral 7:949–960

Carrington DP, Harley SL (1995a) The stability of osumilite in metapelitic granulites. J Metamorph Geol 13:613–625

Carrington DP, Harley SL (1995b) Partial melting and phase relations in high-grade metapelites: an experimental petrogenetic grid in the KFMASH system. Contrib Mineral Petrol 120:270–291

Cesare B (1994) Hercynite as the product of staurolite decomposition in the contact aureole of Vedrette di Ries, eastern Alps, Italy. Contrib Mineral Petrol 116:239–246

Chamberlain CP, Lyons JB (1983) Pressure, temperature, and metamorphic zonation studies of pelitic schists in the Merrimack synclinorium, south-central New Hampshire. Am Mineral 68:530–540

Chopin C (1981) Talc–phengite, a widespread assemblage in high-grade pelitic blueschists of the Western Alps. J Petrol 22:628–650

Chopin C (1983) Magnesiochloritoid, a key-mineral for the petrogenesis of high-grade pelitic blueschists. Bull Mineral 106:715–717

Chopin C (1984) Coesite and pure pyrope in high-grade blueschists of the western Alps: a first record and some consequences. Contrib Mineral Petrol 86:107–118

Chopin C, Schreyer W (1983) Magnesiocarpholite and magnesiochloritoid: two index minerals of pelitic blueschists and their preliminary phase relations in the model system $MgO–Al_2O_3–SiO_2–H_2O$. Am J Sci 283A:72–96

Chopin C, Henry C, Michard A (1991) Geology and petrology of the coesite-bearing terrain, Dora Maira massif, Western Alps. Eur J Mineral 3:263–291

Clarke GL, Powell R, Guirand M (1989) Low-pressure granulite facies metapelitic assemblages and corona textures from MacRobertson Land, east Antarctica: the importance of Fe_2O_3 and TiO_2 in accounting for spinel-bearing assemblages. J Metamorph Geol 7:323–336

Connolly JAD, Cesare B (1993) C–O–H–S fluid composition and oxygen fugacity in graphitic metapelites. J Metamorph Geol 11:379–388

Currie KL, Gittins J (1988) Contrasting sapphirine parageneses from Wilson Lake, Labrador, and their tectonic implications. J Metamorph Geol 6:603–622

Curtis CD (1985) Clay mineral precipitation and transformation during burial diagenesis. Philos Trans R Soc Lond A315:91–105

Daniel CG, Spear FS (1999) The clustered nucleation and growth processes of garnet in regional metamorphic rocks from north-west Connecticut, USA. J Metamorph Geol 17:503–520

Das K, Dasgupta S, Miura H (2001) Stability of osumilite coexisting with spinel solid solution in metapelitic granulites at high oxygen fugacity. Am Mineral 86:1423–1434

Dempster TJ (1985) Garnet zoning and metamorphism of the Barrovian type area, Scotland. Contrib Mineral Petrol 89:30–38

Droop GTR (1985) Alpine metamorphism in the south-east Tauern Window, Austria. 1. P–T variations in space and time. J Metamorph Geol 3:371–402

Droop GTR, Bucher K (1984) Reaction textures and metamorphic evolution of sapphirine-bearing granulites from the Gruf complex, Italian Central Alps. J Petrol 25:766–803

Droop GTR, Harte B (1995) The effect of Mn on the phase relations of medium-grade pelites: constraints from natural assemblage on petrogenetic grid topology. J Petrol 36:1549–1578

Dunoyer de Segonzac G (1970) The transformation of clay minerals during diagenesis and low grade metamorphism: a review. Sedimentology 15:281–346

Earley D III, Stout JH (1991) Cordierite–cummingtonite facies rocks from the Gold Brick District, Colorado. J Petrol 32:1169–1201

Ellis DJ (1987) Origin and evolution of granulites in normal and thickened crusts. Geology 15:167–170

Ellis DJ, Sheraton JW, England RN, Dallwitz WB (1980) Osumilite–sapphirine–quartz–granulites from Enderby Land, Antarctica – mineral assemblages and reactions. Contrib Mineral Petrol 72:123–143

El-Shazly AK, Aley El-Din K (1995) Petrology of Fe–Mg-carpholite-bearing metasediments from NE Oman. J Metamorph Geol 13:379–396

Enami M (1983) Petrology of pelitic schists in the oligoclase–biotite zone of the Sanbagawa metamorphic terrain, Japan: phase equilibria in the highest grade zone of a high-pressure intermediate type of metamorphic belt. J Metamorph Geol 1:141–161

Engi M, Scherrer NC, Burri T (2001) Metamorphic evolution of pelitic rocks of the Monte Rosa nappe: constraints from petrology and single grain monazite age data. Schweiz Mineral Petrogr Mitt 81:305–328

Evans BW, Guidotti CV (1966) The sillimanite–potash feldspar isograd in western Maine, USA. Contrib Mineral Petrol 12:25–62

Evans NH, Speer JA (1984) Low-pressure metamorphism and anatexis of Carolina Slate Belt Phyllites in the contact aureole of the Lilesville pluton, North Carolina, USA. Contrib Mineral Petrol 87:297–309

Fitzsimons ICW (1996) Metapelitic migmatites from Brattstrand Bluffs, East Antarctica – metamorphism, melting and exhumation of the mid crust. J Petrol 37:395–414

Florence FP, Spear FS (1993) Influences of reaction history and chemical diffusion on P–T calculations for staurolite schists from the Littleton Formation, northwestern New Hampshire. Am Mineral 78:345–359

Foster CT Jr (1991) The role of biotite as a catalyst in reaction mechanisms that form sillimanite. Can Mineral 29:943–964

Fransolet A-M, Schreyer W (1984) Sudoite, di/trioctahedral chlorite: a stable low-temperature phase in the system MgO–Al$_2$O$_3$–SiO$_2$–H$_2$O. Contrib Mineral Petrol 86:409–417

Frey M (1987) The reaction-isograd kaolinite + quartz = pyrophyllite + H$_2$O, Helvetic Alps, Switzerland. Schweiz Mineral Petrogr Mitt 67:1–11

Frey M, Teichmüller M, Teichmüller R, Mullis J, Künzi B, Breitschmid A, Gruner U, Schwizer B (1980) Very low-grade metamorphism in external parts of the Central Alps: illite crystallinity, coal rank and fluid inclusion data. Eclogae Geol Helv 73:173–203

Frost BR, Frost CD (1987) CO$_2$, melts, and granulite metamorphism. Nature 327:503–506

Frost BR, Frost CD, Touret JLR (1989) Magmas as a source of heat and fluids in granulite metamorphism. In: Bridgewater D (ed) Fluid movements – element transport, and the composition of the deep crust. Kluwer, Dordrecht, pp 1–18

Giaramita MJ, Day HW (1991) The four-phase AFM assemblage staurolite–aluminum–silicate–biotite–garnet: extra components and implications for staurolite-out isograds. J Petrol 32:1203–1230

Giaramita MJ, Day HW (1992) Buffering in the assemblage staurolite–aluminium silicate–biotite–garnet–chlorite. J Metamorph Geol 9:363–378

Grant JA (1981) Orthoamphibole and orthopyroxene relations in high-grade metamorphism of pelitic rocks. Am J Sci 281:1127–1143

Grant JA (1985) Phase equilibria in low-pressure partial melting of pelitic rocks. Am J Sci 285:409–435

Grew ES (1980) Sapphirine and quartz association from Archean rocks in Enderby Land, Antarctica. Am Mineral 65:821–836

Grew ES (1988) Kornerupine at the Sar-e-Sang, Afghanistan, whiteschist locality: implications for tourmaline–kornerupine distribution in metamorphic rocks. Am Mineral 73:345–357

Guiraud M, Holland T, Powell R (1990) Calculated mineral equilibria in the greenschist–blueschist–eclogite facies in the system Na$_2$O–FeO–MgO–Al$_2$O$_3$–SiO$_2$–H$_2$O: methods, results and geological applications. Contrib Mineral Petrol 104:85–98

Harley SL (1986) A sapphirine–cordierite–garnet–sillimanite granulite from Enderby Land, Antarctica: implications for FMAS petrogenetic grids in the granulite facies. Contrib Mineral Petrol 94:452–460

Harley SL (1989) The origins of granulites: a metamorphic perspective. Geol Mag 126:215–247

Harley SL, Fitzsimons IC (1991) P–T evolution of metapelitic granulites in a polymetamorphic terrane: the Rauer Group, East Antarctica. J Metamorph Geol 9:231–244

Harte B, Hudson NFC (1979) Pelite facies series and the temperatures and pressures of Dalradian metamorphism in E. Scotland. Geological Society, London, pp 323–337

Hesse R, Dalton E (1992) Diagenetic and low-grade metamorphic terranes of Gaspé Peninsula related to the geological structure of the Taconian and Acadian orogenic belts, Quebec Appalachians. J Metamorph Geol 9:775–790

Hirajima T, Compagnoni R (1993) Petrology of a jadeite–quartz/coesite–almandine–phengite fels with retrograde ferro–nyböite from Dora-Maira Massif, Western Alps. Eur J Mineral 5:943–955

Hodges KV, Spear FS (1982) Geothermometry, geobarometry and the Al$_2$SiO$_5$ triple point at Mt Moosilauke, New Hampshire. Am Mineral 67:1118–1134

Holdaway MJ, Mukhopadhyay B (1993) Geothermobarometry in pelitic schists: a rapidly evolving field. Am Mineral 78:681–693

Hollister LS (1966) Garnet zoning: an interpretation based on the Rayleigh fractionation model. Science 154:1647–1651

Kerrick DM (1988) Al$_2$SiO$_5$-bearing segregations in the Lepontine Alps, Switzerland: aluminum mobility in metapelites. Geology 16:636–640

Kisch HJ (1980) Incipient metamorphism of Cambro-Silurian clastic rocks from the Jämtland Supergroup, central Scandinavian Caledonides, western Sweden: illite crystallinity and 'vitrinite' reflectance. J Geol Soc Lond 137:271–288

Lang HM (1991) Quantitative interpretation of within-outcrop variation in metamorphic assemblage in staurolite-kyanite-grade metapelites, Baltimore, Maryland. Can Mineral 29:655–672

Lonker SW (1980) Conditions of metamorphism in high-grade pelites from the Frontenac Axis, Ontario, Canada. Can J Sci 17:1666–1684

Lonker SW (1981) The P–T–X relations of the cordierite–garnet–sillimanite–quartz equilibrium. Am J Sci 281:1056–1090

Loomis TP (1983) Compositional zoning of crystals: a record of growth and reaction history. In: Saxena SK (ed) Kinetics and equilibrium in mineral reactions. Advances in physical geochemistry, vol 3. Springer, Berlin Heidelberg New York, pp 1–60

Loomis TP (1986) Metamorphism of metapelites: calculations of equilibrium assemblages and numerical simulations of the crystallization of garnet. J Metamorph Geol 4:201–229

Mahar EM, Baker JM, Powell R, Holland TJB (1997) The effect of Mn on mineral stability in metapelites. J Metamorph Geol 15:223–238

Mather JD (1970) The biotite isograd and the lower greenschist facies in the Dalradian rocks of Scotland. J Petrol 11:253–275

Miyashiro A, Shido F (1985) Tschermak substitution in low- and middle-grade pelitic schists. J Petrol 26:449–487

Mohr DW, Newton RC (1983) Kyanite–staurolite metamorphism in sulfidic schists of the Anakeesta formation, Great Smoky Mountains, North Carolina. Am J Sci 283:97–134

Motoyoshi Y, Hensen BJ (2001) F-rich phlogopite stability in ultra-high-temperature metapelites from the Napier Complex, East Antarctica. Am Mineral 86:1404–1413

Munz IA (1990) Whiteschists and orthoamphibole–cordierite rocks and the P–T–t path of the Modum Complex, S. Norway. Lithos 24:181–200

Nichols GT, Berry RF, Green DH (1992) Internally consistent gahnitic spinel–cordierite–garnet equilibria in the FMASHZn system: geothermobarometry and applications. Contrib Mineral Petrol 111:362–377

Nyman MW, Pattinson DRM, Ghent ED (1995) Melt extraction during formation of K-feldspar + sillimanite migmatites, West of Revelstoke, British Columbia. J Petrol 36:351–372

Pattison DRM (1987) Variations in Mg/(Mg + Fe), F, and (Fe,Mg)Si = 2Al in pelitic minerals in the Ballachulish thermal aureole, Scotland. Am Mineral 72:255–272

Pattison DRM (2001) Instability of Al_2SiO_5 "triple point" assemblages in muscovite + biotite + quartz-bearing metapelites, with implications. Am Mineral 86:1414–1422

Pattison DRM, Spear FS, Cheney JT (1999) Polymetamorphic origin of muscovite + cordierite + staurolite + biotite assemblages: implications for the metapelitic petrogenetic grids and for P–T paths. J Metamorph Geol 17:685–703

Powell R, Holland T (1990) Calculated mineral equilibria in the pelite system, KFMASH (K_2O–FeO–MgO–Al_2O_3–SiO_2–H_2O). Am Mineral 75:367–380

Reinhardt J (1987) Cordierite–anthophyllite rocks from north-west Queensland, Australia: metamorphosed magnesian pelites. J Metamorph Geol 5:451–472

Robinson D, Warr LN, Bevins RE (1990) The illite 'crystallinity' technique: a critical appraisal of its precision. J Metamorph Geol 8:333–344

Schreyer W (1973) Whiteschist: a high pressure rock and its geological significance. J Geol 81:735–739

Schreyer W (1977) Whiteschists: their compositions and pressure–temperature regimes based on experimental, field and petrographic evidence. Tectonophysics 3:127–144

Schumacher JC, Hollocher KT, Robinson P, Tracy RJ (1990) Progressive metamorphism and melting in central Massachusetts and southwestern New Hampshire, USA. In: Ashworth JR, Brown M (eds) High-temperature metamorphism and crustal anatexis. The Mineralogical Society Series 2. Unwin Hyman, London, pp 198–234

Selverstone J, Spear FS (1985) Metamorphic P–T paths from pelitic schists and greenstones from the south-west Tauern Window, eastern Alps. J Metamorph Geol 3:439–465

Shaw DM (1956) Geochemistry of pelitic rocks. III. Major elements and general geochemistry. Geol Soc Am Bull 67:919–934

Spear FS (1999) Real-time AFM diagrams on your Macintosh. Geol Mater Res 1:1–18

Spear FS, Cheney JT (1989) A petrogenetic grid for pelitic schists in the system SiO_2–Al_2O_3–FeO–MgO–K_2O–H_2O. Contrib Mineral Petrol 101:149–164

Spear FS, Kohn MJ, Florence FP, Menard T (1990) A model for garnet and plagioclase growth in pelitic schists: implications for thermobarometry and P–T path determinations. J Metamorph Geol 8:683–696

Spear FS, Kohn MJ, Paetzold S (1995) Petrology of the regional sillimanite zone west-central New Hampshire, USA, with implications for the development of inverted isograds. Am Mineral 80:361–376

Spear FS, Kohn MJ, Cheney JT (1999) P–T paths from anatectic pelites. Contrib Mineral Petrol 134:17–32

Symmes GH, Ferry JM (1992) The effect of whole-rock MnO content on the stability of garnet in pelitic schists during metamorphism. J Metamorph Geol 10:221–238

Symmes HG, Ferry JM (1995) Metamorphism, fluid flow and partial melting in pelitic rocks from the Onawa Contact Aureole, central Maine, USA. J Petrol 36:587–612

Theye T, Seidel E, Vidal O (1992) Carpholite, sudoite, and chloritoid in low-grade high-pressure metapelites from Crete and the Peloponnese, Greece. Eur J Mineral 4:487–507

Theye T, Chopin C, Grevel KD, Ockenga E (1997) The assemblage diaspore + quartz in metamorphic rocks: a petrological, experimental and thermodynamic study. J Metamorph Geol 15:17–28

Thompson AB (1976) Mineral reactions in pelitic rocks. II. Calculation of some P–T–X$_{(Fe–Mg)}$ phase relations. Am J Sci 276:425–454

Thompson AB (1982) Dehydration melting of pelitic rocks and the generation of H$_2$O-undersaturated granitic liquids. Am J Sci 282:1567–1595

Tinkham DK, Zuluaga CA, Stowell HH (2001) Metapelite phase equilibria modeling in MnNCKFMASH: the effect of variable Al$_2$O$_3$ and MgO/(MgO + FeO) on mineral stability. Geol Mater Res 3:1–42

Torre Dalla M, Livi KJT, Veblen DR, Frey M (1996) White K-mica evolution from phengite to muscovite in shales and shale matrix melange, Diablo Range, California. Contrib Mineral Petrol 123:390–405

Touret J, Dietvorst P (1983) Fluid inclusions in high-grade anatectic metamorphites. J Geol Soc Lond 140:635–649

Tracy RJ, Robinson P (1983) Acadian migmatite types in pelitic rocks of Central Massachusetts. In: Atherton MP, Gribble CD (eds) Migmatites, melting and metamorphism. Shiva, Nantwich, pp 163–173

Tracy RJ, Robinson P (1988) Silicate–sulfide–oxide–fluid reaction in granulite-grade pelitic rocks, central Massachusetts. Am J Sci 288-A:45–74

Vidal O, Goffe B, Theye T (1992) Experimental study of the stability of sudoite and magnesiocarpholite and calculation of a new petrogenetic grid for the system FeO–MgO–Al$_2$O$_3$–SiO$_2$–H$_2$O. J Metamorph Geol 10:603–614

Vidal O, Parra T, Trotet F (2001) A thermodynamic model for FE-MG aluminous chlorite using data from phase equilibrium experiments and natural pelitic assemblages in the 100 to 600 °C, 1 to 15 kb range. Am J Sci 301:557–592

Waters DJ (1991) Hercynite-quartz granulites: phase relations and implications for crustal processes. Eur J Mineral 3:367–386

Weaver BL, Tarney J (1983) Elemental depletion in Archaean granulite facies rocks. In: Atherton MP, Gribble CD (eds) Migmatites, melting and metamorphism. Shiva, Nantwich, pp 250–263

Wenk E (1968) Cordierit im Val Verzasca. Schweiz Mineral Petrogr Mitt 48:455–457

White RW, Powell R, Holland TJB, Worley BA (2000) The effect of TiO$_2$ and Fe$_2$O$_3$ on metapelitic assemblages at greenschist and amphibolite facies conditions: mineral equilibria calculations in the system K$_2$O–FeO–MgO–Al$_2$O$_3$–SiO$_2$–H$_2$O–TiO$_2$–Fe$_2$O$_3$. J Metamorph Geol 18:497–511

Whitney DL, Mechum TA, Kuehner SM, Dilek YR (1996) Progressive metamorphism of pelitic rocks from protolith to granulite facies, Dutchess County, New York, USA: constraints on the timing of fluid infiltration during regional metamorphism. J Metamorph Geol 14:163–181

Xu G, Will TM, Powell R (1994) A calculated petrogenetic grid for the system K$_2$O–FeO–MgO–Al$_2$O$_3$–SiO$_2$–H$_2$O, with particular reference to contact-metamorphosed pelites. J Metamorph Geol 12:99–119

Yardley BWD (1977) The nature and significance of the mechanism of sillimanite growth in the Connemara Schists, Ireland. Contrib Mineral Petrol 65:53–58

Yardley BWD, Leake BE, Farrow CM (1980) The metamorphism of Fe-rich pelites from Connemara, Ireland. J Petrol 21:365–399

CHAPTER 8

Metamorphism of Marls

8.1
General

Marls are carbonate-bearing pelitic sediments covering a large range of composition between "impure" carbonate rocks and "true" pelites. In the Anglo-American literature, this group of sedimentary rocks is better known as argillaceous carbonate rocks, calcareous sediments, calcic pelitic rocks, or calc-pelites. The sediments are widespread and typical of shelf and platform areas.

Unmetamorphosed marls are composed of mixtures of clay minerals (smectite, illite, kaolinite, chlorite, etc.), carbonates (mainly calcite and/or dolomite), quartz, and feldspars in varying proportions. In metamorphosed marls (metamarls), most minerals occurring in metapelites and metacarbonates may be present, plus additional Ca-Al-bearing silicates like Ca-amphiboles, epidote-group minerals, lawsonite, margarite, scapolite, and vesuvianite. Such a large group of possible mineral constituents requires a very complex chemical system: $K_2O-Na_2O-CaO-FeO-MgO-Al_2O_3-SiO_2-H_2O-CO_2$. The approach to an understanding of such a complex system involves the study of simpler subsystems. In this chapter, three such subsystems will be discussed, one dealing with Al-poor and two with Al-rich metamarls. In all three examples, quartz and calcite will be considered to be present in excess, which is true for most low- and medium-temperature metamarls (although quartz and/or calcite may be used up in reactions at high temperature). Therefore, SiO_2 will be treated as an excess component. In addition, a H_2O-CO_2-bearing fluid phase is assumed to have been present during metamorphism. Chemographic relationships of any metamarl system can then be projected from SiO_2, H_2O and CO_2.

8.2
Orogenic Metamorphism of Al-Poor Marls

Al-poor marls are common in many orogenic belts. In the northern Appalachians, as an example, the prograde metamorphism of such rocks has been studied in great detail (e.g., Ferry 1976, 1983a,b, 1992; Hewitt 1973; Zen 1981). In the Vassalboro Formation in Maine, Ferry mapped five mineral zones, each characterized by an index mineral and separated by reaction isograds. With increasing metamorphic grade, the following index minerals

Table 8.1. Reactions in the KCMAS–HC system

Anorthite	$CaAl_2Si_2O_8$
Calcite	$CaCO_3$
Clinochlore	$Mg_5Al_2Si_3O_{10}(OH)_8$
Diopside	$CaMgSi_2O_6$
Dolomite	$CaMg(CO_3)_2$
K-feldspar	$KAlSi_3O_8$
Muscovite	$KAl_3Si_3O_{10}(OH)_2$
Phlogopite	$KMg_3AlSi_3O_{10}(OH)_2$
Quartz	SiO_2
Tremolite	$Ca_2Mg_5Si_8O_{22}(OH)_2$
Zoisite	$Ca_2Al_3Si_3O_{12}(OH)$

All reactions with excess quartz, calcite, and fluid
(1) $Ms + 8\ Dol + 3\ Qtz + 4\ H_2O = Phl + Cln + 8\ Cal + 8\ CO_2$
(2) $5\ Ms + 3\ Cln + 7\ Qtz + 8\ Cal = 5\ Phl + 8\ An + 12\ H_2O + 8\ CO_2$
(3) $Cln + 7\ Qtz + 3\ Cal = Tr + An + 3\ H_2O + 3\ CO_2$
(4) $2\ Zo + CO_2 = 3\ An + Cal + H_2O$
(5) $5\ Phl + 24\ Qtz + 6\ Cal = 5\ Kfs + 3\ Tr + 2\ H_2O + 6\ CO_2$
(6) $Tr + 2\ Qtz + 3\ Cal = 5\ Di + H_2O + 3\ CO_2$
(7) $Cln + 6\ Cal + 4\ CO_2 = 5\ Dol + An + Qtz + 4\ H_2O$
(8) $Phl + 3\ Cal + 3\ CO_2 = Kfs + 3\ Dol + H_2O$
(9) $Ms + 5\ Dol + 3\ Qtz + 3\ H_2O = Kfs + Cln + 5\ Cal + 5\ CO_2$
(10) $5\ Dol + 8\ Qtz + H_2O = Tr + 3\ Cal + 7\ CO_2$
(11) $8\ Kfs + 3\ Cln = 3\ Ms + 5\ Phl + 9\ Qtz + 4\ H_2O$
(12) $5\ Phl + 5\ Cln + 49\ Qtz + 16\ Cal = 5\ Ms + 8\ Tr + 12\ H_2O + 16\ CO_2$
(13) $3\ Cln + 21\ Qtz + 10\ Cal = 3\ Tr + 2\ Zo + 8\ H_2O + 10\ CO_2$
(14) $5\ Ms + 3\ Tr + CO_2 = 5\ Phl + 5\ An + 14\ Qtz + Cal + 3\ H_2O$
(15) $Ms + 2\ Qtz + Cal = Kfs + An + H_2O + CO_2$
(16) $3\ Ms + 6\ Qtz + 4\ Cal = 3\ Kfs + 2\ Zo + 2\ H_2O + 4\ CO_2$

were encountered: ankerite, biotite, Ca-amphibole, zoisite, and diopside. A summary of mineral assemblages and related reactions is given by Yardley (1989, pp. 143–145). Below we shall refer to these reactions and present a simplified model for the progressive metamorphism of Al-poor marls similar to those described by Ferry.

8.2.1
Phase Relationships in the KCMAS–HC System

For the system K_2O–CaO–MgO–Al_2O_3–SiO_2–H_2O–CO_2, phase relationships are considered among anorthite, calcite, clinochlore, diopside, dolomite, K-feldspar, muscovite, phlogopite, quartz, tremolite and zoisite. The chemography of these phases is shown as an inset in Fig. 8.1 projected from $SiO_2 + CaO + H_2O + CO_2$. Reaction equilibria in the KCMAS–HC system with excess quartz, calcite and H_2O–CO_2 are listed in Table 8.1 and depicted in Fig. 8.1 in an isobaric T–X section. The pressure of 3.5 kbar has been chosen in accordance with Ferry (1983a,b), derived from metapelites near the sillimanite isograd. The range of $X_{CO_2} < 0.2$ has been chosen according to Ferry (1983b, Table 2).

The petrogenetic grid in Fig. 8.1 will now be considered in some detail. First, we shall characterize some of the reactions followed by an analysis of the stability fields of some selected phases. Mixed volatile reactions with

Quartz and calcite in excess

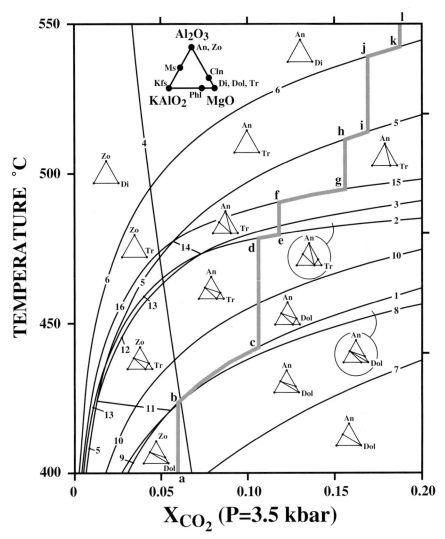

Fig. 8.1. T–X section of the KCMAS–HC system with excess quartz and calcite at $0 < X_{CO_2} < 0.2$ and a constant pressure of 3.5 kbar. Pure end-member mineral compositions are used. The *inset* shows the chemography projected from quartz, calcite, H_2O and CO_2 onto the $KAlO_2$–MgO–Al_2O_3 plane. Compatibility diagrams are shown for some selected divariant fields only. The prograde path a–l is emphasized and discussed in the text (Sect. 8.2.2)

steep (at $X_{CO_2} < 0.05$) to gentle (at $X_{CO_2} > 0.1$) positive slopes are predominant. Reactions (6) and (10) were already encountered in Chapter 6 discussing the metamorphism of dolomite and limestone. Reactions (11) and (14) with a gentle negative slope are dehydration reactions, and reaction (11) was already mentioned in Chapter 7 dealing with the metamorphism of pelites. Reaction (4), with a steep negative slope and crossing the whole diagram of Fig. 8.1, is a special case of a mixed volatile reaction of the type $a + CO_2 = b + H_2O$ (cf. Fig. 3.17). This reaction divides the diagram into zoisite-bearing assemblages to the left and anorthite-bearing assemblages to the right. Because diopside, dolomite and tremolite coincide in the chemographic projection with excess quartz, calcite and fluid (see inset of Fig. 8.1), the T–X stability fields of these three phases do not overlap. Diopside is stable above reaction (6), tremolite is stable between reactions (6) and (10), and dolomite is stable below reaction (10). K-feldspar is stable over the whole diagram because of its single corner position in the chemography. For clinochlore, reaction (7) defines its lower and reactions (3) and (13) its upper stability limits. Phlogopite is stable between reactions (8) and (5). For muscovite, finally, reactions (15) and (16) limit its T–X stability field towards high temperatures and very low X_{CO_2}. The sequence of index minerals observed in the Vassalboro Formation by Ferry, i.e., ankerite, biotite, Ca-amphibole, zoisite, and diopside, is approximated in the KCMAS–HC system by dolomite, phlogopite, tremolite, zoisite, and diopside. In Fig. 8.1, in terms of $T - X_{CO_2}$ values, this means an increase in temperature and, after leaving the tremolite zone, and in order to reach the zoisite zone, a decrease in X_{CO_2} caused by an infiltration of H_2O. This conclusion is in accordance with the findings of Ferry (1976, 1983 a, b). Furthermore, the relative positions of reactions (1)–(6) in T–X space of Fig. 8.1 closely match the sequence of reaction isograds mapped by Ferry (1976). The main difference between the isograd reactions as formulated by Ferry (1976) and reactions (1)–(6) of Table 8.1 concerns plagioclase. Albite and intermediate plagioclase are reactants in reactions (2) and (3), as formulated by Ferry, but the albite component in plagioclase has not been considered in the KCMAS–HC system dealt with here.

In summary, the progressive metamorphism of Al-poor marls described by Ferry (1976, 1983 a, b) can be modeled in the system KCMAS–HC, at least in a qualitative way. A quantitative treatment would require two additional components, FeO and Na_2O, or the use of activity terms.

8.2.2
Prograde Metamorphism in the KCMAS–HC System at Low X_{CO_2}

In this section, the prograde metamorphism of a rock consisting of Ms + Dol with excess Qtz + Cal and an initial fluid composition of $X_{CO_2} = 0.06$ will be considered. The prograde path in $T - X_{CO_2}$ space is shown in Fig. 8.1 by a curve running from **a** to **l** and the changing mineral modal composition is

Fig. 8.2. Modal composition (mol%), fluid composition (X_{CO_2}) and mineral distribution along the prograde path a–l shown in Fig. 8.1

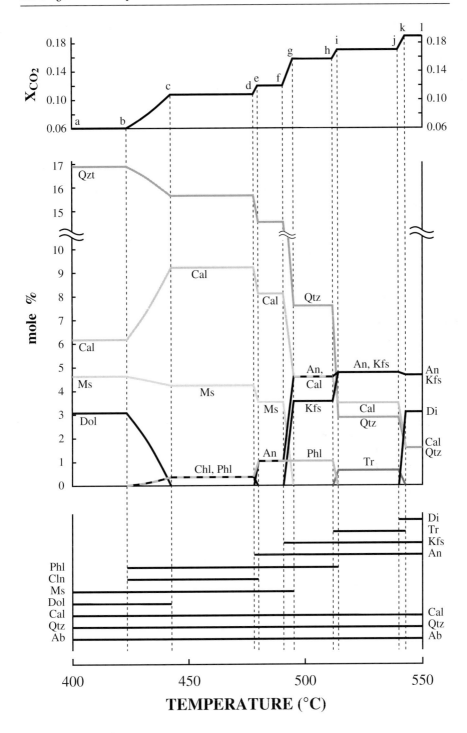

displayed in Fig. 8.2. It is important to note that this calculation, using the computer program THERIAK-DOMINO (de Capitani and Brown 1987; see also Chap. 4), is valid only for a specific bulk composition. The prograde path can be divided into several steps as follows.

a–b. The starting mineral assemblage Dol + Ms + Qtz + Cal is heated up at a constant X_{CO_2} of 0.06.

b–c. Equal amounts (on a mole basis) of clinochlore and phlogopite are formed according to reaction (1) between 424 and 443 °C, while X_{CO_2} increases to 0.107.

c–d. At point **c** all dolomite is used up. The rock is now composed of Phl + Ms + Cln + Qtz + Cal and is heated up at a constant X_{CO_2} of 0.107.

d–e. The first anorthite and additional phlogopite are formed according to reaction (2) between 478 and 480 °C while X_{CO_2} increases to 0.119.

e–f. At point **e** all clinochlore is used up. The rock is now composed of Phl + Ms + An + Qtz + Cal and is heated up at a constant X_{CO_2} of 0.119.

f–g. The first K-feldspar and additional anorthite are formed according to reaction (15) between 491 and 495 °C while X_{CO_2} increases to 0.156.

g–h. At point **g** all muscovite is used up. The rock is now composed of Phl + An + Kfs + Qtz + Cal and is heated up at a constant X_{CO_2} of 0.156.

h–i. The first tremolite and additional K-feldspar are formed according to reaction (5) between 512 and 514 °C while X_{CO_2} increases to 0.167.

i–j. At point **i** all phlogopite is used up. The rock is now composed of Tr + An + Kfs + Qtz + Cal and is heated up at a constant X_{CO_2} of 0.167.

j–k. The first diopside is formed according to reaction (6) between 540 and 543 °C, while X_{CO_2} increases to 0.188.

k–l. At point **k** all tremolite is used up. The rock is now composed of Di + An + Kfs + Qtz + Cal and is heated up at a constant X_{CO_2} of 0.188.

In summary, the model metamorphism of the assemblage Ms + Dol with excess Qtz + Cal in the system KCMAS–HC shows several interesting features. Firstly, X_{CO_2} increases in steps from 0.06 to 0.188 because all acting reactions have a positive slope in $T - X_{CO_2}$ space. Secondly, most reactions (2, 15, 5, 6) take place within a very narrow temperature range of a few °C because of the gentle slope of reaction curves at $X_{CO_2} > 0.1$ and because a relatively large amount of fluid was assumed. In a metamorphic terrain, this would lead to sharp reaction-isograds. Thirdly, with increasing metamorphic grade, sheet silicates (Cln, Ms, and Phl in our example) are replaced by feldspars and chain silicates (Tr and Di).

8.3
Orogenic Metamorphism of Al-Rich Marls

Al-rich marls are present in many orogenic belts. In the Alps, as an example, such rocks are widespread, both in platform sediments of the Helvetic realm and in deep-sea metasediments of the Penninic realm. Their mineralogical transformation has been studied in some detail (e.g., Frey 1978; Bucher et al. 1983; Frank 1983).

Table 8.2. Reactions in the CAS–HC system

Anorthite	$CaAl_2Si_2O_8$
Calcite	$CaCO_3$
Clinozoisite	$Ca_2Al_3Si_3O_{12}(OH)$
Grossular	$Ca_3Al_2Si_3O_{12}$
Kyanite	Al_2SiO_5
Margarite	$CaAl_4Si_2O_{10}(OH)_2$
Pyrophyllite	$Al_2Si_4O_{10}(OH)_2$
Quartz	SiO_2
Wollastonite	$CaSiO_3$

All reactions with excess quartz, calcite, and fluid:
(18) $2\ Prl + Cal = Mrg + 6\ Qtz + H_2O + CO_2$
(19) $Prl = Ky + 3\ Qtz + H_2O$
(20) $3Mrg + 5\ Cal + 6\ Qtz = 4\ Czo + H_2O + 5\ CO_2$
(21) $Mrg + CO_2 = 2\ Ky + Cal + H_2O$
(22) $Mrg + 2\ Qtz + Cal = 2\ An + H_2O + CO_2$
(23) $2\ Czo + CO_2 = 3\ An + Cal + H_2O$
(24) $2\ Czo + 3\ Qtz + 5\ Cal = 3\ Grs + H_2O + 5\ CO_2$
(25) $Ky + Qtz + Cal = An + CO_2$
(26) $An + Qtz + 2\ Cal = Grs + 2\ CO_2$
(27) $Qtz + Cal = Wo + CO_2$

8.3.1
Phase Relationships in the CAS–HC System

The system $CaO–Al_2O_3–SiO_2–H_2O–CO_2$ permits description of phase relationships among anorthite, calcite, clinozoisite, grossular, kyanite, margarite,
pyrophyllite, quartz, and wollastonite. The chemography of these phases is
shown as an inset in Fig. 8.3 as projected from $H_2O + CO_2$. Note the presence
of several collinearities among three phases in this system, e.g., Ky–Prl–Qtz,
Ky–Mrg–Cal, An–Czo–Cal. Reaction equilibria in the CAS–HC system with
excess quartz, calcite and $H_2O–CO_2$ are listed in Table 8.2 and depicted in
Fig. 8.3 in a polybaric T–X section along the Ky geotherm. According to the
phase rule (cf. Chap. 3.8.1), each divariant assemblage consists of one single
phase (in addition to Qtz + Cal + Fluid).

In the CAS–HC system, the assemblage Prl–Qtz–Cal represents the characteristic assemblage in Al-rich marls at the onset of metamorphism. The fluid
composition will be H_2O-rich because sheet silicates have not yet liberated
CO_2 by reacting with carbonates, and the interstitial water is therefore not
yet diluted by CO_2.

Margarite will form during prograde metamorphism according to reaction
(1) at temperatures below 390 °C, and this temperature will be lower than
350 °C for very H_2O-rich fluid compositions $X_{CO_2} < 0.02$. However, as a consequence of the steep equilibrium position of reaction (1) at $X_{CO_2} < 0.05$ (Fig.
8.3), a noticeable amount of margarite will be produced between 375 and
390 °C only. Therefore, the first occurrence of margarite should represent a
mappable reaction isograd at low greenschist facies conditions. Assuming a
low pyrophyllite modal content and closed system behavior, reaction (1) will
consume all pyrophyllite before reaching the invariant point involving

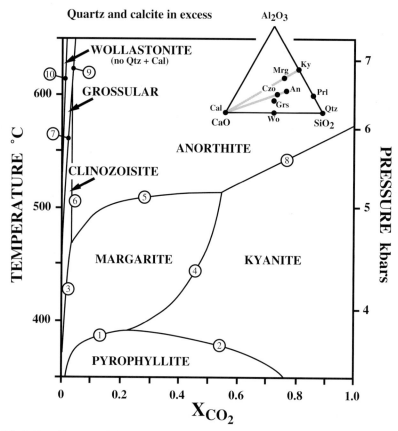

Fig. 8.3. P–T–X diagram of the CAS–HC system with excess quartz and calcite. The vertical axis represents an orogenic geotherm characteristic for kyanite-type terrains. The inset shows the chemography projected from H_2O and CO_2 onto the $CaO–Al_2O_3–SiO_2$ plane. Two collinearities corresponding to reactions (4) and (6) are indicated

Prl + Mrg + Ky (+ Qtz + Cal). The metamarl will now enter the divariant margarite field, the assemblage Mrg + Qtz + Cal being characteristic for the greenschist facies. It should be pointed out that, at such low metamorphic grade, the Ca–Al mica margarite will be very fine-grained and may be easily missed in thin sections, and X-ray diffraction work is needed (see, e.g., Frey 1978, p. 110).

Next the rock will meet reaction (5) between 470 and 510 °C, depending on fluid composition, producing anorthite components in plagioclase. As discussed in Section 8.3.2, however, in margarite-bearing rocks, plagioclase will be formed already at considerably lower temperatures if Na is considered as an additional component.

Clinozoisite is formed by reaction (3), but the steep course of this reaction in P–T–X space and the corresponding small change in X_{CO_2} allows only for very minor modal clinozoisite. At the invariant point involving Czo + Mrg +

An ($+$ Qtz + Cal), the anorthite component in plagioclase is produced at the expense of all other solid phases present. Depending on whether clinozoisite or margarite would be used up first, reaction boundary (5) or (6) would be followed, respectively. Because of the small amount of clinozoisite formed by reaction (3), it is conceivable to assume that this phase will be used up first. Following reaction (5), more anorthite component in plagioclase is produced. Another and more effective way of clinozoisite formation is by interaction of a metamarl with an externally derived H_2O-rich fluid driving reaction (3) or (6).

Grossular and wollastonite do not form in orogenic metamorphic rocks under closed-system conditions. According to Fig. 8.3, these minerals may be generated by reactions (7) or (9) in the case of grossular and by reaction (10) in the case of wollastonite by interaction with a H_2O-rich fluid. Such metamarls (calc-silicate rocks with Grt + Wo + Cpx + Pl + Cal in regional metamorphic settings) have been described, e.g., by Trommsdorff (1968) and Gordon and Greenwood (1971).

In summary, in orogenic metamorphic terrains, relatively Al-rich but quartz- and calcite-bearing metamarls (Fig. 8.3) will show two index minerals with prograde metamorphism: pyrophyllite under subgreenschist and margarite under greenschist facies conditions. Grossular- and wollastonite-bearing metamarls are diagnostic for interaction with an externally derived H_2O-rich fluid.

8.3.2
Phase Relationships in the KNCAS–HC System

Addition of K_2O and Na_2O to the CAS–HC system discussed above allows discussion of the phase relations among four extra phases: muscovite, paragonite, K-feldspar and albite. The chemography for the system KNCAS–HC is shown as an inset in Fig. 8.4. Grs and Wo have been omitted for simplicity, and all remaining phases considered in the CAS–HC system are coincident in the Al_2O_3 corner (projection from excess Qtz + Cal + H_2O + CO_2). The six additional equilibria are listed in Table 8.3 and depicted in Fig. 8.4. Phase relationships in this figure indicate the following maximum thermal stability for pure end member white mica in the presence of excess quartz and calcite: 515 °C for margarite, 540 °C for paragonite, and 590 °C for muscovite. These temperature values are valid only for the kyanite-type paths. Note that equilibria (5), (12), and (15) are all mixed volatile reactions with identical coefficients for all the phases involved, except for feldspars: 2 mol An are produced from Mrg + 2 Qtz + Cal, 1 mol each of Ab and An from Pg + 2 Qtz + Cal, and 1 mol each of Kfs and An from Ms + 2 Qtz + Cal.

The petrogenetic grid of Fig. 8.4 is not yet applicable to natural rocks because of solid solution of white micas and feldspars. For simplicity, only the effect of plagioclase solid solution will be quantified below. Anorthite-involving reactions from the grid of Fig. 8.4 may be contoured for different An contents in the plagioclase as depicted in Fig. 8.5. The stability field of Pl + Mrg + Qtz + Cal is emphasized in gray tone. Its low-temperature boundary is fixed by reaction (5) contoured for a plagioclase composition of An$_{30}$,

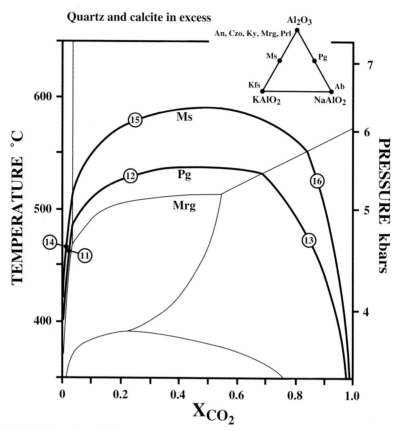

Fig. 8.4. P–T–X section of the KNCAS–HC system with excess quartz and calcite. Pure end-member mineral compositions are used. The vertical axis represents a path typical of kyanite-type terrains. Note the maximum thermal stabilities for margarite, paragonite and muscovite. The inset shows the chemography projected from quartz, calcite, H_2O and CO_2 onto the $KAlO_2$–$NaAlO_2$–Al_2O_3 plane

Table 8.3. Reactions in the KNCAS–HC system (in addition to reactions of Table 8.2)

Mineral	Formula
Muscovite	$KAl_3Si_3O_{10}(OH)_2$
Paragonite	$NaAl_3Si_3O_{10}(OH)_2$
K-feldspar	$KAlSi_3O_8$
Albite	$NaAlSi_3O_8$

All reactions with excess quartz, calcite, and fluid:
(28) $3\ Pg + 6\ Qtz + 4\ Cal = 3\ Ab + 2\ Czo + 2\ H_2O + 4\ CO_2$
(29) $Pg + 2\ Qtz + Cal = Ab + An + H_2O + CO_2$
(30) $Pg + Qtz = Ab + Ky + H_2O$
(31) $3\ Ms + 6\ Qtz + 4\ Cal = 3\ Kfs + 2\ Czo + 2\ H_2O + 4\ CO_2$
(32) $Ms + 2\ Qtz + Cal = Kfs + An + H_2O + CO_2$
(33) $Ms + Qz = Kfs + Ky + H_2O$

Quartz and calcite in excess

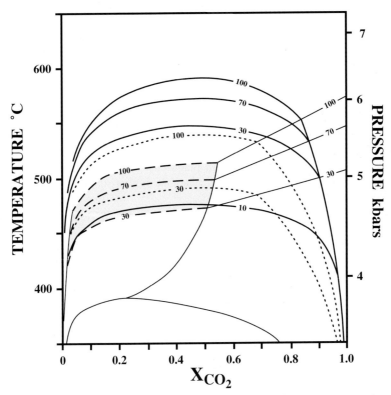

Fig. 8.5. P–T–X section of the KNCAS–HC system with excess quartz and calcite. The vertical axis represents a path characteristic of kyanite-type terrains. Polybaric divariant equilibria for a particular composition of plagioclase, An_x, are shown as *solid lines* (muscovite-involving reactions), *stippled lines* (paragonite-involving reactions) or *dashed lines* (margarite-involving reactions). *Numbers* refer to mol% An content in the plagioclase. Most clinozoisite-involving equilibria from Fig. 8.4 have been omitted for clarity. The stability field for the assemblage Mrg + Pl (+ Qtz + Cal) is emphasized with *gray shading*

based on field observations by Frey and Orville (1974). The effect of margarite-paragonite solid solution would shift the gray-shaded field to higher temperatures, but would also expand it (because of increasing solubility of the paragonite component in margarite with increasing temperature). Compared with Fig. 8.4, the stability field of Mrg + Qtz + Cal in Fig. 8.5 is reduced considerably, with an upper temperature limit of about 470 °C.

In contrast to margarite-bearing rocks, albite does occur in paragonite-bearing metamarls of the greenschist facies (e.g., Ferry 1992). According to Fig. 8.5, the equilibrium curve according to reaction (12) for Pl (An_{30}) is stable up to about 490 °C, that is, about 40 °C lower than the maximum thermal stability of Pl + Pg + Qtz + Cal. An interesting feature is noticed for the breakdown of the assemblage Pg + Qtz + Cal in the presence of intermediate plagioclase composition (about An_{50}–An_{70}). The assemblage Pg + Qtz + Cal is now breaking down at a lower temperature than the assemblage Mrg + Qtz + Cal (compare with Fig. 8.4), and reaction (12) is then replaced by the following reaction:

$$2\,Pg + 2\,Qtz + Cal = 2\,Ab + Mrg + H_2O + CO_2 \tag{17}$$

For a given plagioclase composition, contours of reactions (5) and (17) are separated by a few degrees only. As an example, for a plagioclase composition of An_{70} and for fluid compositions of $X_{CO_2} = 0.2 - 0.5$, equilibrium (17) is located 5 °C below equilibrium (5), but for graphic reasons this is not shown in Fig. 8.5.

The production of K-feldspar and plagioclase from Ms + Qtz + Cal for plagioclase compositions > An_{30} takes place at higher temperatures than the breakdown of Pg + Qtz + Cal (Fig. 8.5).

Prograde metamorphism of Al-rich marls will be discussed separately for low and very low X_{CO_2}.

8.3.2.1
Prograde Metamorphism at Low X_{CO_2}

Let us now consider the prograde metamorphism of a rock consisting of Ms + Pg + Prl + Qtz + Cal and an initial fluid composition of $X_{CO_2} = 0.1$. The beginning of the metamorphic evolution will be identical to that described for the CAS–HC system in Fig. 8.3, i. e., formation of margarite according to reaction (1) (Table 8.2). The resulting mineral assemblage with Ms + Pg + Mrg + Qtz + Cal is characteristic for the lower greenschist facies, and such assemblages with three coexisting white micas have been described, e. g., by Frey (1978). Whether a small amount of sodic plagioclase is already formed from Pg + Qtz + Cal at this grade of metamorphism is not documented in the literature. Between 450 and 470 °C, the first plagioclase of composition An_{30} is produced by reaction (5). Within the stability field of margarite + plagioclase + quartz + calcite (gray-shaded area in Fig. 8.5), phase relationships are complex and depend, besides P–T–X conditions, on the modal rock composition considered, including the amount of fluid. Margarite and paragonite will react, together with quartz and calcite, to form plagioclase according to reactions (5) and (12), respectively. On the other hand, reaction (17) will produce some additional margarite. Concomitantly, the fluid composition, X_{CO_2}, will slightly increase. If during prograde metamorphism the upper limit of the gray-shaded area in Fig. 8.5 is reached, the net result will be the consumption of all margarite. For most marly compositions, also all paragonite will be used up at this stage. However, for rocks rich in Al and Na, remaining Pg + Qtz + Cal will react to An-rich plagioclase above the gray-shaded area in Fig. 8.5, between 515 and 540 °C. From this it follows that the following mineral assemblages will be encountered in the middle and upper greenschist facies: Ms + Mrg + Pl + Qtz + Cal±Pg and Ms + Pg + Pl + Qtz + Cal (for Al- and Na-rich compositions only). After the final breakdown of margarite and paragonite in the presence of quartz and calcite, at temperatures up to 590 °C, the mineral assemblage in Al-rich metamarls will be Ms + Pl + Kfs + Qtz + Cal. Finally, at even higher temperatures, after the final breakdown of muscovite in the presence of quartz and calcite, the assemblage Pl + Kfs + Qtz + Cal will remain.

According to the contouring of reactions (5), (12), (15) and (17) in Fig. 8.5, at a given temperature and pressure, these reactions will simultaneously produce plagioclase of different An content. Provided that these differences in plagioclase composition are not eliminated by diffusion processes, then Al-rich marly rocks of the greenschist and amphibolite facies will display a range of plagioclase An content within a single thin section. Such a situation has actually been observed in nature (e.g., Frank 1983). However, such varying plagioclase composition may also be due to miscibility gaps within the plagioclase feldspar series.

8.3.2.2
Prograde Metamorphism at Very Low X_{CO_2}

Phase relationships for the system KNCAS–HC involving the phases albite, anorthite, clinozoisite, grossular, K-feldspar, margarite, muscovite, paragonite, pyrophyllite with excess quartz and calcite are shown in Fig. 8.6. Calculation of this petrogenetic grid was performed using pure end-member mineral compositions, except for clinozoisite, where an activity of 0.64 was computed based on mineral composition data of Frank (1983). These clinozoisites contain about 6 wt% of Fe_2O_3. Note that Fig. 8.6 is similar to the water-rich portion of Fig. 8.4, but with an enlarged stability field for clinozoisite [reaction (6)] is now at X_{CO_2} of ca. 0.13 instead of 0.035 in Fig. 8.4 [and with addition of reaction (7)].

Figure 8.7 shows phase relationships as in Fig. 8.6, complemented by contours for anorthite-involving reactions as discussed earlier for Fig. 8.5. The stability field of the assemblage Pl + Mrg + Qtz + Cal is emphasized by gray tone and is limited by reaction (5): its high- and low-temperature boundaries are given by the An_{100} and An_{30} contours, respectively. As already discussed above for Fig. 8.5, the assemblage Pg + Qtz + Cal in the presence of intermediate plagioclase composition breaks down by reaction (17), and not by reaction (12). This is indicated in Fig. 8.7 for the An_{70} contours of margarite- and paragonite-involving reactions, which meet at an invariant point at about 470 °C and $X_{CO_2} = 0.11$. The configuration around this invariant point is shown in the inset of Fig. 8.7.

The prograde metamorphism of a rock consisting of Ms + Pg + Prl + Qtz + Cal and an initial fluid composition of $X_{CO_2} = 0.02$ will be considered next. The prograde path in $T - X_{CO_2}$ space is shown in Fig. 8.7 by a curve running from **a** to **k** and the changing mineral modal composition is displayed in Fig. 8.8. Calculations were performed with the computer program THERIAK-DOMINO (de Capitani; see Chap. 4). Note that these calculations are valid only for a specific bulk composition. The prograde path can be divided into several steps as follows.

Quartz and calcite in excess

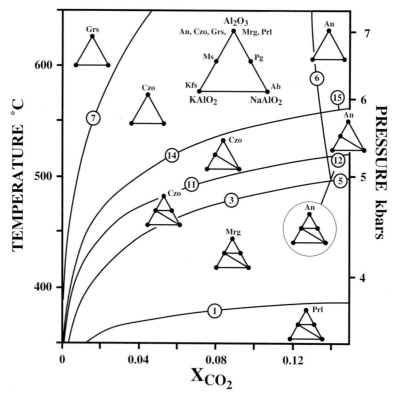

Fig. 8.6. P–T–X phase relationships for the KNCAS–HC system with excess quartz and calcite at $0 < X_{CO_2} < 0.15$. Pure end-member mineral compositions are used, except for clinozoisite ($a_{Czo} = 0.64$). The vertical axis represents a kyanite-type path. The inset shows the chemography projected from quartz, calcite, H_2O and CO_2 onto the $KAlO_2$–$NaAlO_2$–Al_2O_3 plane. Compatibility diagrams are given for each divariant field

a–b. The starting assemblage Ms + Pg + Prl + Qtz + Cal is heated up at a constant X_{CO_2} of 0.020.

b–c. Margarite is formed, according to reaction (1) (Table 8.2), between 359 and 364 °C and X_{CO_2} increases to 0.025.

c–d. At point **c** all pyrophyllite is used up. The rock is now composed of Ms + Pg + Mrg + Qtz + Cal and is heated up at a constant X_{CO_2} of 0.025. This mineral assemblage is characteristic for the lower greenschist facies.

d–e. Clinozoisite is formed for the first time according to reaction (3) between 425 and 436 °C and X_{CO_2} increases to 0.034.

e–f. At point **e** all margarite is used up and the resulting mineral assemblage Ms + Pg + Czo + Qtz + Cal is heated up at a constant X_{CO_2} of 0.034.

Quartz and calcite in excess

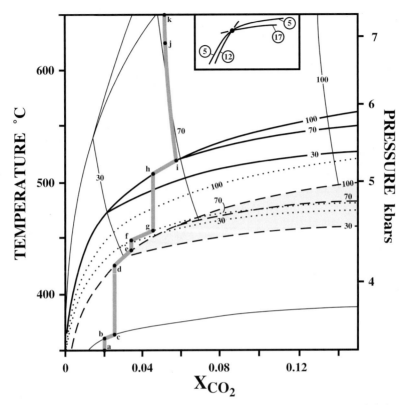

Fig. 8.7. P–T–X diagram of the KNCAS–HC system with excess quartz and calcite at $0 < X_{CO_2} < 0.15$. The vertical axis represents a path of kyanite-type terrains. Pure end-member mineral compositions are used, except for clinozoisite ($a_{Czo} = 0.64$) and plagioclase. Polybaric divariant equilibria for a particular composition of plagioclase, An_x, are shown as *solid lines* (muscovite-involving reactions), *stippled lines* (paragonite-involving reactions) or *dashed lines* (margarite-involving reactions). Numbers refer to mol% An-content in the plagioclase. The stability field for the assemblage $Mrg + Pl$ ($+ Qtz + Cal$) is shown in *gray*. The prograde path from a to k is emphasized and discussed in the text

f–g. Plagioclase of composition An_{33-36} is formed according to reaction (11) between 450 and 458 °C and X_{CO_2} increases to 0.045.

g–h. At point **g** all paragonite is used up. The mineral assemblage $Ms + Pl + Czo + Qtz + Cal$ is present between 458 and 507 °C and is characteristic for the upper greenschist facies. In this temperature interval, reactions (6), (11) and (12) cause minor changes in the mode and mineral composition (note that these three reactions are linearly dependent). Reactions (11) and (12) are responsible for a decrease in the paragonite component in muscovite, and all three reactions produce additional plagioclase. The very small increase in X_{CO_2} is not visible in Figs. 8.7 and 8.8.

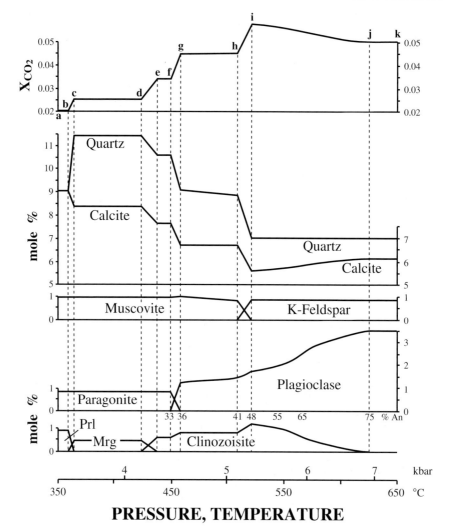

Fig. 8.8. Modal composition (mol%), fluid composition (X_{CO_2}) and plagioclase composition (An %) along the prograde path a–k shown in Fig. 8.7

h–i.   K-feldspar and additional clinozoisite are formed by reaction (14) between 507 and 520 °C and X_{CO_2} increases to 0.058. Some additional plagioclase is produced as in the previous step.

i–j.   At point **i** all muscovite is used up. The mineral assemblage Kfs + Pl + Czo + Qtz + Cal is now present and is characteristic for the lower and middle amphibolite facies. Additional plagioclase is produced by reaction (6); the An content of plagioclase increases from 48 to 75 mol%, and X_{CO_2} decreases slightly.

j–k. At point **k** at 625 °C all clinozoisite is used up and the resulting mineral assemblage Kfs + Pl + Qtz + Cal is heated up at a constant X_{CO_2} of 0.051 without further reaction.

In summary, phase relationships of relatively Al-rich metamarls with excess quartz and calcite are complex. Nevertheless, several diagnostic assemblages (always with excess Qtz + Cal) are found along a kyanite-type path of regional metamorphism. The assemblage Ms + Pg + Prl is characteristic for sub-greenschist facies conditions. The assemblage Ms + Pg + Mrg is characteristic for the lower greenschist facies, Ms + Mrg + Pl and Ms + Czo + Pl for the upper greenschist facies. The assemblages Ms + Pl + Kfs and Czo + Pl + Kfs are characteristic for the lower and middle amphibolite facies, Pl + Kfs for the upper amphibolite facies. The first plagioclase has an An content of ca. 30 mol%, and the An content increases with metamorphic grade. In short, sheet silicate-rich assemblages containing two or three white micas at low grade are replaced by feldspar-rich assemblages at high grade.

References

Aines RD, Rossman GR (1984) The hydrous component in garnets: pyralspites. Am Mineral 69:1116–1126

Bucher K (1976) Occurence and chemistry of xanthophyllite in roof pendants of the Bergell granite, Sondrio, N Italy. Schweiz Mineral Petrogr Mitt 56:413–426

Bucher K, Frank E, Frey M (1983) A model for the progressive regional metamorphism of margarite-bearing rocks in the Central Alps. Am J Sci 283A:370–395

Castelli D (1991) Eclogitic metamorphism in carbonate rocks: the example of impure marbles from the Sesia-Lanzo Zone, Italian Western Alps. J Metamorph Geol 9:61–77

Connolly JAD, Memmi I, Trommsdorff V, Franceschelli M, Ricci CA (1994) Forward modeling of calc-silicate microinclusions and fluid evolution in a graphitic metapelite, northeastern Sardinia. Am Mineral 79:960–972

Dachs E (1990) Geothermobarometry in metasediments of the southern Grossvenediger area (Tauern Window, Austria). J Metamorph Geol 8:217–230

Dasgupta S (1993) Contrasting mineral parageneses in high-temperature calc-silicate granulites: examples from the eastern Ghats, India. J Metamorph Geol 11:193–202

De Capitani C, Brown TH (1987) The computation of chemical equilibrium in complex systems containing non-ideal solutions. Geochim Cosmochim Acta 51:2639–2652

Ferry JM (1976) Metamorphism of calcareous sediments in the Waterville-Vassalboro area, south-central Maine: mineral reactions and graphical analysis. Am J Sci 276:841–882

Ferry JM (1981) Petrology of graphitic sulfide-rich schists from south-central Maine: an example of desulfidation during prograde regional metamorphism. Am Mineral 66:908–930

Ferry JM (1983 a) Applications of the reaction progress variable in metamorphic petrology. J Petrol 24:343–376

Ferry JM (1983 b) Mineral reactions and element migration during metamorphism of calcareous sediments from the Vassalboro Formation, south-central Maine. Am Mineral 68:334–354

Ferry JM (1983 c) Regional metamorphism of the Vassalboro Formation, south-central Maine, USA: a case study of the role of fluid in metamorphic petrogenesis. J Geol Soc Lond 140:551–576

Ferry JM (1984) A biotite isograd in south-central Maine, USA: mineral reactions, fluid transfer, and heat transfer. J Petrol 25:871–893

Ferry JM (1987) Metamorphic hydrology at 13-km depth and 400–500 °C. Am Mineral 72:39–58

Ferry JM (1988) Contrasting mechanisms of fluid flow through adjacent stratigraphic units during regional metamorphism, south-central Maine, USA. Contrib Mineral Petrol 98:1–12

Ferry JM (1992) Regional metamorphism of the Waits River Formation, eastern Vermont: delineation of a new type of giant metamorphic hydrothermal system. J Petrol 33:45–94

Ferry JM (1994) Overview of the petrologic record of fluid flow during regional metamorphism in northern New England. Am J Sci 294:905–988

Frank E (1983) Alpine metamorphism of calcareous rocks along a cross-section in the Central Alps: occurrence and breakdown of muscovite, margarite and paragonite. Schweiz Mineral Petrogr Mitt 63:37–93

Frey M (1969) Die Metamorphose des Keupers vom Tafeljura bis zum Lukmanier-Gebiet. (Veränderungen tonig-mergeliger Gesteine vom Bereich der Diagenese bis zur Staurolith-Zone). Beitr Geol Karte Schweiz NF 131:160 S

Frey M (1978) Progressive low-grade metamorphism of a black shale formation, Central Swiss Alps, with special reference to pyrophyllite and margarite bearing assemblages. J Petrol 19:93–135

Frey M, Orville PM (1974) Plagioclase in margarite-bearing rocks. Am J Sci 274:31–47

Ghent ED, Stout MZ (2000) Mineral equilibria in quartz leucoamphibolites (quartz-garnet-plagioclase-hornblende calc-silicates) from southeastern British Columbia, Canada. Can Mineral 38:233–244

Gordon TM, Greenwood HJ (1971) The stability of grossularite in H_2O-CO_2 mixtures. Am Mineral 56:1674–1688

Griffin WL, Styles MT (1976) A projection for analyses of mineral assemblages in calc-pelitic metamorphic rocks. Norsk Geol Tidsskrift 56:203–209

Harley SL, Buick IS (1992) Wollastonite-scapolite assemblages as indicators of granulite pressure-temperature-fluid histories: the Rauer Group, East Antarctica. J Petrol 33:693–728

Hewitt DA (1973) The metamorphism of micaceous limestones from south-central Connecticut. Am J Sci 273-A:444–469

Hoschek G (1980a) Phase relations of a simplified marly rock system with application to the Western Hohe Tauern (Austria). Contrib Mineral Petrol 73:53–68

Hoschek G (1980b) The effect of Fe-Mg substitution on phase relations in marly rocks of the Western Hohe Tauern (Austria). Contrib Mineral Petrol 75:123–128

Hoschek G (1984) Alpine metamorphism of calcareous metasediments in the western Hohe Tauern, Tyrol: mineral equilibria in COHS fluids. Contrib Mineral Petrol 87:129–137

Jacobs G, Kerrick D (1981) Devolatilization equilibria in H_2O-CO_2 and H_2O-CO_2-NaCl fluids: an experimental and thermodynamic evaluation at elevated pressures and temperatures. Am Mineral 66:1135–1153

Jamtveit B, Bucher K, Stijfhoorn DE (1992) Contact metamorphism of layered metasediments in the Oslo rift. I. Buffering, infiltration and the mechanisms of mass transport. J Petrol 33:377–422

Joesten R (1974) Local equilibrium and metasomatic growth of zoned calc-silicate nodules from a contact aureole, Christmas Mountains, Big Bend region, Texas. Am J Sci 274:876–901

Kohn MJ, Spear FS (1993) Phase equilibria of margarite-bearing schists and chloritoid + hornblende rocks from western New Hampshire, USA. J Petrol 34:631–651

Labotka TC (1987) The garnet + hornblende isograd in calcic schists from an andalusite-type regional metamorphic terrain, Panamint Mountains, California. J Petrol 28:323–254

Labotka TC (1995) Evidence for immiscibility in Ti-rich garnet in a calc-silicate hornfels from northeastern Minnesota. Am Mineral 80:1026–1030

Labotka TC, Nabelek PI, Papike JJ, Hover-Granath VC, Laul JC (1988) Effects of contact metamorphism on the chemistry of calcareous rocks in the Big Horse Limestone Member, Notch Peak, Utah. Am Mineral 73:1095–1110

Léger A, Ferry JM (1991) Highly aluminous hornblende from low-pressure metacarbonates and a preliminary thermodynamic model for the Al content of calcic amphibole. Am Mineral 76:1002–1017

Letargo CMR, Lamb WM (1993) P-T-X conditions of calc-silicate formation: evidence from fluid inclusions and phase equilibria; Llano Uplift, central Texas, USA. J Metamorph Geol 11:89–100

Lopez Sanchez-Vizcaino V, Connolly JAD, Gomez-Pugnaire MT (1997) Metamorphism and phase relations in carbonate rocks from the Nevado-Filabride Complex (Cordilleras Beticas, Spain): application of the Ttn + Rt + Cal + Qtz + Gr buffer. Contrib Mineral Petrol 126:292–302

Mathavan V, Fernando GWAR (2001) Reactions and textures in grossular-wollastonite-scapolite calc-silicate granulites from Maligawila, Sri Lanka: evidence for high-temperature isobaric cooling in the metasediments of the Highland Complex. Lithos 59:217–232

Melson WG (1966) Phase equilibria in calc-silicate hornfels, Lewis and Clark County, Montana. Am Mineral 51:402–421

Menard T, Spear FS (1993) Metamorphism of calcic pelitic schists, Strafford Dome, Vermont: compositional zoning and reaction history. J Petrol 34:977–1005

Misch PM (1964) Stable association wollastonite-anorthite and other calc-silicate assemblages in amphibolite-facies crystalline schists of Nanga Parbat, northwest Himalayas. Beitr Mineral Petrogr (later Contrib Mineral Petrol) 10:315–356

Motoyoshi Y, Thost DE, Hensen BJ (1991) Reaction textures in calc-silicate granulites from the Bolingen Islands, Prydz Bay, East Antarctica: implications for the retrograde P–T path. J Metamorph Geol 9:293–300

Nesbitt BE, Essene EJ (1983) Metamorphic volatile equilibria in a portion of the southern Blue Ridge Province. Am J Sci 283:135–165

Orville PM (1969) A model for metamorphic differentiation origin of thin-layered amphibolites. Am J Sci 267:64–86

Selverstone J, Munoz JL (1987) Fluid heterogeneities and hornblende stability in interlayered graphitic and nongraphitic schists (Tauern Window, Eastern Alps). Contrib Mineral Petrol 96:426–440

Selverstone J, Spear FS, Frank G, Morteani G (1984) High-pressure metamorphism in the SW Tauern Window, Austria: P–T paths from hornblende–kyanite–staurolite schists. J Petrol 25:501–531

Sillanpää J (1986) Mineral chemistry study of progressive metamorphism in calcareous schists from Ankarvattnet, Swedish Caledonides. Lithos 19:141–152

Svensen H, Jamtveit B (1998) Contact metamorphism of shales and limestones from the Grua area, the Oslo Rift, Norway: a phase–petrological study. Norsk Geol Tidsskrift 78:81–98

Tanner PWG (1976): Progressive regional metamorphism of thin calcareous bands from the Moinian rocks of NW Scotland. J Petrol 17:100–134

Thompson AB (1975) Mineral reactions in a calc-mica schist from Gassetts, Vermont, USA. Contrib Mineral Petrol 53:105–127

Thompson PH (1973) Mineral zones and isograds in "impure" calcareous rocks, an alternative means of evaluating metamorphic grade. Contrib Mineral Petrol 42:63–80

Trommsdorff V (1968) Mineralreaktionen mit Wollastonit in einem Kalksilikatfels der alpinen Disthenzone (Claro, Tessin). Schweiz Mineral Petrogr Mitt 48:655–666

Ulmer P (1982) Monticellite-clintonite bearing assemblages at the southern border of the Adamello-Massif. R Soc Ital Mineral Petrol 38:617–628

Valley JW, Peacor DR, Bowman JR, Essene EJ, Allard MJ (1985) Crystal chemistry of a Mg-vesuvianite and implications of phase equilibria in the system CaO–MgO–Al$_2$O$_3$–SiO$_2$–H$_2$O–CO$_2$. J Metamorph Geol 3:137–154

Will TM, Powell R, Holland T, Guiraud M (1990) Calculated greenschist facies mineral equilibria in the system CaO–FeO–MgO–Al$_2$O$_3$–SiO$_2$–CO$_2$–H$_2$O. Contrib Mineral Petrol 104:353–368

Yardley BWD (1989) An introduction to metamorphic petrology. Longman, Edinburgh, 248 pp

Zen E-A (1981) A study of progressive regional metamorphism of pelitic schists from the Taconic allochthon of southwestern Massachusetts and its bearing on the geologic history of the area. US Geological Survey Professional Paper 1113. US Government Printing Office, Washington, DC, 128 pp

CHAPTER 9

Metamorphism of Mafic Rocks

9.1
Mafic Rocks

Metamorphic mafic rocks (e.g., mafic schists and gneisses, amphibolites) are derived from mafic igneous rocks, mainly basalts and andesites, and of lesser importance, gabbros (Chap. 2). Metamorphic assemblages found in mafic rocks are used to define the intensity of metamorphism in the metamorphic facies concept (Chap. 4).

Extrusive mafic igneous rocks, basalts and andesites comprise by far the largest amount of mafic rocks and greatly outweigh their plutonic equivalents (Tables 2.1 and 2.2). Basalts and andesites occur as massive lava flows, pillow lavas, hyaloclastic breccias, tuff layers, sills, and dykes. Basaltic rocks constitute a major portion of the oceanic crust and most basalts appear to have been subjected to ocean-floor metamorphism immediately after formation at a spreading ridge. When transported to a continental margin via plate motion, oceanic mafic rocks are again recrystallized at or near convergent plate junctions; the alteration in mineralogy depends on whether the oceanic crust was subducted under a continental plate or was obducted onto continental crust. On the other hand, andesitic rocks and other related calc-alkaline volcanics as well as associated graywacke are the dominant lithologies within island arcs and Pacific-type continental margins. These rocks are subjected to alteration by hydrothermal fluids as evidenced by present-day geothermal activity in many island arc districts or are transformed during burial and orogenic metamorphism.

Metamorphosed mafic rocks are very susceptible to changes in temperature and pressure. This is the reason why most names of individual metamorphic facies are derived from mineral assemblages of this rock group, e.g., greenschist, amphibolite, granulite, blueschist, and eclogite (cf. Chaps. 2 and 4). In addition, mafic rocks that are metamorphosed under very weak conditions below the greenschist and blueschist facies often show systematic variations in mineralogy that permits a further subdivision into characteristic metamorphic zones. All these distinct low-grade zones may be given separate metamorphic facies names if one wishes. However, we prefer to use the expression subgreenschist facies for very-low-grade conditions of incipient metamorphism.

9.1.1
Hydration of Igneous Mafic Rocks

Basalts and gabbros have solidus temperatures in the order of 1200 °C. Consequently, hydrates are not typical members of the solidus mineralogy of basalts and other mafic rocks. At the onset of metamorphism, mafic igneous rocks and pyroclastics are, therefore, in their least hydrated state as opposed to "wet" sedimentary rocks that start metamorphism in their maximum hydrated state. Because newly formed minerals in metamafics at low temperature are predominantly hydrous phases, access of water is absolutely essential to initiate metamorphism. Otherwise, igneous rocks will persist unchanged in metamorphic terrains. Partial or complete hydration of mafic rocks may occur during ocean-floor metamorphism in connection with hydrothermal activity in island arcs or during orogenic metamorphism where deformation facilitates the influx of water. In metamorphism of mafic igneous rocks it is not self-evident that the condition $P_{total} = P_{H_2O}$ is continuously maintained during the prograde reaction history. On the other hand, worldwide experience with mafic rocks shows that partial persistence of igneous minerals and microstructures is widespread in subgreenschist facies rocks and it is typically absent from the greenschist facies upward. Primary igneous mesostructures such as magmatic layering and pillow structures may be preserved even in eclogite and granulite facies terrains. Coarse-grained gabbroic rocks have the best chance to conserve primary igneous minerals up to high-grade metamorphic conditions. Gabbro bodies often escape pervasive internal deformation and this, in turn, prevents access of water, hampers recrystallization and hinders hydration of the igneous minerals.

As pointed out in Chapter 3, dehydration reactions are strongly endothermic. Consequently, hydration of basalt is exothermic and releases large amounts of heat. The heat of reaction released by replacing the basalt assemblage Cpx + Pl by a collection of low-temperature hydrates such as prehnite + chlorite + zeolites could raise the temperature by as much as 100 °C (in a heat-conserving system). Another interesting aspect of exothermic reactions is their self-accelerating and feedback nature. Once initiated, they will proceed as long as water is available. The increasing temperature will initially make the reactions go faster, but it will also bring the reaction closer to its equilibrium where it eventually will stop. Also note that the reactions that partially or completely hydrate igneous mafic rocks are metastable reactions and the reactions tend to run far from equilibrium. Consider, for instance, the reaction diopside + anorthite + $H_2O \Rightarrow$ chlorite + prehnite ± quartz that was mentioned above. In the presence of water at low temperature, this reaction will always run to completion and the product and reactant assemblage will never reach reaction equilibrium. The unstable or metastable nature of the hydration reactions that replace high-T anhydrous igneous assemblages by low-T hydrate assemblages has the consequence that low-grade mafic rocks often show disequilibrium assemblages. Unreacted high-T igneous assemblages may occur together with various generations of low-T assemblages of the subgreenschist facies in intimate spatial association. The hydration process commonly leads to the development of zoned metasomatic structures. Examples are: networks of veins and concentric shells of mineral zones that reflect progressive hydration of basaltic pillow lavas. The highly permeable inter-pillow zones serve as aquifers for hydration water and the hydration process progresses towards the pillow centers. The nature of the product assemblages depends not only on the pressure and temperature during the hydration process, but also on the chemical composition of the hydration fluid. CO_2 in the fluid plays a particularly important role. CO_2-rich aqueous fluids may result in altered basalts with modally abundant carbonate minerals (calcite, ankerite). CO_2-rich fluids also tend to favor the formation of less hydrous assemblages compared with pure H_2O fluids. The cation composition of aqueous fluids that have already reacted with large volumes of basalt is dramatically different from that of seawater. Fluid that moves fast through permeable fractured rocks will have a different composition than fluid that trickles slowly through porous basalt. Fluid-rock interaction typically also leads to extensive redistribution of chemical elements in piles of basaltic pillow lavas. This may still be witnessed in high-grade terrains. In the Saas-Zermatt ophiolite complex of the Central Alps, for example, mesoscopic pillow lava structures are often preserved with pillow cores of 25 kbar eclogite and inter-pillow material of nearly monomineralic glaucophane. Finally, the alteration products of the fluid-basalt interaction process will, of course, reflect these compositional differences in the hydration fluid.

Heterogeneous hydration results also in features such as incipient to extensive development, sporadic distribution, and selective growth of low-grade minerals in vesicles and fractures, and the topotaxic growth of these minerals after igneous plagioclase, clinopyroxene, olivine, hornblende, opaques, and volcanic glass. From the viewpoint of igneous petrologists, these minerals are often referred to as "secondary minerals". Depending on the effective bulk composition of local domains of the rock, different associations of secondary minerals may develop in vesicles, veins and irregular patches replacing primary phases even within a single thin section. Furthermore, relic igneous phases such as plagioclase and clinopyroxene are common. From the viewpoint of metamorphic petrologists the low-grade hydrates and carbonates, the "secondary minerals", constitute the stable low-grade assemblage and the starting material of prograde metamorphism. In the conceptually simplest case, no relic igneous minerals remain in the mafic rock and its igneous assemblage has been completely replaced by a stable low-T assemblage in its maximum hydrated state.

9.1.2
Chemical and Mineralogical Composition of Mafic Rocks

The characteristic composition of alkali-olivine basalt is listed in Table 2.1 and of MOR basalt in Table 2.3. Mid-ocean ridge basalt (MORB) that is produced along ocean spreading centers is by far the most common type of igneous mafic rocks. It may serve as a reference composition for all mafic rocks discussed in this chapter. Mafic rocks are characterized by SiO_2 contents of about 45–60 wt% and are also relatively rich in MgO, FeO, CaO and Al_2O_3. It is general custom in petrology to call igneous rocks with 45–52 wt% of SiO_2 basic, and their metamorphic derivatives are then called metabasic rocks or, in short, metabasites. Metabasalts represent the most commonly encountered group of metabasites. Andesitic rocks, on the other hand, contain higher SiO_2, Al_2O_3, and alkalis, but lower MgO, FeO, and CaO than basaltic compositions (e.g., Carmichael 1989); basaltic andesites and andesites belong to the intermediate igneous rocks, defined by SiO_2 contents of 52–63 wt%.

Igneous mafic rocks contain appreciable amounts of at least the following eight oxide components: SiO_2, TiO_2, Al_2O_3, Fe_2O_3, FeO, MgO, CaO, and Na_2O. K_2O, H_2O, and CO_2 may also be present in small amounts. These components are stored in relatively few different minerals. The mineralogical inventory of the mafic protolith comprises the major constituents plagioclase and clinopyroxene. In addition, quartz, orthopyroxene, olivine and nepheline can be present in various associations with Cpx + Pl. Plagioclase and clinopyroxene are the prime and most common minerals of most mafic rocks and many gabbros and basalts contain more than 90–95 vol% of these two minerals. Other basalts and gabbros may be composed of Pl + Cpx + Opx + Qtz or Pl + Cpx + Ol + Ne; troctolites contain plagioclase + olivine only; anorthosites more than 90 vol% plagioclase, and so on. However, all rocks are always combinations of a few different mineral species. A large variety of minor and

Table 9.1. Minerals and compositions in metabasaltic rocks

Nesosilicates		Sheet silicates	
Garnet	$(Fe,Mg,Ca)_3(Al,Fe)_2Si_3O_{12}$	Muscovite	$KAl_3Si_3O_{10}(OH)_2$
		Celadonite	$KMgAlSi_4O_{10}(OH)_2$
Sorosilicates		Paragonite	$NaAl_3Si_3O_{10}(OH)_2$
Kyanite	Al_2SiO_5	Phlogopite	$KMg_3[AlSi_3O_{10}](OH)_2$
Zoisite	$Ca_2Al_3Si_3O_{12}(OH)$	Biotite	$K(Mg,Fe,Al,Ti)_3$
Epidote	$Ca_2FeAl_2Si_3O_{12}(OH)$		$[(Al,Si)_3O_{10}](OH,F,Cl)_2$
Pumpellyite	$Ca_4Mg_1Al_5Si_6O_{23}(OH)_3 \cdot 2\ H_2O$	Clinochlore	$Mg_5Al_2Si_3O_{10}(OH)_8$
Vesuvianite	$Ca_{19}Mg_2Al_{11}Si_{18}O_{69}(OH)_9$	Chlorite	$(Fe,Mg)_5Al_2Si_3O_{10}(OH)_8$
Lawsonite	$CaAl_2Si_2O_7(OH)_2 \cdot H_2O$	Prehnite	$Ca_2Al_2Si_3O_{10}(OH)_2$
Chloritoid	$Mg_1Al_2Si_1O_6(OH)_2$		
		Tectosilicates	
Pyroxenes		Quartz	SiO_2
Diopsid	$CaMgSi_2O_6$	Anorthite	$CaAl_2Si_2O_8$
Jadeite	$NaAlSi_2O_6$	Albite	$NaAlSi_3O_8$
Hyperstene	$(Mg,Fe)_2Si_2O_6$	Analcite	$NaAlSi_2O_6 \cdot H_2O$
Omphacite	$(Ca,Na)(Mg,Fe,Al)Si_2O_6$	Scapolite	$Ca_4(Al_2Si_2O_8)_3(CO_3, SO_4, Cl_2)$
Amphiboles		**Zeolites**	
Tremolite	$Ca_2Mg_5Si_8O_{22}(OH)_2$	Laumontite	$CaAl_2Si_4O_{12} \cdot 4\ H_2O$
Actinolite	$Ca_2(Fe,Mg)_5Si_8O_{22}(OH)_2$	Heulandite	$CaAl_2Si_7O_{18} \cdot 6\ H_2O$
Glaucophane	$Na_2(Fe,Mg)_3(Al)_2Si_8O_{22}(OH)_2$	Stilbite	$CaAl_2Si_7O_{18} \cdot 7\ H_2O$
Barroisite	$(Ca,Na)_2(Fe,Mg,Al)_5$	Wairakite	$CaAl_2Si_4O_{12} \cdot 2\ H_2O$
	$Si_8O_{22}(OH)_2$		
Tschermakite	$Ca_2(Fe,Mg)_3(Al)_2$	**Carbonates**	
	$(Al_2Si_6O_{22}(OH)_2$	Calcite	$CaCO_3$
Hornblende	$(Na,K)Ca_2(Fe,Mg,Al)_5$	Aragonite	$CaCO_3$
	$(Si,Al)_8O_{22}(OH,F,Cl)_2$	Dolomite	$Ca(Mg,Fe)(CO_3)_2$

accessory minerals including ilmenite, magnetite, spinel, garnet, and even hornblende and biotite can be found in igneous mafic rocks. The latter two hydrates are often poor in OH-groups as a result of oxidation and/or halogen substitution.

The complex chemical bulk rock composition of basalts will be redistributed from the few igneous mineral species into a variety of new minerals during metamorphism. The high calcium content of basalts that is stored in Cpx and Pl results in the formation of numerous separate calcium-bearing metamorphic minerals. This is the basic difference to metapelitic rocks (Chap. 7) where CaO is very low and does not form separate calcic phases but rather is found as a minor component in ordinary Fe-Mg minerals such as garnet. The most important minerals found in metamafic rocks are listed in Table 9.1 as a refresher.

9.1.3
Chemographic Relationships and the ACF Projection

The composition space that must be considered in an analysis of phase relationships in metamafic rocks is rather complex and requires eight or more system components. This follows from above and from the chemical analysis of MOR basalt in Table 2.3. A graphical analysis of phase relationships in metamafic rocks is often accomplished by means of the ACF projection. The principles of the ACF projection were introduced in Chapter 2.5.3 and the general chemography and a sample ACF diagram are shown in Fig. 2.10. The ACF diagram is used, in general, for the display and analysis of phase relationships that involve calcic minerals. Because of its importance for the subsequent treatment of metamafic rocks, some important features of the ACF projection will be briefly reviewed.

The following discussion makes extensive use of Fig. 9.1a that is a pseudo-3D display of the NACF tetrahedron, and Fig. 9.1b, a conventional ACF diagram. The ACF diagram represents, as usual, a mole fraction triangle and displays the three composition variables Al_2O_2, CaO and FeO. Any mineral that is composed of only these three components can be directly shown on an ACF diagram (e.g., hercynite). Any other mineral composition is projected onto the ACF triangle or is used as a projection point. In the latter case, the mineral must be present in all assemblages in order to deduce meaningful phase relationships. ACF diagrams are projections from SiO_2 and this means that at any given P–T condition the stable polymorph of SiO_2 (e.g., quartz) must be present in all assemblages. This restriction imposed by the conventional ACF diagram often causes problems when dealing with metamorphic silica-poor igneous mafic rocks such as troctolites, nepheline basalts, olivine basalts and olivine gabbros. However, other meaningful diagrams may be designed for such quartz-free metamafics following the suggestions given in Chapter 2.5. ACF diagrams are also projections from H_2O and CO_2 in order to permit the display of hydrates and carbonates. This means that a fluid phase of a specified composition must be present (or the chemical potentials of H_2O and CO_2 must be defined otherwise) under the

ACF-projection

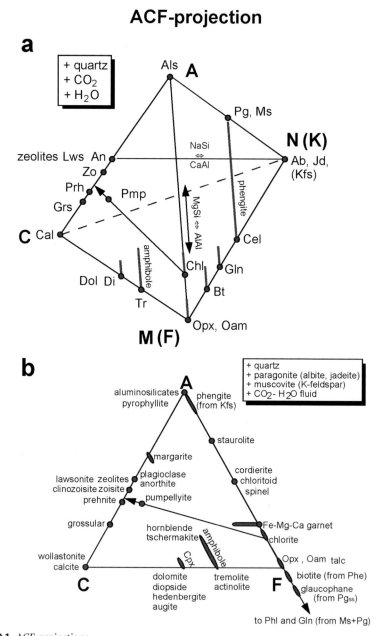

Fig. 9.1. ACF projections

condition of the diagram. This again may cause pain, particularly at low metamorphic grades, because the universal presence of a fluid phase is not unavoidably obvious in anhydrous igneous protolith rocks (see discussion above). Note that pumpellyite (Pmp) is on the ACF surface in Fig. 9.1 a and its composition projects between zoisite and prehnite seen from the clinochlore composition.

ACF diagrams also represent projections parallel to the $MgFe_{-1}$ exchange vector. This is a rigid and lamentable feature of ACF diagrams. All complex relationships, continuous and discontinuous reactions arising from variations in the Fe/Mg ratio of ferro-magnesian minerals cannot be properly analyzed and understood with the aid of such diagrams. The AF binaries on ACF and AFM diagrams are identical. Consequently, many of the relationships discussed in Chapter 7 on metapelites are also valid here. The AF binary is expanded to the AFM triangle in Chapter 7 and, in order to include Ca-minerals, to the ACF triangle in this chapter. An ACFM tetrahedron would permit a rather thorough analysis of phase relationships in metamafic rocks. However, it is very inconvenient to work with three-dimensional composition phase diagrams.

The $MgFe_{-1}$ projection on ACF diagrams has the consequence that crossing tie-line relationships among ferro-magnesian silicates (with contrasting X_{Fe}) and Ca-minerals do not necessarily represent a discontinuous reaction relationship but rather span a composition volume in the ACFM space. An example (Fig. 9.1b): the four minerals garnet, hornblende, kyanite, and zoisite can be related to the reaction 2 tschermakite + 1 kyanite = 2 garnet + 2 zoisite + 1 quartz + 1 H_2O. This important reaction in high-pressure amphibolites is not discontinuous, as may erroneously be concluded from Fig. 9.1 b. The four minerals actually occupy corners of a composition volume in ACFM space and may occur as a stable assemblage over a certain P–T interval. The reaction formulated above is thus in reality a continuous reaction. Another effect of the $MgFe_{-1}$ projection is that Fe and Mg end members of a given mineral species project to identical positions (and of course any intermediate X_{Fe} composition of that mineral too, e.g., pyrope and almandine). The consequences of the $MgFe_{-1}$ projection on ACF diagrams must always be kept in mind.

About 3 wt% of Na_2O is present in typical basalts and sodium-bearing minerals are important in metamafics (plagioclase, amphibole, mica, pyroxene; Table 2.3). If one wants to represent sodic phases on ACF diagrams, one may expand the ACF triangle of Fig. 9.1 a (that actually is an ACM figure, but this is equivalent, see above) to a tetrahedron with $NaAlO_2$ as an additional corner. In this corner we will find the projected compositions of albite, jadeite and analcite (Qtz, H_2O projection). In a similar fashion, K-feldspar will project to a $KAlO_2$ corner of an analogous K tetrahedron. End-member paragonite and muscovite project onto the AN and AK binary, respectively. All plagioclase compositions project onto the ACN ternary. The compositional variation in plagioclase is mainly related to the $NaSiAl_{-1}Ca_{-1}$ exchange that connects albite and anorthite in Fig. 9.1a. The same exchange component can also be found in, for example, amphibole, pyroxene and margarite.

End-member celadonite, glaucophane and phlogopite (biotite) project onto the MN and MK binary, respectively. The micas and amphibole show strong P-T- and assemblage-dependent compositional variations along the $MgSiAl_{-1}Al_{-1}$-exchange direction. This tschermak exchange is parallel to the AM-(AF) binary. Phengites project onto the AMN (AFN) ternary and the solid solution series connects Ms and Cel. Sodic phases are often depicted on ACF diagrams by projecting from the $NaAlO_2$ apex of the ACFN tetrahedron onto the ACF triangle. This means then that Ab, Jd, or Anl (whatever is stable at the pressure and the temperature of the diagram) is present as an extra mineral or that the potentials of these phase components are fixed. It can be seen in Fig. 9.1a that pure albite cannot be shown on ACF diagrams, but that any Ca-bearing plagioclase will project to the An point. This means that the consequences of the plagioclase composition on phase relationships in metamafic rocks cannot be analyzed on ACF diagrams. Ms and Pa will project to the A-apex of ACF diagrams, whereas phengites will occupy the entire AF binary when projected from Kfs (Ab). However, Kfs is very rare in metamafics and consequently Kfs projections are not much used. In analogy to AFM diagrams, one may project from the white micas rather than from the feldspar components, particularly for blueschist facies rocks. In this case, the feldspars do not project onto the ACF plane with the exception of anorthite; phengite does not project onto the ACF plane either, whereas glaucophane projects onto the AF binary with a negative A-coordinate, like biotite projected from muscovite (Fig. 9.1a,b). In metamafic rocks, just one of the minerals, K-mica, phengite or biotite, is generally present. This makes it possible to ignore the small amount of potassium that is stored in just one extra minor phase. One may also choose to project from albite (jadeite) and muscovite. The consequences of the countless alternatives can be understood by carefully studying Fig. 9.1a. Also note that not all possible compositional variations of minerals are shown in Fig. 9.1a (e.g., Jd–Di solution). Whatever projection you prefer for your own work, it is important to specify the projections on any of your phase diagrams carefully.

Another complication when dealing with metamafic rocks is the fact that they tend to be much more oxidized than, for example, metapelites. Redox reactions that involve phase components with ferric iron are common and important. Notably, in garnet (andradite), amphibole (e.g., riebeckite, crossite) and pyroxene (e.g., acmite), a considerable amount of the total iron is usually present in the trivalent state. However, a considerable substitution of Fe_2O_3 for Al_2O_3 can also be found in most of the low-grade Ca–Al hydrosilicates. The presence of epidote in rocks always signifies the presence of Fe^{3+} in the total iron. Magnetite is a widespread oxide phase in metamafics. A separate treatment of redox reactions will not be given here, however. One of the effects of variable Fe^{3+}/Fe^{2+} ratios in minerals is a further increase of the variance of the considered assemblages. Consequently, the assemblages occur over a wider P–T range compared to the situation where all iron is present as Fe^{2+}. One may also construct ACF diagrams for a fixed oxygen activity. For instance, the coexistence of hematite and magnetite in a rock at P and T will fix a_{O_2} (as shown in Chap. 3) and the conditions will be rather oxidizing. Thus, a significant amount of the total iron of the rock will be present as Fe^{3+}. The effects on the minerals and mineral compositions can be related, to a large extent, to the $Fe^{3+}Al_{-1}$ exchange. This exchange makes epidote from clinozoisite, magnetite from hercynite and ribeckite from glaucophane (for example). By considering ACF projections as projections parallel to the $Fe^{3+}Al_{-1}$ exchange vector, one may represent important minerals in metamafics such as epidote, grandite garnet, magnetite and crossite. Note that magnetite does not project to the F-apex but to the same projection point as hercynite and Mg–Al-spinel.

Carbonates are widespread and abundant in low-grade metamafics. Fe-bearing calcite and ankerite (dolomite) are predominant. Carbonates are also often present in high-grade rocks. Carbonates can be displayed on ACF diagrams as explained above. However, any reactions that involve carbonates are in general mixed volatile reactions and must be analyzed accordingly (see Chaps. 3, 6 and 8). The presence of much CO_2 in an aqueous fluid phase also has effects on pure dehydration reactions that do not involve carbonate minerals (Chap. 3). Compared with pure H_2O fluids, in CO_2-rich fluids hydrates (e.g., chlorite, amphibole) may also break down at much lower temperatures in rocks that do not even contain carbonates.

Basalts also contain TiO_2 in the % range. Titanium is mainly shelved in one major Ti-phase, i.e., titanite, ilmenite, or rutile. If two of these minerals are present in the rock, Ti-balanced reactions may be useful for P-T estimates. An example: Ky + 3 Il + 2 Qtz = 3 Rt + Grt. Ti is also found in considerable amounts in rock-forming silicates (biotite, amphiboles, garnet, etc.) with unavoidable consequences for solution properties and equilibrium conditions of reactions.

In conclusion, the composition of metamafic rocks is rather complex and the minerals that typically occur in the assemblages show extensive chemical variation along several exchange directions. This makes comprehensive graphical analysis of phase relationships in metamorphic mafic rocks a difficult task. The complex chemical variation of solid-solution minerals can be simplified for graphical analysis by projecting parallel to some of the exchange components. The choice of projection depends entirely on the problem one wants to solve and the kind of rocks one is working with. For many metamafic rocks it has turned out that a very advantageous and powerful projection can be made from SiO_2, $NaAlO_2$, H_2O and CO_2, parallel to $MgFe_{-1}$ and $Fe^{3+}Al_{-1}$ onto the Al_2O_3-CaO-FeO mole fraction triangle. This ACF projection will be used below.

The chemical complexity of mafic rocks makes it difficult to discuss phase relationships by means of chemical subsystems and comprehensive petrogenetic grids and phase diagrams in P-T space (as, for example, in previous chapters). Note, however, the complexity of mafic rocks can be quantitatively analyzed but this must be done for each given suite of rocks individually. The presentation below is thus mainly a discussion of the ACF system and MORB composition with some important reactions discussed separately where necessary.

9.2
Overview of the Metamorphism of Mafic Rocks

The best overview of the metamorphism of mafic rocks can be gained from the metamorphic facies scheme shown in Fig. 4.2. The characteristic assemblages in metabasalts are used for the definition of metamorphic facies and serve as a reference frame for all other rock compositions. The assemblages of metamafic rocks that are diagnostic for the different facies are given in Chapter 4. From Fig. 4.2 it is evident that basalt undergoing prograde metamorphism along a Ky- or Sil-type path will first show diagnostic assemblages of the subgreenschist facies; later it will become a greenschist then an amphibolite and finally end up as a mafic granulite. High-pressure low-temperature (HPLT) metamorphism turns basalt first into blueschist and then into eclogite. Any geologic process that brings basalts to great depth (>50 km) will result in the formation of eclogite. In contact metamorphism, basalts are metamorphosed to mafic hornfelses. Partial melting of H_2O-saturated metamafics starts at a temperature that is significantly higher than in metagranitoids and metapelites.

Prograde metamorphism of mafic rocks produces sequences of mineral zones that can be compared with mineral zones defined by minerals in metapelites. Figure 9.2 shows sequences of minerals formed in mafic and pelitic rocks by prograde metamorphism from northern Michigan as an example. Metapelites contain muscovite and quartz throughout and the sequence is of the low-pressure type with andalusite and sillimanite as aluminosilicates. The staurolite zone along the Sil geotherm is narrow, as expected from Fig. 7.3. The corresponding prograde mineral zonation in metabasites shows a series of features that are very characteristic.

- There are very few different mineral species present in metamafic rocks. In the greenschist facies metamafic rocks contain: albite + chlorite + actinolite + epidote ⇒ greenschist. In the amphibolite facies the minerals are: plagioclase (>An_{17}) + hornblende ± biotite ± epidote ⇒ amphibolite.
- Most minerals occur over many of the mineral zones defined by metapelites.
- The characteristic prograde changes in metabasites pertain to the composition of plagioclase and amphibole.
- Plagioclase systematically changes its composition from albite at low grade to more calcic plagioclase (andesine in the example of Fig. 9.2). The transition from albite to oligoclase is abrupt and marks a sharp mappable boundary in the field. The discontinuous nature is caused by a miscibility gap in the plagioclase system. This discontinuity can be used to define the greenschist – amphibolite facies boundary. Along a Ky-type path the oligoclase boundary coincides with the staurolite zone boundary that marks the beginning of the amphibolite facies in metapelites. In low-pressure metamorphism (e.g., Fig. 9.2) staurolite occurs for the first time inside the amphibolite facies (Fig. 7.3).
- Amphibole systematically changes its composition from actinolite at low grade to alkali- and aluminum-bearing hornblende at high grade (the

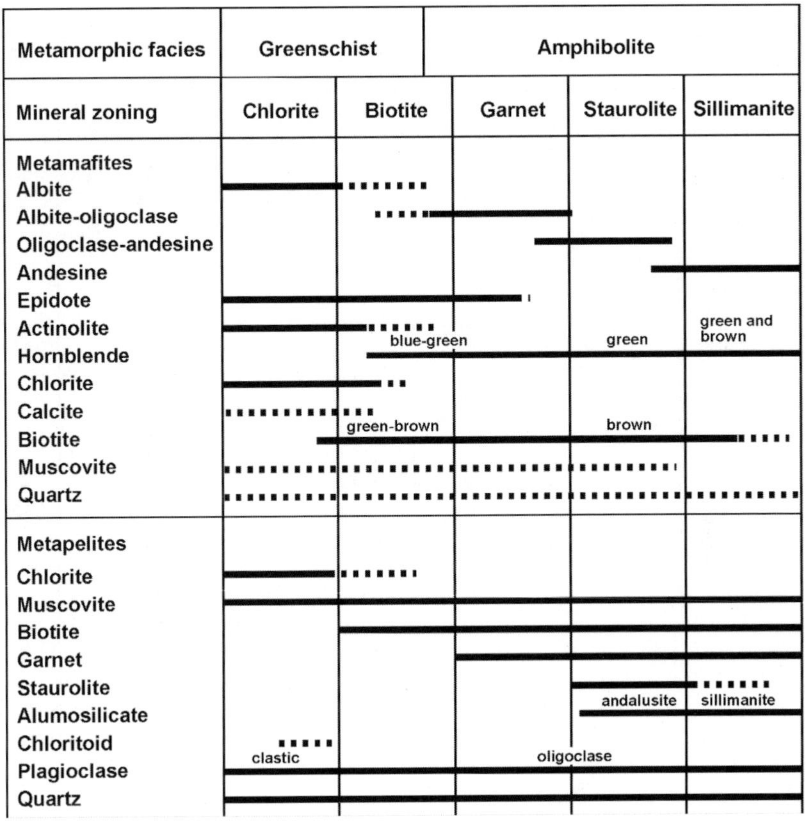

Metamorphic facies	Greenschist		Amphibolite		
Mineral zoning	Chlorite	Biotite	Garnet	Staurolite	Sillimanite
Metamafites					
Albite					
Albite-oligoclase					
Oligoclase-andesine					
Andesine					
Epidote					
Actinolite		blue-green		green	green and brown
Hornblende					
Chlorite					
Calcite		green-brown		brown	
Biotite					
Muscovite					
Quartz					
Metapelites					
Chlorite					
Muscovite					
Biotite					
Garnet					
Staurolite				andalusite	sillimanite
Alumosilicate					
Chloritoid	clastic		oligoclase		
Plagioclase					
Quartz					

Fig. 9.2. Progressive mineral changes in northern Michigan. (James 1955)

hornblende color information in Fig. 9.2 relates to the changing mineral composition).

- Quartz is only occasionally present and, hence, ACF diagrams must be used with care.
- Biotite is present as an extra K-bearing mineral from the greenschist to the upper amphibolite facies. The mineral systematically changes its composition during prograde metamorphism.
- Calcite is present in low-grade rocks but is used up by mixed volatile reactions in prograde metamorphism.

A comprehensive representation of the effects of metamorphism on mafic rocks is shown in Fig. 9.3. The figure depicts the characteristic assemblages in metabasalts in the form of ACF diagrams representative for the respective position in P–T space. The ACF diagrams are arranged along three typical paths of metamorphism and the 9 °C/km geotherm limits the geologically accessible P–T space. The aluminosilicate phase diagram is given for reference.

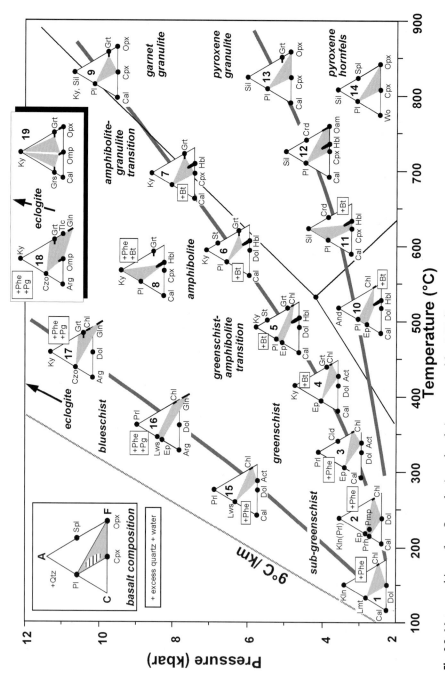

Fig. 9.3. Metamorphism of mafic rocks (metabasalts) represented by ACF diagrams

The inset in the upper left corner shows the composition of typical mid-ocean ridge basalt ($Pl + Cpx \pm Opx \pm Qtz$). Many metabasalts are expected to lie compositionally within the horizontally ruled field. On the ACF diagrams of Fig. 9.3, the assemblages that represent the approximate composition of MORB have been shaded. Metabasalts that have been strongly metasomatized during ocean-floor metamorphism may project to other, often more Ca-rich, compositions in the ACF diagram. In the following presentation, the details of the phase relations in each of the represented ACF chemographies will be discussed separately.

9.3
Subgreenschist Facies Metamorphism

9.3.1
General Aspects and a Field Example

ACF chemographies #1 and #2 (Fig. 9.3) represent mineral assemblages typical of the subgreenschist facies. At very low grade, the characteristic assemblage includes albite + chlorite + carbonate + a variety of zeolites. On the ACF chemography #1 zeolite is typified by laumontite but any of the four zeolites listed in Table 9.1 would project to the same point. In addition, clay (e. g., smectite, vermiculite) and "white mica" (illite, sericite) may be present as well as kaolinite. At slightly higher grade (ACF chemography #2), the zeolites are replaced by the minerals prehnite, pumpellyite and epidote in a number of different combinations. Still albite, chlorite and carbonates are major minerals in prehnite-pumpellyite rocks. Pyrophyllite may be present instead of kaolinite. Zeolite-bearing metamafics are representative for temperatures below about 150–200 °C; metabasalts with prehnite and pumpellyite are most characteristic for the temperature range 150–300 °C.

The zoneography of minerals in metamafic rocks from the Tanzawa mountains in central Japan is shown in Fig. 9.4. The metamorphic terrain has been subdivided into five zones ranging from lower subgreenschist facies to amphibolite facies. The terrain is of the low-pressure sillimanite type. The first two zones are characterized by the presence of zeolites in the assemblages: low-T zeolites in zone I and high-T zeolites in zone II. The zone I zeolites are typically stilbite and heulandite; zone II zeolites are laumontite and wairakite. This is understandable by reference to Table 9.1 where one can see that zone I zeolites clearly contain more crystal water (zeolite water) than zeolites of zone II. The transition of zone I to zone II is therefore related to ordinary dehydration reactions. For the same reason, it is also clear that wairakite is typically the last zeolite in prograde metamorphism. Analcite occurs in both zones sporadically. Mixed-layer clays are typically present in the first two zones. Plagioclase is present in all zones; its composition is that of albite in the first four zones. The protolith is SiO_2-rich, and quartz is present in excess throughout (ACF diagrams). No zeolites survive in zone III, which is characterized by three new metamorphic minerals: pumpellyite, prehnite and epidote. Also chlorite has transformed into a well-defined ferro-magnesian

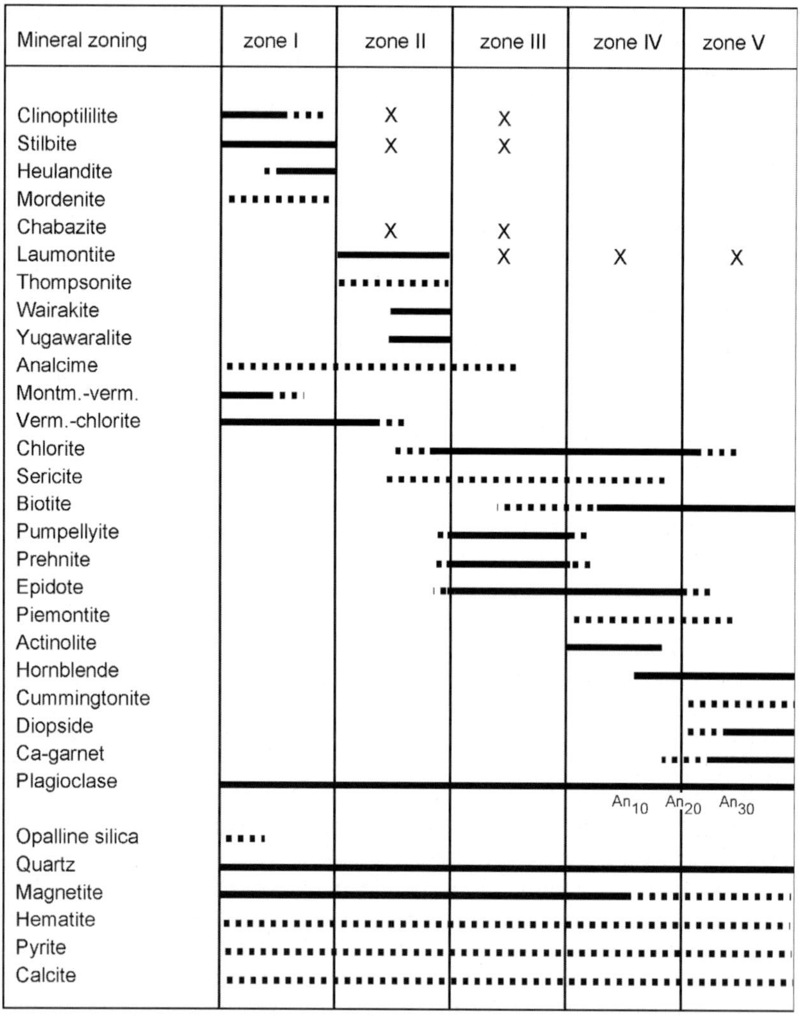

Mineral zoning	zone I	zone II	zone III	zone IV	zone V
Clinoptililite	▬▬ ···	X	X		
Stilbite	▬▬▬	X	X		
Heulandite	· ▬▬				
Mordenite	·········				
Chabazite		X	X		
Laumontite		▬▬▬	X	X	X
Thompsonite		··········			
Wairakite		▬▬			
Yugawaralite		▬▬			
Analcime	········|··········|······			
Montm.-verm.	▬▬ ··· ı				
Verm.-chlorite		··			
Chlorite		··· ▬▬▬▬|▬▬▬▬▬▬▬|▬▬▬▬▬▬▬|▬▬ ···		
Sericite		·······|·········|···········		
Biotite			ı ·······|·· ▬▬▬▬|▬▬▬▬▬	
Pumpellyite			· ▬▬▬▬ ·		
Prehnite			· ▬▬▬▬ ··		
Epidote			ı· ▬▬▬▬|▬▬▬▬▬|▬▬▬▬ ··	
Piemontite				············|····	
Actinolite				▬▬▬▬	
Hornblende				▬▬▬|▬▬▬▬	
Cummingtonite					·········ı
Diopside					··· ▬▬▬
Ca-garnet				·· |·· ▬▬▬	
Plagioclase				An₁₀ An₂₀ An₃₀	
Opalline silica	···· ı				
Quartz	▬▬▬▬|▬▬▬▬▬|▬▬▬▬▬|▬▬▬▬▬|▬▬▬▬				
Magnetite	▬▬▬▬|▬▬▬▬▬|▬▬▬▬▬|▬▬▬ ···|···········				
Hematite	·········|··········|··········|··········|···········				
Pyrite	·········|··········|··········|··········|···········				
Calcite	·········|··········|··········|··········|···········				

Fig. 9.4. Occurrence of some metamorphic minerals in Tanzawa Mountains, central Japan (Seki et al. 1969). Mineral zonation in the Tanzawa terrain, Japan. *Zone I* Stilbite (clinoptilo-lite)-vermiculite; *zone II* laumontite–"mixed layer"–chlorite; *zone III* pumpellyite–prehnite–chlorite; *zone IV* actinolite–greenschist; *zone V* amphibolite. X only in veins

sheet silicate. Sericite is present sporadically. Several zeolite- and chlorite-consuming reactions are responsible for the generation of the new meta-morphic minerals Prh, Pmp, and Ep. In zone IV, typical greenschist facies assemblages replace the zone III mineralogy. In particular, pumpellyite and prehnite disappear and the characteristic greenschist facies mineral association chlorite + epidote + actinolite + albite±quartz becomes dominant. In zone V, the minerals are diagnostic for the amphibolite facies. Calcic plagioclase and hornblende are the dominant minerals but biotite, garnet and Cpx may appear as well. In low-pressure terrains such as the Tanzawa Mountains, cummingtonite or other Fe–Mg–amphiboles typically occur together with calcic amphiboles (hornblende).

9.3.2
Metamorphism in the CASH and NCMASH Systems

Quantitative phase relationships in the simple CASH system in low-grade metamorphism are shown in Fig. 9.5. Some of the mineral stability fields are highlighted on the figure. The reaction stoichiometries can be found in Table 9.2. The predicted sequence of Ca-zeolites in progressively metamorphosed metamafic rocks is: stilbite, heulandite, laumontite and wairakite. This sequence is clearly consistent with the sequence of observed mineral zones in the Tanzawa Mountains shown in Fig. 9.4. It also follows from Fig. 9.5 that lawsonite favorably forms in terrains that are characterized by high-pressure low-temperature geotherms. In fact, lawsonite is an important mineral in blueschist terrains.

Figure 9.6 depicts phase relations in the more complex NCMASH system involving the following 14 minerals: actinolite, plagioclase, chlorite, epidote, garnet, heulandite, laumontite, lawsonite, paragonite, prehnite, pumpellyite, quartz, stilbite, and wairakite. The relationships in Fig. 9.6 are based on thermodynamic data for Mg end-member phase components in the NCMASH system. The deviations of *average* natural mineral compositions from the model end-member compositions have been accounted for by incorporating appropriate activity terms. However, amphiboles, chlorite, epidote, and pumpellyite from low-grade metabasites have large variations in their chemical composition, depending on mineral assemblage, rock composition, and metamorphic grade. It is therefore important to realize that the petrogenetic grid in Fig. 9.6 is valid only for an average metabasite (meta-MORB). The effect of variable mineral composition on the P–T stability fields of various index minerals may be considerable and will be evaluated separately.

Solid solution in zeolites may also be important, especially the "plagioclase" substitution $NaSiCa_{-1}Al_{-1}$. However, rather poorly constrained thermodynamic data for zeolites and limited chemical data do not justify including activity terms for zeolites.

Figure 9.6 is a projection from albite + chlorite (of fixed average composition) + quartz + water, because these minerals are ubiquitous in low-grade metabasites. At very low temperatures (T < 200 °C), however, this represents a simplification because smectite may be present, either as a discrete phase or as interstratified layers in chlorite. Thus it is fortunate that chlorite is not involved in reactions limiting the stability fields of zeolites.

Below follows a brief discussion of the typical P–T conditions under which some diagnostic index minerals may form and subsequently decompose in prograde metamorphism of metamafic rocks. The discussion is based on Figs. 9.5 and 9.6. Reaction stoichiometries are given in Table 9.2.

Analcite is commonly found in quartz-bearing diagenetic and low-grade metamorphic rocks. The analcite-quartz assemblage, along with stilbite +

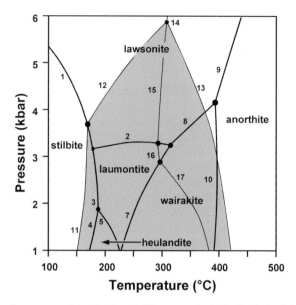

Fig. 9.5. Low-grade metamorphism in the CASH system. Prehnite field *shaded*

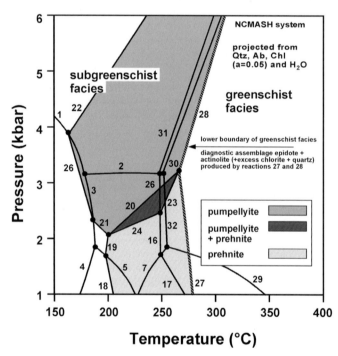

Fig. 9.6. Low-grade metamorphism in the NCMASH system. Prehnite, pumpellyite and prehnite + pumpellyite fields *shaded*

Table 9.2. Reaction stoichiometries of reactions shown in the figures in this chapter and reactions discussed in the text

Subgreenschist facies
CASH system (Fig. 9.5)
(1) $Stb = Lws + 5\ Qtz + 5\ H_2O$
(2) $Lmt = Lws + 2\ Qtz + 2\ H_2O$
(3) $Stb = Lmt + 3\ Qtz + 3\ H_2O$
(4) $Stb = Hul + H_2O$
(5) $Hul = Lmt + 3\ Qtz + 2\ H_2O$
(6) $Hul = Wa + 3\ Qtz + 4\ H_2O$
(7) $Lmt = Wa + 2\ H_2O$
(8) $Wa = Lws + 2\ Qtz$
(9) $Lws = An + 2\ H_2O$
(10) $Wa = An + 2\ Qtz + 2\ H_2O$
(11) $Stb + Grs = 2\ Prh + 4\ Qtz + 5\ H_2O$
(12) $2\ Prh = Lws + Grs + Qtz$
(13) $5\ Prh = 2\ Zo + 2\ Grs + 3\ Qtz + 4\ H_2O$
(14) $5\ Lws + Grs = 4\ Zo + Qtz + 8\ H_2O$
(15) $2\ Lws + Prh = 2\ Zo + Qtz + 4\ H_2O$
(16) $Prh + 2\ Lmt = 2\ Zo + 5\ Qtz + 8\ H_2O$
(17) $Prh + 2\ Wa = 2\ Zo + 5\ Qtz + 4\ H_2O$

NCMASH system (Fig. 9.6)
(18) $4\ Hul + Tr = 3\ Prh + Chl + 24\ Qtz + 18\ H_2O$
(19) $4\ Lmt + Tr = 3\ Prh + Chl + 12\ Qtz + 10\ H_2O$
(20) $20\ Pmp + 3\ Tr + 6\ Qtz = 43\ Prh + 7\ Chl + 2\ H_2O$
(21) $86\ Lmt + 17\ Tr = 30\ Pmp + 11\ Chl + 267\ Qtz + 212\ H_2O$
(22) $86\ Lws + 17\ Tr = 30\ Pmp + 11\ Chl + 95\ Qtz + 40\ H_2O$
(23) $5\ Pmp + 3\ Qtz = 7\ Prh + 3\ Zo + Chl + 5\ H_2O$
(24) $6\ Lmt + 17\ Prh + 2\ Chl = 10\ Pmp + 21\ Qtz + 14\ H_2O$
(25) $14\ Lmt + 5\ Pmp = 17\ Zo + Chl + 32\ Qtz + 61\ H_2O$
(26) $86\ Stb + 17\ Tr = 30\ Pmp + 11\ Chl + 525\ Qtz + 470\ H_2O$

Subgreenschist to greenschist facies transition (Fig. 9.6)
Essential reactions
(27) $5\ Prh + Chl + 2\ Qtz = 4\ Zo + Tr + 6\ H_2O$
(28) $25\ Pmp + 2\ Chl + 29\ Qtz = 7\ Tr + 43\ Zo + 67\ H_2O$

Additional reactions
(29) $4\ Wa + Ab = Pg + 2\ Zo + 10\ Qtz + 6\ H_2O$
(30) $4\ Lws + Ab = Pg + 2\ Zo + 2\ Qtz + 6\ H_2O$
(31) $14\ Lws + 5\ Pmp = 17\ Zo + Chl + 4\ Qtz + 33\ H_2O$
(32) $4\ Lmt + Ab = Pg + 2\ Zo + 10\ Qtz + 14\ H_2O$

Reactions involving carbonates (examples)
(33) $3\ Chl + 10\ Cal + 21\ Qtz = 2\ Zo + 3\ Tr + 10\ CO_2 + 8\ H_2O$
(34) $15\ Pmp + 9\ Qtz + 4\ CO_2 = 4\ Cal + 25\ Zo + 3\ Tr + 37\ H_2O$

Greenschist facies
CMASH reactions (Fig. 9.7)
(35) $2\ Zo + 5\ Prl = 4\ Mrg + 18\ Qtz + 2\ H_2O$
(36) $Mrg + 2\ Qtz + 2\ Zo = 5\ An + 2\ H_2O$
(37) $Mrg + Qtz = An + And + H_2O$
(38) $4\ Mrg + 3\ Qtz = 2\ Zo + 5\ Ky + 3\ H_2O$
(39) $2\ Chl + 2\ Zo + 2\ Qtz = 2\ Tr + 5\ Ky + 7\ H_2O$
(40) $6\ Zo + 7\ Qtz + Chl = 10\ An + Tr + 6\ H_2O$
(41) $2\ An + Chl + 4\ Qtz = Tr + 3\ Sil + 3\ H_2O$

Example of reaction producing tschermak component
(42) $7\ Chl + 14\ Qtz + 12\ Zo = 12\ Tr + 25\ TS\ (Al_1Al_1Si_{-1}Mg_{-1}) + 22\ H_2O$

Table 9.2 (continued)

Mica-involving reactions
(43) 16 Tr + 25 Ms = 25 Phl + 16 Zo + 1 Chl + 77 Qtz + 4 H_2O
(44) 4 Tr + 6 Chl + 25 Cel = 25 Phl + 4 Zo + 63 Qtz + 26 H_2O
(45) 1 Chl + 4 Cel = 3 Phl + 1 Ms + 7 Qtz + 4 H_2O
(46) 2 Chl + 2 Zo + 5 Ab + 4 Qtz = 2 Tr + 5 Pg + 2 H_2O

Carbonate-involving reactions (examples)
(47) 1 Ms + 3 Qtz + 8 Dol + 4 H_2O = 1 Phl + 1 Chl + 8 Cal + 8 CO_2
(48) 9 Tr + 2 Cal + 15 Ms = 15 Phl + 10 Zo + 42 Qtz + 2 CO_2 + 4 H_2O
(49) 19 Cal + 3 Chl + 11 CO_2 = 15 Dol + 2 Zo + 3 Qtz + 11 H_2O

Greenschist-amphibolite facies transition
(50) 4 Chl + 18 Zo + 21 Qtz = 5 Ts-amphibole + 26 An + 20 H_2O
(51) 7 Chl + 13 Tr + 12 Zo + 14 Qtz = 25 Ts-amphibole + 22 H_2O
(52) Ab + Tr = Ed + 4 Qtz
(53) 12 Zo + 15 Chl + 18 Qtz = 8 Grs + 25 Prp + 66 H_2O

Amphibolite facies
(54) 4 Tr + 3 An = 3 Prp + 11 Di + 7 Qtz + 4 H_2O
(55) 1 Ts-amphibole + 6 Zo + 3 Qtz = 10 An + 4 Di + 4 H_2O
(56) 3 Ts-amphibole + 7 Ky = 6 An + 4 Prp + 4 Qtz + 3 H_2O
(57) 7 Ts-amphibolie + 7 Sil + 4 Qtz = 14 An + 4 Ath + 3 H_2O
(58) 7 Tr = 3 Ath + 14 Di + 4 Qtz + 4 H_2O

Amphibolite-granulite facies transition
(59) Tr = 2 Cpx + 3 Opx + Qtz + H_2O
(60) Ts = Cpx + 3 Opx + An + H_2O
(61) Tr + 7 Grs = 27 Cpx + Prp + 6 An + H_2O
(62) Tr + Grs = 4 Cpx + Opx + An + H_2O

Granulite facies
(63) 4 En + 1 An = 1 Di + 1 Qtz + 1 Prp

Blueschist facies (Fig. 9.9)
(64) Tr + 10 Ab + 2 Chl = 2 Lws + 5 Gln
(65) 6 Tr + 50 Ab + 9 Chl = 25 Gln + 6 Zo + 7 Qtz + 14 H_2O
(66) 13 Ab + 3 Chl + 1 Qtz = 5 Gln + 3 Pg + 4 H_2O
(67) 12 Lws + 1 Gln = 2 Pg + 1 Prp + 6 Zo + 5 Qtz + 20 H_2O
(68) 4 Pg + 9 Chl + 16 Qtz = 2 Gln + 13 Prp + 38 H_2O
(69) 10 Pg + 3 Chl + 14 Qtz = 5 Gln + 13 Ky + 17 H_2O

Blueschist–eclogite facies transition (Figs. 9.9 and 9.10)
(70) Jd + Qtz = Ab
(71) 1 Gln + 1 Pg = 1 Prp + 3 Jd + 2 Qtz + 2 H_2O

Granulite- and amphibolite–eclogite facies transition
(72) CaTS + Qtz = An
(73) 2 Zo + Ky + Qtz = 4 An + H_2O
(74) 1 Grs + 2 Ky + 1 Qtz = 3 An
(75) 4 Tr + 3 An = 3 Prp + 11 Di + 7 Qtz + 4 H_2O
(76) 4 En + 1 An = 1 Prp + 1 Di + 1 Qtz
(77) 3 Di + 3 An = 1 Prp + 2 Grs + 3 Qtz
(78) 1 An + 2 Di = 2 En + 1 Grs + 1 Qtz

Eclogite facies
(79) Pg = Jd + Ky + H_2O
(80) 6 Zo + 4 Prp + 7 Qtz = 13 Ky + 12 Di + 3 H_2O
(81) Tlc + Ky = 2 En + Mg-TS + 2 Qtz + H_2O
(82) Grs + Prp + 2 Qtz = 3 Di + 2 Ky
(83) Grs + 6 Opx = 3 Di + Prp

quartz, can be used as an indicator of the initial stage of zeolite facies metamorphism. In the NASH model system, analcite decomposes in the presence of quartz to albite according to the reaction: $Anl + Qtz = Ab + H_2O$ (the reaction is not listed in Table 9.2). The equilibrium conditions of the analcite breakdown reaction very nearly coincide with the three reactions that limit the stilbite field towards higher temperature in Fig. 9.5 (reactions 1, 2, and 3). In order to avoid overcrowding of the figure, the analcite reaction has not been shown in Fig. 9.5 explicitly. Natural analcite, however, often shows extensive solid solution with wairakite, thus expanding the stability field of analcite to higher temperature. This has the consequence that the P–T fields of analcite and heulandite overlap.

Stilbite, a widespread zeolite mineral, dehydrates to heulandite, laumontite or lawsonite with increasing temperature and successively higher pressures in the CASH model system (Fig. 9.5). In nature, however, stilbite is sometimes associated with heulandite and/or laumontite, as, e.g., in geothermal areas in Iceland. Although little is known about the composition of stilbite in low-grade metamorphic rocks, this feature may result from solid-solution effects.

Heulandite is one of the most common zeolites besides analcite and laumontite. In Fig. 9.5, its stability field is located at temperatures around 200 °C and at pressures below 2 kbar, and limited by the stability fields of three other zeolites, i.e., stilbite towards low-T (reaction 4), wairakite towards high-T (reaction 6; not shown in Fig. 9.5, reaction is stable below 1 kbar pressure), and laumontite towards high P–T (reaction 5). Heulandite and laumontite seem to be equally abundant in ocean-floor thermal, hydrothermal, and burial metamorphism. Laumontite, but not heulandite, has been described from subduction zone metamorphism, e.g., from the Franciscan Complex of California and the Sanbagawa Belt of Japan (for a summary, see Liou et al. 1987). This indicates that heulandite is a low-pressure zeolite consistent with the topology shown in Fig. 9.5.

Natural heulandite commonly contains Na or K or both substituted for Ca, and its chemical composition may extend to those of clinoptilolite and alkaliclinoptilolite. This effect, together with other variables (e.g., pore-fluid chemistry), may explain the considerable overlap between the stability regions of heulandite and other zeolites.

Laumontite is widespread in ocean-floor, burial, and subduction zone metamorphism. In Fig. 9.5, its field of stable occurrence is limited by five univariant reactions. With increasing temperature, laumontite is formed from heulandite at lower pressure and from stilbite at higher pressure. The upper pressure limit of laumontite with respect to lawsonite is about 3 kbar [reaction (2)], and the upper temperature limit with respect to wairakite [reaction (7)] and epidote (zoisite; reaction 16) is about 230–260 °C. Compositionally, laumontite is near the ideal composition, $CaAl_2Si_4O_{12} \cdot 4H_2O$, but solid solution in heulandite and wairakite will allow the coexistence of Lmt + Hul and Lmt + Wa.

The presence of laumontite is often taken to indicate low-grade metamorphic conditions, and thus its lower thermal stability limit is of interest. In the CASH model system, this limit is about 180 °C. From field evidence, however, it has been inferred that laumontite may have formed at temperatures as low as 50–100 °C (e.g., Boles and Coombs 1975).

Wairakite is restricted to areas of relatively high geothermal gradients, including geothermal systems and areas with intrusive igneous bodies. In the CASH model system, wairakite is produced from heulandite or laumontite at temperatures between 220 and 260 °C. Wairakite has the highest thermal stability of any zeolite and, in the presence of excess albite + quartz, its upper thermal stability is between 260 and 380 °C. As mentioned earlier, wairakite shows considerable solid solution with analcite and this may explain the natural occurrence of Wa + Anl and Wa + Lmt (+ Ab + Qtz).

Lawsonite is one of the most definitive of the blueschist suite of minerals.

Pumpellyite is a very common mineral in metamafics and is found in various associations in low-grade metamafic rocks. The shaded pumpellyite field in Fig. 9.6 shows the P–T range of the assemblage Pmp + Qtz + Chl + H_2O for average pumpellyite and chlorite composition. The reactions forming and consuming pumpellyite are listed in Table 9.2. In rocks that depart from the average meta-MORB, pumpellyite may occur over a much larger or much smaller P–T interval. However, for all rock compositions, the center of the pumpellyite fields clusters around 220 °C±20° and 4±2 kbar. Most pumpellyite-bearing rocks probably formed close to 200–250 °C and at pressures of 2–3 kbar.

Prehnite occurs widespread in low-grade metamafic rocks as a part of the matrix assemblage but also very often in late veins and open fracture spaces. The shaded prehnite fields in Figs. 9.5 and 9.6 clearly suggest that the mineral is diagnostic for low-pressure conditions. The upper pressure limit of prehnite is about 5 kbar but a more typical value is probably around 3 kbar.

The prehnite + pumpellyite assemblage is shown as a dark shaded field in Fig. 9.6. The assemblage may be found in meta-MORB's at 200–250 °C at pressures of 2–3 kbar in association with quartz and chlorite (of the average composition). Depending on rock composition (and the resulting average mineral compositions at low grade), the two minerals (Pmp + Prh) may not have a common field of occurrence at all. However, most metamafics may be able to generate the Prh + Pmp assemblage. The assemblage has, however, under any circumstances an extremely narrow P–T field where it may be stable. The cumulative conditions for the co-occurrence of Pmp + Prh + Chl + Chl is about 200–280 °C and 1–4 kbar. In many metamorphic terrains Pmp + Prh can be found in rocks of the subgreenschist facies. This is not surprising because of the natural variation of bulk composition of mafic rocks and the geotherms followed by prograde metamorphism will almost certainly pass through the mentioned cumulative P–T field of Pmp + Prh. The higher-grade portion of the subgreenschist facies is therefore characterized by assemblages that involve: Pmp, Prh, Wa, Ep, Chl, Ab, Pg and Qtz.

9.3.3
Transition to the Greenschist Facies

The transition to the greenschist facies is marked by the first appearance of the diagnostic assemblage actinolite + epidote in the presence of chlorite (of average composition), albite and quartz. The assemblage is produced from prehnite decomposition at low pressures (< 3 kbar) and pumpellyite decomposition at higher pressure (> 3 kbar). Both reactions [reactions (27) and (28), Table 9.2] produce the typical greenschist facies mineralogy at a temperature of about 280±30 °C at pressures below 6 kbar. Along the characteristic Ky- and Sil-type path, the first occurrence of actinolite + epidote + chlorite + albite + quartz defines the beginning of the greenschist facies. The assemblage is shown on chemography #3 in Fig. 9.3. Chemography #3 also suggests that the greenschist facies assemblage may also form from dolomite- or calcite-involving reactions (an example is reaction 33). As carbonates are present in many low-grade mafic rocks, these mixed volatile reactions are important in removing carbonate minerals from metamafic rocks at an early stage of prograde metamorphism and connecting to the greenschist facies. Typically, these reactions consume chlorite and carbonate and produce epidote + actinolite. The reactions are of the maximum-type mixed volatile reactions (see Chap. 3). Note, however, that pumpellyite- and prehnite-consuming reactions *produce* rather than consume carbonates. Reaction (34) is an example reaction that limits pumpellyite in CO_2-bearing fluids and produces the greenschist facies assemblage. This means that, in the presence of CO_2-rich fluids, greenschist facies assemblages may appear at lower temperature compared with pure H_2O fluids.

At high pressures (e.g., along a low-T subduction geotherm), the subgreenschist facies assemblages are replaced by assemblages that are diagnostic for the blueschist facies. These rocks contain glaucophane, lawsonite, paragonite, epidote, and other minerals in various assemblages. The blueschist facies assemblages and blueschist metamorphism will be discussed below (see Sect. 9.7).

9.4
Greenschist Facies Metamorphism

9.4.1
Introduction

Greenschist facies mafic rocks are, as the name suggests, green schists. The green color of greenschists results from the modal dominance of green minerals in the rocks notably chlorite, actinolite and epidote. The most characteristic assemblage found in greenschists is: chlorite + actinolite + epidote + albite ± quartz. The term greenschist is exclusively reserved for a schistose chlorite-rich rock derived from mafic igneous rocks that were metamorphosed under greenschist facies conditions (hence contains the diagnostic assemblage: Chl + Act + Ep + Ab ± Qtz). Note, however, pumpellyite + epidote + chlorite schists of the subgreenschist facies are green schists as well, although not in greenschist facies. Serpentinites may be green schists of the greenschist facies but they are not greenschists.

Chemographies #3 and #4 (Fig. 9.3) represent the typical range of assemblages in the greenschist facies. In addition to the most important minerals named above, some further minerals may be found in greenschists in minor amounts. Their occurrence mostly depends on the bulk composition of the protolith and the details of the hydration history prior to greenschist facies metamorphism. Greenschists commonly lost all relic structures from previous metamorphic and magmatic stages. It is typical and ordinary to find mineralogically and structurally perfectly equilibrated greenschists.

9.4.2
Mineralogical Changes Within the Greenschist Facies

9.4.2.1
Reactions in the CMASH System

The difference between chemography #3 and #4 (Fig. 9.3) is the presence of pyrophyllite and phengite in the lower greenschist facies and of kyanite and biotite in the upper greenschist facies. The phase relationships in the ASH system that covers the changes in the A-apex of ACF diagrams has been discussed in Chapter 7 and they are displayed in Fig. 7.2. The reactions in the ASH system also represent the backbone of Fig. 9.7 that depicts some phase relationships in the CMASH system (using the same average mineral compositions as in Fig. 9.6, unless indicated otherwise). The reaction stoichiometry of reactions shown in Fig. 9.7 is listed in Table 9.2. Note, however, that not all equilibria shown are listed in Table 9.2; in particular, the details of the phase relations in the sillimanite field of Fig. 9.7 have been omitted because we are concerned with the greenschist facies here.

It is evident from Fig. 9.7 that the first reaction relevant for mafic rocks in greenschist facies is reaction (35). It replaces epidote + pyrophyllite by margarite in quartz-bearing rocks. The shaded margarite field in Fig. 9.7 is created by the stable reaction (35) and it is terminated by other reactions that will be discussed below. Margarite is not shown in Fig. 9.3 (ACF figure). Figure 9.7 suggests that large central portions of the greenschist facies are within the stability field of margarite. However, margarite in greenschists is not very common. The reason for this apparent mismatch is that most metamafic rocks fall compositionally into the shaded Chl–Ep–Act triangle in Fig. 9.3. However, if bulk composition is more aluminous than the epidote-chlorite tie line, then the stable assemblages may be read from Fig. 9.7. The sequence is (for Qtz-saturated rocks following a Ky-type path): Prl + Ep $\Rightarrow$ Mrg + Ep $\Rightarrow$ Ky + Ep.

Figure 9.7 also suggests that the assemblage epidote + kyanite is diagnostic for relatively high pressures (>about 5 kbar). The Ep + Ky assemblage is produced by reaction (38). Other reactions that terminate the margarite + quartz assemblage in greenschists are reactions (36) and (37). The assemblage actinolite (amphibole) + kyanite commences with the higher greenschist facies. It

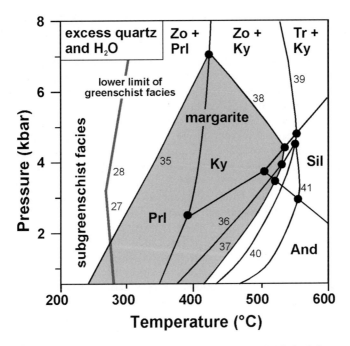

Fig. 9.7. Phase relationships in the greenschist facies. Margarite field *shaded*

is diagnostic for pressures typically above 5–6 kbar. The composition of amphibole that occurs stable with kyanite is, however, far removed from common actinolite and reaction (39), that produces the Act + Ky assemblage, runs close to the amphibolite facies boundary.

Inside the Chl–Ep–Act triangle of Fig. 9.3, there are also reactions that modify the typical assemblage of the greenschist facies. The processes are best explained by using reaction (42) as an example. The reaction describes the production of tschermak component that is taken up by amphibole and chlorite. Both minerals systematically change their Al content across the greenschist facies. The effect can be shown on ACF diagrams by a displacement of the Chl + Ep + Act triangle along the TS vector. Note, however, concurrent rotations of FM relationships cannot be displayed on ACF diagrams. In general, chlorite and actinolite become more aluminous across the greenschist facies. AF relationships are complex and can be evaluated along the lines described in Chapter 7 (metapelites). Redox reactions may change the composition of epidote and other Fe-bearing minerals.

9.4.2.2
Reactions Including Micas

Muscovite (in general K-white mica, sericite, phengite) is the most common K-bearing mineral in low-grade metamafic rocks. This is also true for micas in the lower greenschist facies. However, in the middle of the greenschist facies (~400 °C), biotite appears for the first time in metamafics replacing K-white mica. This is shown in Fig. 9.3 (chemographies #3 and #4). Biotite formation can be modeled by reaction (43). It involves all greenschist minerals and it transfers the Ms component to the Phl component. The equilibrium of reaction (43) is close to 600 °C in the pure KCMASH system and it is rather insensitive to pressure. For real mineral compositions of greenschist facies rocks, biotite formation from reaction (43) takes place around 400 °C ($a_{Phl} = 0.05$; $a_{Ms} = 0.2$; all others see Fig. 9.6). The first biotite that appears in prograde metamorphism of mafic rocks is often green under the microscope. This is an indication of high Fe^{3+} in low-grade biotite and demonstrates the importance of redox reactions in low-grade metamafics (so be aware of that when calculating phase relationships involving biotite).

Biotite production is a continuous process. The biotite-producing reaction (44) is similar to reaction (43). However, it consumes the celadonite component in K-white mica in order to produce biotite. The reaction leaves behind a white mica that is closer to muscovite end-member composition. This is in accordance with field evidence that K-white mica becomes progressively depleted in Cel component and enriched in Ms component during prograde metamorphism along a Ky-type path. The effect can also be understood by investigating reaction (45), where celadonite component is consumed in white mica and biotite + muscovite components are formed in prograde metamorphism.

Again, biotite forms in metamafics at about 400–450 °C. K-white mica is typically celadonitic (phengite, sericite) in low-grade rocks; it becomes more muscovitic (muscovite) at the upper end of the greenschist facies. However, the small amount of potassium in mafic rocks is usually bound in K-white mica below 400 °C and in biotite at temperatures above 400 °C, and there is not much overlap of the two micas.

Paragonite occurs in greenschists at high pressures. The equilibrium conditions of reaction (46) are extremely sensitive to small compositional variations in the protolith and the resulting metamorphic minerals. The typical actinolite + paragonite assemblage forms, however, at pressures above about 6 kbar, and this pressure is insensitive to temperature.

Garnet may form in mafic schist from reactions very similar to the reactions that form garnet in pelitic schist. The first garnet is manganiferous and contains little grossular component. Garnet appears in the uppermost part of the greenschist facies for the first time, but garnet is not really typical in mafic rocks until the amphibolite facies.

Stilpnomelane is a characteristic mineral in many low-grade mafic schists and, if present, is diagnostic for lower greenschist facies conditions (or blueschist facies). The brown color of this pleochroic mineral often resembles biotite under the microscope. Stilpnomelane is replaced by green biotite at around 400 °C. Mafic rocks containing both stilpnomelane and biotite occur over a narrow temperature interval near 400 °C.

Carbonate minerals may still be present in greenschists. Both dolomite (ankerite) and calcite can be present and the carbonates participate in many mixed volatile reactions that involve the characteristic greenschist facies silicates, Chl, Ep, and Act. The reactions may also involve micas such as reaction (47) that replaces K-white mica by biotite in Dol- and Cal-bearing rocks. Another example is reaction (48), where the assemblage actinolite + muscovite + calcite is replaced by epidote + biotite. Reaction (49) involves two carbonates in rocks containing epidote + actinolite. Some of the carbonate-involving reactions are also relevant for marbles (CMS–HC system) or for marly rocks. These have been discussed in Chapters 6 and 8, respectively, and we will not present an extensive discussion of phase equilibria in carbonate-bearing greenschists here. There is, however, a great potential for useful information contained in carbonate-bearing greenschists.

9.4.3
Greenschist-Amphibolite Facies Transition

The greenschist facies assemblage experiences two basic changes as metamorphic temperatures approach about 500 °C:
- Albite disappears and it is replaced by plagioclase. The composition of the first Ca-bearing plagioclase is typically An_{17} (oligoclase).
- Amphibole becomes capable of taking up increasing amounts of aluminum and alkalis. Actinolite disappears and it is replaced by alkali-bearing aluminous hornblende.

The combined transformation results in the replacement of the Ab + Act pair by the Pl + Hbl pair, i.e., in other words, replacement of greenschist by amphibolite. In orogenic metamorphism along a Ky-type path, the greenschist–amphibolite facies transition occurs at temperatures of about 500 °C (5 kbar).

Anorthite component is produced by a series of continuous reactions. However, albite present in greenschist does not continuously change its composition along the albite–anorthite binary. At the conditions of the upper greenschist facies the plagioclase solid-solution series is not continuous but rather shows several miscibility gaps. The first Ca-bearing plagioclase that forms in this manner is an oligoclase and its typical composition is An_{17}. Because of the abrupt appearance of oligoclase, due to the miscibility gap in the plagioclase series, the first appearance of plagioclase with An_{17} can be mapped in the field as an oligoclase-in isograd in mafic schists. This isograd defines the beginning of the amphibolite facies; it separates greenschist facies from amphibolite facies terrains.

The systematic compositional changes in the amphibole solid-solution series are generally more continuous in nature (although miscibility gaps do exist there as well). The amphibole composition changes basically by: (1) taking up tschermak component produced by a series of continuous reactions, (2) incorporation of edenite component produced by albite-consuming reactions, and (3) the unavoidable FM exchange. Other effects are related to Ti incorporation and redox reactions.

The resulting amphibole is a green tschermakitic to pargasitic hornblende that is now present together with plagioclase. The two minerals constitute, by definition, an amphibolite and define the amphibolite facies. Chemography #5 (Fig. 9.3) shows this new situation; hornblende has changed its composition along the TS exchange direction and plagioclase shows up on ACF projections. In the beginning, chlorite and epidote may still by present in amphibolites.

The most important mineralogical changes at the greenschist-amphibolite transition can be related to reaction (50). The reaction consumes epidote and chlorite from the greenschist assemblage and produces anorthite component of plagioclase and tschermak component of amphibole. The reaction will eventually consume all epidote or all chlorite in the rock, leaving behind a chlorite- or epidote-bearing amphibolite. All three FM minerals of the greenschist facies assemblage are consumed by reaction (51). It produces tschermak component and consumes actinolite component of the amphibole. Reaction (52) describes the formation of the edenite component of the amphibolite facies hornblende. The combined effect of the three reactions (51), (52) and (53) is that the chlorite + epidote assemblage gradually disappears, plagioclase becomes increasingly calcic and amphibole systematically changes its composition from actinolite to alkali- and aluminum-bearing green hornblende.

Garnet may also appear at the transition to the amphibolite facies. Its formation is accomplished by similar reactions as in metapelitic rocks (Chap. 7). Garnet grows initially mostly at the expense of chlorite. Chlorite decomposition contributes the bulk of the almandine and pyrope components of garnet. Epidote-consuming reactions produce much of the grossular and andradite components found in garnets in metamafic rocks. As in metapelitic rocks, low-grade garnets are generally rich in manganese and the strong fractionation of Mn into garnet makes the mineral appear in mafic schists at temperatures as low as 450 °C. Reaction (54) consumes chlorite and epidote and produces both grossular and Fe–Mg-garnet components, which results in the typical ternary Ca–Fe–Mg-garnets found in mafic rocks.

9.5
Amphibolite Facies Metamorphism

9.5.1
Introduction

The amphibolite facies is characterized by chemographies #6, #7 and #8 (Fig. 9.3). The minerals plagioclase and hornblende make up the bulk volume of amphibolites. All other minerals that can be present, such as quartz, epidote, muscovite, biotite, garnet, and Cpx, are modally subordinate. Calcite can be found in some amphibolites. Mineralogical changes within the amphibolite facies mostly result from reactions that run continuously over a wide P–T range. The main effect of these continuous reactions is seen in systematic variations in the composition of the two master minerals, plagioclase and hornblende. The continuous reactions also cause epidote–clinozoisite to decrease in modal amount and eventually they disappear. Muscovite, if it survived from the greenschist facies, also gradually disappears. Garnet persists and becomes modally more important up-grade (garnet amphibolites); clinopyroxene appears at higher temperatures in the amphibolite facies.

In principle, given a MORB composition and the composition of plagioclase and hornblende, the P–T conditions of equilibration are uniquely defined. However, at present, experimental data do not permit a rigorous treatment of plagioclase-hornblende relationships. In particular, solution properties of amphiboles are still poorly known and few end-member phase components are well constrained. Surprisingly, the low-temperature behavior of the plagioclase system is also not quantitatively known (e.g., quantitative thermodynamic description of the various miscibility gaps and structural transitions along the Ab–An binary at low temperature).

9.5.2
Mineralogical Changes Within the Amphibolite Facies

In prograde orogenic metamorphism of metabasalts, the rocks contain hornblende and plagioclase at the beginning of the amphibolite facies (~500 °C), as explained in Section 9.4.3. In addition, amphibolites may still contain some epidote and/or chlorite that have not been completely consumed by reactions that produced anorthite and tschermak components. Biotite may be present as well. The same continuous reactions that produced the amphibolite facies mineralogy initially continue to consume chlorite and epidote within the lower part of the amphibolite facies. Eventually, chlorite completely disappears at about 550 °C and epidote is not typically found in amphibolites that were metamorphosed to 600 °C. Some of the epidote- and chlorite-consuming reactions produce garnet that, in general, becomes modally more important with increasing grade. Chemography #6 is characteristic for the central portion of the amphibolite facies around 600 °C. Here, amphibolites contain plagioclase (typically andesine) and green hornblende ± garnet ± biotite. At still higher temperatures, clinopyroxene appears in amphibolites (chemography #7). Cpx of the diopside–hedenbergite series usually appears around 650 °C (along the Ky path) and a reaction that produces it is listed as reaction (54). The reaction simultaneously produces Cpx and garnet; it is typical of relatively high pressures. However, Cpx is also often found in amphibolites lacking garnet. A typical assemblage is Hbl + Pl + Cpx + Bt. Reaction (55) produces diopside and anorthite components from amphibole and epidote or zoisite.

This extremely important reaction has several significant effects: (1) it continuously consumes epidote or clinozoisite that may still be present in mid-amphibolite facies rocks and eventually eliminates them; (2) the reaction consumes amphibole component, a process typical for the higher amphibolite facies; (3) the reaction produces Cpx that appears in higher-grade amphibolites; and (4), last but not least, the reaction produces even more anorthite component that is incorporated in plagioclase in high-grade amphibolites. Plagioclase in high-grade amphibolites becomes progressively more calcic and in the upper amphibolite facies andesine–labradorite compositions are typical of plagioclase in amphibolites (the range is An_{30}–An_{70}). Note, however, that phase relations in the plagioclase system are complex and little understood, so if you find a bytownite or anorthite (An_{95} plagioclase) in an upper amphibolite facies terrain, this is no reason to be confused.

The first appearance of clinopyroxene in amphibolites can be used to define the boundary to the upper amphibolite facies. Reactions in the upper amphibolite facies, such as reactions (54) and (55), begin to break down amphibole components and replace them with pyroxene components. This is a continuous process, however, and the first appearance of Cpx in mafic rocks is usually not a sharp isograd in the field. Nevertheless, the "Cpx-in" transition zone marks the beginning of the upper amphibolite facies and a representative temperature is about 650 °C. In water-saturated environments, metamafic rocks show the first structural evidence in the field of local partial melt formation, migmatization and appearance of quartzo-feldspathic seams, patches, veins and similar mobilisate structures in the upper amphibolite facies. The processes remove quartz, plagioclase and biotite from the rocks and transfer them to a melt phase.

At higher pressures than that of the ordinary Ky-geotherm (Fig. 9.3), amphibolites may contain the diagnostic kyanite + hornblende assemblage. The Ky + Hbl pair may have formed by reaction (39) earlier in the course of prograde metamorphism (see Fig. 9.7). The link to the more common plagioclase + garnet assemblage found in amphibolites is given by reaction (56). The reaction replaces the garnet + plagioclase tie line of chemographies #6 and #7 by the kyanite + hornblende tie line of chemography #8. The continuous dehydration reaction runs typically between chemographies #6 and #8 in Fig. 9.3 for typical metabasalt compositions. Therefore, kyanite-bearing amphibolites are diagnostic of high-pressure amphibolite facies (pressures typically greater than 7 kbar). Remember, however, that the Ky–Hbl–Grt–Pl assemblage is not normally coplanar (as one may erroneously conclude from ACF figures such as Fig. 9.3) but rather defines a phase volume that occurs over a relatively wide P–T interval.

9.5.3
Low-Pressure Series Amphibolites

At low metamorphic grades, mafic rocks that were metamorphosed along a Ky-type and a Sil-type path, respectively, are very similar in mineralogy. The most significant difference may be found in the composition of chlorite and, especially, K-white mica. The celadonite component of K-white mica is relatively sensitive to pressure variations and it is controlled by processes such as reaction (45). The tschermak component in mica (and in principle also chlorite or biotite) can be used to monitor pressure conditions in low-grade metamorphism.

The transition to the amphibolite facies at low pressures is similar to the one described in Section 9.4.3. In extremely Al-rich rocks, andalusite may appear instead of kyanite (chemography #10). However, andalusite + amphibole is not very common. The phase relationships shown in Fig. 9.7 suggest that the transition to the amphibolite facies occurs at slightly lower temperatures in low-pressure metamorphism. This is reasonable because the reactions that produce amphibolite from greenschist are ordinary continuous dehydration reactions. The lower boundary of the amphibolite facies along a Sil-path is typically at about 450 °C (at 3 kbar). Reactions (36), (37), (40) and the andalusite equivalent of reaction (41) pass through the andalusite field in Fig. 9.3. The reactions generally produce anorthite and amphibole components from greenschist facies mineralogy. At 550 °C (Figs. 9.3 and 9.7), epidote and chlorite typically disappeared from low-pressure amphibolites and the characteristic assemblage is hornblende + plagioclase ± andalusite ± biotite.

Compared with orogenic metamorphism along a Ky-path, clinopyroxene forms at significantly lower temperatures in low-pressure amphibolites (600 °C or lower). Chemography #11 is characteristic of amphibolites that formed in the range 3–4 kbar. It shows the usual hornblende + plagioclase assemblage together with clinopyroxene and biotite. Sillimanite may be present in Al-rich amphibolites and the Sil + Hbl pair forms a stable, though rare, assemblage. However, garnet is scarce or even absent in low-pressure amphibolites.

Chemography #12 is representative for amphibolites in the upper amphibolite facies at low pressures (Sil path). The most significant feature is the presence of ferro-magnesian amphiboles in addition to the calcic amphiboles (hornblende). The types of Fe–Mg amphiboles that occur together with hornblende include anthophyllite, gedrite and cummingtonite. In some amphibolites, as many as three different amphibole species may be present such as hornblende, gedrite and anthophyllite. The abbreviation "Oam" for orthoamphibole in Fig. 9.3 includes all Fe–Mg amphiboles and also gedrite. Phase relationships in such multi-amphibole rocks can be very complex. Miscibility gaps in various amphibole series as well as structural changes in amphiboles complicate the picture. However, such low-pressure amphibolites possess a great potential for detailed analysis of relationships among minerals and assemblages and for the reconstruction of the reaction history of low-pressure amphibolites. Fe–Mg-amphiboles may form by a number of different conceivable mechanisms. The first and obvious one is reaction (57) that links chemography #11 to chemography #12. The reaction breaks down the hornblende + sillimanite assemblage and produces anorthite component that increases the An content of plagioclase and also produces anthophyllite component that is a major component in any Fe–Mg-amphibole. As metamorphic grade increases, reaction (58) becomes more important in producing anthophyllite component. The reaction consumes calcic amphibole component and replaces it with clinopyroxene and Fe–Mg-amphibole. The reaction equilibrium of reaction (58) in metabasaltic rocks involves phase components of the typical assemblage: $Hbl + Oam + Cpx + Pl \pm Qtz$. As a general rule of thumb, the assemblage is characteristic for the temperature range between 650 and 750 °C along the Sil geotherm.

9.5.4
Amphibolite–Granulite Facies Transition

Much of the quartz that has been produced in the subgreenschist facies to greenschist facies metamorphism has been consumed by continuous reactions in the amphibolite facies. Many amphibolites are quartz-free (use ACF diagrams with care!). Most of the water bound in hydrous minerals has been released by reactions during prograde metamorphism. Metabasaltic rocks that have been progressively metamorphosed to 700 °C contain $Pl + Hbl \pm Cpx \pm Grt \pm Bt$ (Ky-path) and $Pl + Hbl \pm Cpx \pm Oam \pm Bt$ (Sil path). Amphibole is the last remaining hydrous phase and further addition of heat to the rock by tectono-thermal processes will ultimately destroy the amphibole. The transition from a hydrous amphibolite facies assemblage to a completely anhydrous granulite facies assemblage is gradual and takes place over a temperature interval of at least 200 °C (from about 650–850 °C). The first clear and unequivocal indication that granulite facies conditions are reached is the appearance of orthopyroxene in Cpx-bearing quartz-free rocks. Granulite facies is obvious if Opx turns up in quartz-bearing amphibolites. Orthopyroxene is most common in low-pressure mafic granulites. At higher pressures, the typical anhydrous granulite facies assemblage is $Pl + Cpx + Grt \pm Qtz$. The reactions that eventually produce this assemblage are continuous in nature and the temperature interval is very wide. Hbl gradually decreases in modal amount, leaving behind the anhydrous granulite facies assemblage. Note, however, that $Pl + Cpx + Grt$ may also occur in the amphibolite facies and that the assemblage as such is by no means diagnostic for granulite facies conditions. If pressure conditions are too high for Opx to form, then the ultimate granulite facies condition is reached when the last crystal of amphibole disappeared from the rock. Reactions (59)–(62) consume the major amphibole components tremolite and tschermakite and produce pyroxene, garnet and anorthite components and, hence, represent the transition of the amphibolite to the granulite facies. Reaction (60), in particular, is of great importance. It decomposes hornblende and makes the two-pyroxene assemblage that is diagnostic for the granulite facies in quartz-absent mafic rocks. It produces, in addition, anorthite component that ultimately leads to the calcic plagioclase found in mafic granulites.

What happens to the biotite? Biotite is a subordinate hydrous phase in high-grade amphibolites. The small total amount of potassium stored in biotite will be taken up by pyroxene and ternary feldspar or it is used to form a Kfs–Ab–Qtz melt in the migmatite-producing process in high-grade metamorphism of mafic rocks. The total amount of water that can be stored in high-grade amphibolites is very small (in the order of 0.4 wt% H_2O or less). Consequently, mafic rocks are less susceptible to partial melting compared with their metapelitic or metagranitoid counterparts.

9.6
Granulite Facies and Mafic Granulites

Granulite facies metamorphism of mafic rocks is represented by chemographies #9 and #13 in Fig. 9.3. At high pressures the characteristic assemblage is: plagioclase + clinopyroxene (augite) + garnet. At lower pressures the typical granulite facies assemblage is: plagioclase + clinopyroxene (augite) + orthopyroxene (hypersthene). The two assemblages are linked by the important reaction (63). The reaction separates a field with pyroxene granulites at pressures below about 5–7 kbar from a field with garnet granulites at pressures above 5–7 kbar and below the eclogite field at very high pressures (at 800 °C). The continuous nature of reaction (63) generates a wide overlap zone with both garnet and Opx present in Pl + Cpx rocks. If strictly concerned with meta-MORB's, however, the overlap is smaller and the boundary occurs close to 6–7 kbar at 800 °C.

Because high-grade brown hornblende is an extremely stable mineral, very high temperatures are necessary to destroy the last hydrous phase in mafic rocks. Completely anhydrous mafic granulites form normally above 850 °C or at even higher temperatures. Dehydration of amphibole, on the other hand, is often aided by interaction of the rocks with a foreign fluid phase that is poor in H_2O. Alternatively, dehydration of amphibole may be triggered by partial melting and removal of H_2O in a melt phase; see also Chapter 7 (metapelitic granulites), where some aspects of granulite facies metamorphism are discussed. Figure 9.8 shows some of the relationships that are important in high-grade and granulite facies metamorphism of mafic rocks. Figure 9.8 is a schematic temperature versus fluid composition diagram at an approximate total pressure of 6 kbar. Along the 750 °C isotherm, three different assemblage fields intersect. If H_2O-rich fluids are present (section A), the mafic rocks undergo partial melting and mafic migmatites are characteristic for such conditions. Intermediate fluid compositions are consistent with amphibolite facies rocks with the typical mineralogy: Pl + Hbl + Cpx (section B). Migmatite structures are absent. Fluids low in H_2O coexist with granulite facies rocks at the same temperature and the characteristic assemblage is: Pl + Cpx + Opx ± Grt (section C). Note that, in real rocks, the transition zone between the granulite and amphibolite facies assemblages (Fig. 9.8) covers a wide range of fluid composition. The important message of Fig. 9.8 is, however, that at the same temperature (e.g., 750 °C) mafic migmatites, amphibolites and mafic granulites may be found intimately associated in a metamorphic complex or terrain depending on the composition of the fluid phase present in the volume of rocks under consideration. The situation may be complicated by variations of fluid composition over time at a particular site. Also note that the first melt in mafic rocks forms if pure H_2O fluids are present (wet solidus in Fig. 9.8). With increasing temperature, migmatites form with fluids that contain decreasing amounts of water. The amphibolite facies assemblage has a temperature maximum in Fig. 9.8. The actual value of the maximum temperature for amphibolite can be correlated with the amphibole-out isograd that is at about 900 °C (at 6 kbar) in pure

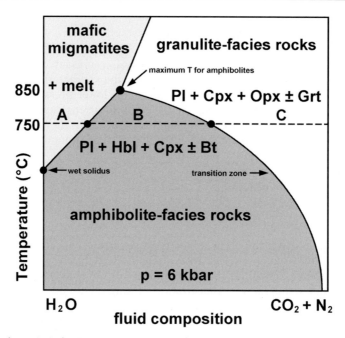

Fig. 9.8. Schematic isobaric temperature versus fluid composition diagram showing amphibolite–granulite facies relationships

H_2O fluids and may be closer to 850 °C in mixed volatile fluids at the same pressure.

Although completely anhydrous mafic granulites are widespread (e.g., in the Jotun nappe and the Bergen arcs of the Scandinavian Caledonides), many mafic granulites still contain prograde hornblende (as opposed to retrograde post-granulite Hbl). The transitional nature of the amphibolite to granulite facies transition that occurs over a wide P–T range requires a definition of granulite facies in mafic rocks. In low-pressure terrains this is a simple task: the first appearance of orthopyroxene in mafic rocks (quartz-free or quartz-bearing does not matter) marks the onset of granulite facies conditions. Mafic rocks will, at this stage, always contain clinopyroxene as a major mineral. The assemblage orthopyroxene + clinopyroxene will then be diagnostic for two-pyroxene granulites. The case is more difficult at high pressures where Opx does not form in mafic rocks and the typical assemblage Pl + Cpx + Grt may also be present in the amphibolite facies or hornblende persists in mafic rocks to very high temperatures (1000 °C at 10 kbar). Obviously, in this case, the definition of an amphibolite–granulite facies *boundary* is impossible on the basis of mafic rocks alone (see comments in Chap. 4 on metamorphic grade).

In addition to the minerals discussed so far, mafic granulites may also contain more exotic minerals such as sapphirine or scapolite in various assemblages. Also, hercynitic spinel is commonly present in mafic granulites and its presence is diagnostic for low-pressure conditions (<4 kbar).

At very low pressures, such as represented by chemography #14, pyroxene hornfelses form by high-temperature contact metamorphism of mafic rocks. Ultimately, the mineralogy of the protolith basalt is expected to form in ultra-high-temperature contact metamorphism of mafic rocks. The assemblage $Pl + Cpx + Opx \pm Spl$ (chemography #14) is then identical to the basalt mineralogy shown in the upper left corner of Fig. 9.3. The upper boundary of granulite facies metamorphism of mafic rocks is given by the dry liquidus for basalt ($\sim 1200\,^{\circ}C$).

Geological causes of granulite facies metamorphism at low pressures are commonly attributed to intrusions of mafic (gabbroic/basaltic) or charnockitic magma into the continental crust (see references and further reading of this chapter and Chap. 7). Large volumes of dry or CO_2-rich hot magma from the mantle or lower crust serve as heat source for extensive high-temperature metamorphism at shallower crustal levels. The dry or CO_2-rich nature of the magma facilitates dehydration of the crustal rocks. There is little tectonic activity associated with this type of granulite facies metamorphism. After heating to high temperature the rocks cool essentially isobarically at the crustal level at which they reside. Many granulite facies terrains are characterized by isobaric cooling paths (and perhaps a counter-clockwise P–T loop; see also Chap. 3). In general, the exhumation of granulite facies terrains requires a later contractional orogenic event that involves stacking and tilting of crustal slices and, e.g., the formation of fold nappe structures that ultimately permit the erosion surface to intersect with granulites of the middle or lower crust. If water becomes available during slow cooling, the high-grade assemblages may be completely erased and granulites may lose their memory of the granulite facies event. In particular, if Opx disappears from mafic granulites during amphibolitization, it is very hard to tell that the rock ever went through a granulite facies episode.

9.7
Blueschist Facies Metamorphism

9.7.1
Introduction

Blueschists are rocks that contain a significant amount of blue alkali-amphibole with a very high proportion of glaucophane end-member component. Such rocks are, in an outcrop or hand specimen, blue in color. Note, however, that pure glaucophane is colorless under the microscope. If a rock contains a blue pleochroic amphibole in thin section it means that it is not a pure glaucophane and appreciable amounts of ribeckite and other sodic ferro-ferric amphibole components are important and the overall composition of the amphibole is crossite (amphibole name recently disapproved by IMA, however, I like it!). Note also that Cl-rich hornblende may have a blue color under the microscope. However, if mafic rocks of basaltic bulk composition

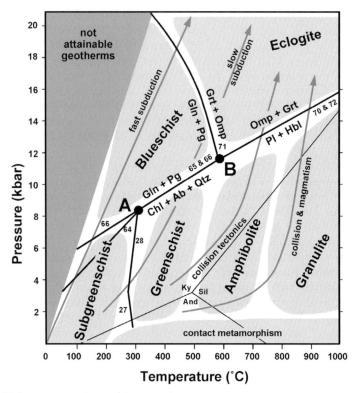

Fig. 9.9. High-pressure metamorphism overview

contain sodic amphibole, the rocks are likely to have been metamorphosed under the conditions of the blueschist facies.

The general P–T field of the blueschist facies is delineated in Fig. 4.2. The blueschist facies is represented in Fig. 9.3 by chemographies #16 and #17. The lower-pressure version of the blueschist facies #16 is characterized by glaucophane (sodic amphibole) + lawsonite + chlorite assemblages. In the higher-grade blueschist facies #17, typical metamafics contain various assemblages among glaucophane (sodic amphibole), zoisite (clinozoisite-epidote), garnet, paragonite, phengite, chlorite, talc, kyanite, rutile, ankerite, and other minerals.

A detailed rendering of the blueschist field is shown in Fig. 9.9. It is bounded towards the subgreenschist and greenschist facies by reactions (64), (65) and (66). Its high-pressure boundary is given by reaction (71). Blueschist facies terrains are worldwide associated with subduction zone metamorphism that typically occurs along destructive plate margins where oceanic crust (basalt) is recycled to the mantle under continental lithosphere. Two typical geotherms related to subduction tectonics are depicted in Fig. 9.9. The slow subduction geotherm only marginally passes through the blueschist field and eclogite assemblages are produced at pressures as low as 13–

14 kbar. If subduction is fast, prominent blueschist facies terrains may develop and eclogite formation does not occur until pressures of 18–20 kbar are reached (corresponding to subduction depths of 50–60 km). Because there are natural limits to attainable subduction velocities in global tectonic processes, the P–T conditions in the upper left wedge in Fig. 9.9 are not accessible conditions for rock metamorphism.

9.7.2
Reactions and Assemblages

As discussed above (Fig. 9.9), reactions (27) and (28) mark the boundary between the subgreenschist facies and the greenschist facies. The boundary between the subgreenschist facies and the blueschist facies is given by reaction (64). This reaction replaces the low-grade assemblage actinolite + chlorite + albite by the high-pressure assemblage glaucophane + lawsonite. The reaction is H_2O-conserving, and as such independent of water pressure or the composition of the fluid phase. Equilibrium of reaction (64) is shown in Fig. 9.9 for average mineral compositions found in low-grade equivalents of basaltic rocks.

Point A in Fig. 9.9 marks an important P–T region where blueschist, greenschist and subgreenschist facies assemblages meet. It is at about 8 kbar and 300 °C. The P–T region is defined by the approximate intersection of reactions (22), (28), (30), (31), (64), (66) and (70). Note, however, that point A is not an invariant point in Schreinemakers' sense but rather a relatively narrow P–T range of locations of various invariant points generated by the intersection of continuous reactions in the NCMASH system.

The blueschist facies is separated from the greenschist facies by reactions (65) and (66). The stable equilibrium of these reactions is shown in Fig. 9.9. Reaction (65) consumes the low-grade assemblage chlorite + actinolite + albite and replaces it by the most typical blueschist facies assemblage glaucophane + epidote (zoisite). Reaction (65) connects the points A and B in Fig. 9.9, respectively. It nearly coincides with another very important reaction, reaction (66). The reaction replaces the greenschist facies assemblage chlorite + albite by the characteristic blueschist assemblage glaucophane + paragonite. The combined effect of reactions (65) and (66) is that the chlorite + albite + actinolite assemblage, that is diagnostic for low-grade metamafic rocks, disappears and the blueschist assemblage glaucophane + epidote + paragonite is formed. If for any reason some albite should still be present in the rocks after all the chlorite has been used up, that albite will be destroyed by reaction (70). The reaction boundary of reaction (66) roughly coincides with the diagonally ruled boundary that separates a plagioclase present from a plagioclase absent region (Fig. 9.10). Like in the eclogite facies, blueschists do not contain stable albite (plagioclase). In most low-grade mafic rocks, however, chlorite is much more abundant than albite. Consequently, reactions (65) and (66) will normally destroy all albite.

The glaucophane + paragonite assemblage is replaced by omphacite + garnet as a result of reaction (71). The reaction terminates the blueschist field

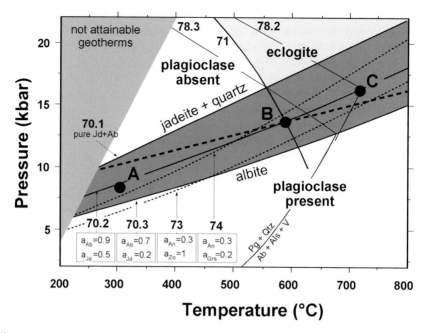

Fig. 9.10. Selected mineral equilibria in high-pressure metabasites. □=plagioclase present-absent transition zone and coexistence of Na-pyroxene, plagioclase and quartz. The *heavy dashed line* is the approximate high-pressure boundary for plagioclase in mafic rocks. On the high-T side of reaction (71) and at the high-P side of the *dashed line*=field of eclogites. $a_{H_2O}=1$; $a_{Pg}=0.9$ in reaction (78), other activities in reaction (78) correspond to those of reaction (70), $a_{Prp}=0.3$ and $a_{Jd}=0.4$ in reaction (71)

towards higher pressures where it is grading into the eclogite field with the diagnostic omphacite + garnet assemblage. The sodium present in mafic igneous rocks is stored at low grade in albite feldspar, in the blueschist facies in alkali amphibole (glaucophane, crossite) and paragonite, and in the eclogite facies in sodic pyroxene (omphacite, jadeite). Reaction (71) has a negative slope on a P–T diagram (Figs. 9.9 and 9.10) and terminates at point B towards lower pressure where it becomes metastable relative to feldspar-involving reactions. The negative slope of reaction (71) means that the most favorable conditions for the development of extensive blueschist facies terrains are met in tectonic settings of very rapid subduction of cold (old) oceanic crust.

Inside the blueschist facies field, some additional reactions may modify the assemblages. Most important is reaction (67). It terminates the lawsonite + glaucophane assemblage that is represented by chemography #16 and garnet shows up in blueschists towards higher pressure (chemography #17). Another important garnet producer in the blueschist facies is reaction (68). Reaction (69) generates the relatively rare glaucophane + kyanite assemblage that is diagnostic for very low-T high-P glaucophane schists.

The continuously running reactions in the blueschist facies field result in a systematic, also gradual, change in the mineralogical and modal composition of metabasaltic rocks during prograde metamorphism. Let us follow a metabasaltic rock with a subgreenschist facies assemblage that undergoes subduction at a rate intermediate to the slow and fast subduction geotherms in Fig. 9.9. Such a geotherm will cover a temperature interval in the blueschist facies field from 250–500 °C and a pressure range from 7–16 kbar. At pressures between 6–8 kbar, lawsonite + alkali amphibole will begin to replace the low-grade chlorite + actinolite + albite assemblage. The alkali amphibole is initially fairly Fe-rich, containing high proportions of ribeckite component. As pressure further increases, albite gradually disappears and amphibole becomes increasingly sodic and depleted in actinolite component. Sodium gradually replaces calcium on the M4-position of the amphibole structure by an exchange mechanism that can be formulated as the glaucophane exchange component: $Na^{M4}AlCa^{M4}_{-1}Mg_{-1}$. The exchange leaves the A-site of the amphibole structure untouched. The continuous reactions produced the lawsonite + crossite + chlorite + paragonite ± phengite assemblage that is most characteristic for pressures around 10 kbar. At still higher pressures, lawsonite gradually disappears, alkali amphibole becomes increasingly enriched in glaucophane component, zoisite and epidote become important members of the assemblages, and chlorite continuously decreases in modal amount. The characteristic assemblage for this intermediate blueschist facies stage is: glaucophane + epidote + chlorite + paragonite ± phengite. Typical pressures for that assemblage are in the range 12–14 kbar (along the geotherm we are presently looking at). With increasing pressure, garnet becomes a member of the assemblage. Chlorite further decreases in modal amount giving rise to increasing amounts of garnet. High-pressure blueschists most typically contain: glaucophane + epidote (clinozoisite, zoisite) + garnet + paragonite ± phengite. Other minerals that might be present in high-P low-T blueschists include Mg-chloritoid, talc and kyanite. For a discussion of systematic compositional changes in the only important potassic mineral in blueschists, i.e., phengite, with increasing pressure you should consult Chapter 7.8. Also note that blueschists often contain carbonate minerals that give rise to mixed volatile reactions that are not discussed here. Carbonates found in blueschists include calcite and its high-pressure polymorph aragonite, ankerite, dolomite, breunerite and magnesite. The origin of carbonates in blueschists is related to either ocean-floor metamorphism of basalt and metasomatism associated with it or they are derived from carbonate sediment deposited concurrently with the extrusion of pillow basalt lavas on the ocean floor.

The blueschist facies assemblages are gradually replaced with the typical eclogite facies assemblage omphacite + garnet at pressures greater than about 14–16 kbar. Because high-pressure blueschists typically contain garnet at these conditions, it is the appearance of omphacite (sodic pyroxene) that marks the transition to the eclogite facies. The boundary is rather gradual and there is a wide pressure range where glaucophane + paragonite + epidote + garnet + omphacite may occur in a stable association. Strictly speaking, this assemblage is diagnostic for the eclogite facies. Along an intermediate high-P low-T geotherm, typical eclogite assemblages are formed at about 16–18 kbar and a corresponding temperature of about 500 °C.

9.8
Eclogite Facies Metamorphism

9.8.1
Eclogites

Eclogites are metamorphic mafic rocks that contain the stable mineral pair **garnet and omphacite** in significant modal amounts. Eclogites are free of plagioclase. Eclogites are very dense rocks with a density even greater than some ultramafic mantle rocks ($\varrho_{\text{eclogite}} > 3300 \text{ kg m}^{-3}$). Because of the high density of eclogites, their origin can be readily related to very high-pressure conditions during formation. Omphacite + garnet is the diagnostic assemblage in metabasaltic rocks that recrystallized above about 12–14 kbar pressure outside the stability field of plagioclase. Typical eclogites form at pressures of 18–22 kbar and higher. Many of the eclogites found in collisional mountain belts such as the Alps or the Scandinavian Caledonides still display clear compositional and structural evidence for being derived from basaltic lavas. In the Zermatt region of the Central Alps, for example, preserved MORB compositions and pillow lava structures may be found in eclogites that formed at pressures greater than 20 kbar corresponding to a subduction depth of 60 km. This, in turn, reveals compelling evidence that surface rocks (basaltic lava) can be transported by active tectonic processes to depths of 60 km below the surface (or in the case of coesite-bearing eclogites to > 100 km).

The P–T field of the eclogite facies shown in Fig. 9.9 can be accessed in prograde metamorphism from all three neighboring facies fields, blueschist, amphibolite and granulite facies, depending on the tectonic setting. In subduction tectonics, eclogites originate from blueschists, as outlined above. These types of subduction-related eclogites often register very high pressures of formation.

In a setting of continent-continent collision, the continental crust often attains twice its normal thickness and the deeper parts of the double-crust are exposed to pressures in the range 12–24 kbar. Any mafic rocks of suitable composition in the deeper part of the double-crust could be transformed to eclogite (Fig. 9.9). An example of such a setting is the collision of the Eurasian and the Indian plate and the resulting double-crust formation of the Tibetan plateau. A Ky-type geotherm, typical of collision tectonics, is shown in Fig. 9.9. It can be seen that eclogites in such a setting will be created from amphibolites. Eclogites that form in such a setting of crustal stacking and nappe formation do not record extremely high pressures of formation. Pressures in the range 14–18 kbar are most characteristic.

If collision is accompanied by substantial magmatic heat transfer to the crust, eclogites may form from granulites and the "hot" geotherm shown in Fig. 9.9 may be more appropriate in this situation. Mantle-generated basaltic magmas often sample pieces of rocks along the conduits during ascent to the surface. The rock fragments typically found as xenolith in lavas of basalt volcanoes include ultramafic rocks from the mantle, granulites and eclogites.

The association is characteristic of abnormally hot geotherms in conjunction with collision and magmatism, initial stages of extension or magmatism alone. In such settings, eclogites may have been created from mafic granulites or directly crystallized from basaltic magma at great depth. The pressure range of high-temperature eclogites is very large; however, typical pressures are in the same range as for the eclogites formed from amphibolites. Basaltic magmas may, on the other hand, also collect some eclogite xenoliths from great depth in the mantle and transport them to the surface. The ultimate origin of these deep eclogite samples is ambiguous, however. They also may represent pieces of recycled former oceanic crust.

From the comments above it follows that eclogites may form in a wide range of tectonic settings and the eclogite facies comprises the widest region of temperature and pressure conditions of any of the metamorphic facies fields. The temperature ranges from about 400 to 1000 °C. It may be reasonable to distinguish three general types of eclogites depending on geological setting and temperature of formation:

1. Low-temperature eclogites are related to subduction tectonics and form from blueschists (high-pressure low-temperature eclogites = **LT eclogites**).
2. Intermediate-temperature eclogites form in continent collision settings from amphibolites (medium-temperature eclogites = **MT eclogites**).
3. High-temperature eclogites form in settings of collision or extension where the geotherm is abnormally "hot" due to magmatic heat transfer from the mantle (e. g., basalt magma underplating), from mafic granulites or crystallize directly as eclogites from mafic magma (high-temperature eclogites = **HT eclogites**).

The three different genetic types of eclogite are also characterized by typical mineral associations because of dramatically different temperatures of formation and because of widely varying H_2O pressures associated with their formation. While the LT eclogites that formed in subduction zones often contain modally large amounts of hydrate minerals, HT eclogites often hold "dry" assemblages.

In dealing with high-pressure rocks, important issues also include the kind of mechanism by which they return to the surface, what kind of detailed P–T path they follow on their way to the surface, and, most importantly, what kind of modifications the rocks experience as they are returned to the surface. For example, if an eclogite formed by a subduction mechanism and equilibrated at 650 °C and 20 kbar, it is crucial for the fate of the eclogite assemblages whether the rocks reach the surface via the blueschist field (cooling and decompression occur simultaneously), via the amphibolite and greenschist fields (first decompression then cooling), or via the granulite field (first heating and decompression then cooling). Most eclogites display some mineralogical impact of reactions along the return path to the surface. The access of H_2O along that path is, of course, essential whether or not the high-P assemblages are conserved or extensive retrogression occurs. On the other hand, as will be shown below, LT eclogites often contain abundant hydrate minerals at peak pressure. Such rocks will undergo extensive dehydration reactions that will eventually eliminate the eclogite assemblage if decompression occurs without concurrent cooling.

9.8.2
Reactions and Assemblages

Reactions that connect the blueschist facies to the **low-temperature eclogite facies** were discussed in Section 9.7.2. The most typical assemblages in LT eclogites involve the following minerals (bold = compulsory minerals; ± = optional minerals): **garnet + omphacite** ± zoisite ± chloritoid ± phengite ± paragonite ± glaucophane ± quartz ± kyanite ± talc ± rutile ± dolomite.

Omphacite is a sodic high-pressure clinopyroxene that is composed of the major phase components jadeite, acmite, diopside and hedenbergite (a solid solution of: $NaAlSi_2O_6$–$NaFe^{3+}Si_2O_6$–$CaMgSi_2O_6$–$CaFeSi_2O_6$). The M2 site of the pyroxene structure is filled with about 50% Na and 50% Ca in typical omphacite. LT eclogites often contain pure jadeite together with omphacite. Omphacite contains commonly minor amounts of chromium giving the mineral a grass-green color. This is particularly the case in metagabbros. Note that Cr-diopside in garnet-peridotite also shows a characteristic grass-green color. However, because ultramafic rocks contain very little sodium, the jadeite component in Cpx of high-pressure ultramafic rocks is generally low.

Garnet in eclogite is a solid solution of the major phase components almandine, pyrope and grossular (a solid solution of: $Fe_3Al_2Si_3O_{12}$–$Mg_3Al_2Si_3O_{12}$–$Ca_3Al_2Si_3O_{12}$). However, due to its extreme refractory nature, garnet in eclogite can be inherited from earlier stages of metamorphism or be relics from the protolith (e.g., garnet-granulite).

Chloritoid in LT eclogites is very magnesium-rich and X_{Mg} may be as high as 0.5. Chloritoid and paragonite are diagnostic minerals in LT eclogites. The two minerals are removed by various reactions from eclogites at higher temperatures and are typically absent in MT- and HT eclogites.

Phengite and other sheet silicates in LT eclogites derived from gabbros also often contain small amounts of chromium giving them a green color (Cr-phengite = fuchsite).

9.8.2.1
Amphibolite and Granulite to Eclogite Facies Transition

Amphibolites and granulites contain plagioclase feldspar that is rich in anorthite component. Reaching the eclogite facies from the amphibolite facies or the granulite facies, the reactions that destroy feldspar at high pressures must also decompose the anorthite component of plagioclase.

Reaction (72) produces Ca-tschermak component that is taken up by omphacite. The equilibrium of this reaction is nearly independent of temperature and it can be used for reliable pressure estimates. Reaction (73) replaces anorthite by zoisite and kyanite. This reaction is the most important anorthite-consuming reaction in high-pressure metamorphism of both MT and

HT eclogites. The reaction is a hydration reaction that consumes H_2O as well as anorthite component. In the absence of water, anorthite will not decompose to zoisite. However, if water is not available as solvent, then reaction kinetics are so slow and transport distances are so small that albite and other low-pressure phases or phase components will also survive the high pressures. For example, in an olivine gabbro from the Alps (Allalin gabbro, see below), the primary igneous minerals survive pressures in excess of 20 kbar at 650–700 °C without forming eclogite except where water had access to the rocks. Another example is from the Bergen Arcs (Norwegian Caledonides), a basement nappe complex that experienced P ~20 kbar and T ~700 °C during Caledonian metamorphism. It was found that the mineralogy of Precambrian mafic granulite is perfectly preserved over large areas except along an anastomosing network of shears where the access of water permitted the formation of the stable rock under the Caledonian metamorphic conditions, i.e., MT eclogite.

The equilibrium conditions of reaction (73) are shown in Fig. 9.10. It runs roughly parallel to the albite breakdown reaction (70). Reaction (74) decomposes anorthite component and produces grossular component in eclogite garnet together with kyanite and quartz (Fig. 9.10). The slope of the reaction is similar to those of reactions (70) and (73).

Reaction (70) was used in Chapter 3 to illustrate some general principles of chemical reactions in rocks. In high-pressure mafic rocks it is the most important of all reactions. It limits a plagioclase-present from a plagioclase-absent region of the P–T space (Fig. 9.10). If all three minerals are present (quartz, plagioclase and sodic pyroxene), the equilibrium conditions of reaction (70) can be used for pressure estimates (if the temperature can be estimated from a thermometer, e.g., Fe–Mg distribution between garnet and omphacite). If quartz is not present, a maximum pressure can be estimated from the pyroxene-feldspar pair. If feldspar is not present (the normal case in eclogites), the omphacite + quartz assemblage can be used to estimate a minimum pressure for eclogite formation by assuming unit activity for the phase component albite in the equilibrium constant expression for reaction (70). Other important reactions of the amphibolite (or granulite) to eclogite transition zone must consume calcic amphibole or pyroxene. An example is reaction (75) that destroys tremolite and anorthite component (hornblende + plagioclase = amphibolite) and produces garnet + pyroxene. Reactions (76)–(78) are examples of anorthite-consuming reactions that are relevant for the transition from the granulite to the HT eclogite facies.

As a result of the reactions that produce **medium-temperature eclogites (from amphibolites)** the typical mineralogy found in meta-MORB is: **garnet + omphacite ± zoisite** (clinozoisite) ± phengite ± kyanite ± amphibole (Na-Ca amphibole) ± quartz (at very high pressure coesite) ± rutile.

High-temperature eclogites (from granulites) contain typically: **garnet + omphacite ± kyanite ± orthopyroxene ± amphibole ± quartz (at very high pressure coesite) ± rutile.**

9.8.2.2
Reactions in Eclogites

Prograde metamorphic mineral reactions that affect eclogites are basically ordinary dehydration reactions. The reactions transform assemblages involving
hydrous minerals of LT eclogites and substitute them with less hydrous assemblages of MT eclogites and ultimately HT eclogites. Examples are the
paragonite breakdown reaction (79), the zoisite breakdown reaction (80), and
the replacement of talc + kyanite by orthopyroxene [reaction (81)]. Other reactions gradually remove chlorite from quartz-free MT eclogites, some of
them producing Opx. Reaction (80) replaces zoisite and garnet and produces
kyanite + Cpx. This continuous reaction relates MT to HT eclogites. The reaction ultimately leads to the change from chemography #18 to #19 shown in
Fig. 9.3.

Other important reactions in HT eclogites are reactions (82) and (83). The
last reaction is important in the relatively rare Opx-bearing eclogites that
occur, e.g., in the Western Gneiss region of the Scandinavian Caledonides.
Phengite is the only K-bearing phase in LT eclogites. At higher temperatures,
the mica continuously decomposes under production of K-amphibole component. The K-bearing Na–Ca-amphiboles found in high-grade eclogites are
very stable and remain the prime K-carrier in HT eclogites.

9.8.3
Eclogite Facies in Nonbasaltic Mafic Rocks

Mafic or basic rocks cover a very wide range in bulk composition. Typical
eclogites with garnet and omphacite as the dominant minerals usually develop from basaltic protoliths. The above discussion of eclogites refers to
MORB-type mafic rocks. However, many gabbros are compositionally dissim-
ilar to MORB. In particular, gabbros are commonly (but not exclusively; e.g.,
ferro-gabbros) much more Mg-rich compared with MORB. It can be expected
that high-pressure equivalents of troctolites and olivine-gabbros may develop
mineral assemblages that are quite different from normal eclogite facies garnet + omphacite rocks. In fact, Mg-rich olivine gabbro metamorphosed at
25 kbar and 650 °C may not contain garnet + omphacite at all as a peak assemblage. The low X_{Fe} of the protolith prevents the formation of garnet. An
example of LT eclogite metamorphism of an olivine gabbro from the Alps
will be presented in Section 9.8.3.1. Such mafic rocks were metamorphosed
under eclogite facies conditions; however, they are not eclogites. We also recommend avoiding confusing rock names like pelitic eclogites or pelitic
blueschists. The proper expressions for such kind of rocks are metapelites in
blueschist facies or pelitic gneisses in eclogite facies (see Chap. 7.7).

Anorthosites, also basic rocks, cannot be converted into eclogites by high-
pressure metamorphism because the rocks contain more than 90–95% plagioclase. Plagioclase of the composition An_{70} will be converted into a jadeite
+ zoisite + kyanite + quartz assemblage in the LT- and MT eclogite facies.

9.8.3.1
Blueschist and Eclogite Facies Metamorphism of an Olivine Gabbro:
a Case History from the Central Alps

The 2-km gabbro lens forms the Allalinhorn between Zermatt and Saas and is referred to as the Allalin gabbro in the literature (Bearth 1967; Chinner and Dixon 1973; Meyer 1983 a,b). The gabbro belongs to a typical ophiolite complex that once formed the oceanic lithosphere of the Mesozoic Tethys ocean. The ophiolites make up a part of the South Pennine nappes of the Alps. During early Alpine orogeny, the ophiolites were subducted to great depth and metamorphosed under LT eclogite facies conditions. Pillow basalts were turned first into ordinary LT eclogites, later retrogressed to blueschists, and finally were overprinted by the Tertiary metamorphism that has reached the greenschist facies in the Zermatt area. Abundant eclogite garnet represents the only relic mineral from the eclogite stage in locally strongly retrogressed eclogites. The garnet-bearing greenschists of the Zermatt-Saas ophiolites demonstrate the very refractory nature of garnet.

The Allalin gabbro is an Mg-rich olivine gabbro with irregular layers of troctolite. The succession of minerals that formed in the gabbro through its entire P–T history is given in Fig. 9.11. The data shown in Fig. 9.11 are based on Meyer (1983 a). The P–T diagram in the upper left corner of Fig. 9.11 shows the path followed by the gabbro from the magmatic crystallization, via the LT eclogite facies stage to the present-day erosion surface. The metamorphic history of the gabbro can be subdivided into various stages that will be briefly explained. (A) Crystallization of the igneous assemblage. In undeformed domains within the gabbro lens, where water never had access to the rock, igneous minerals are often well preserved. (B) Onset of subduction of the partly cooled gabbro (or of the completely cooled gabbro that has been partly altered by oceanic metamorphism: Barnicoat 1995; Barnicoat and Cartwright 1997). Associated with stage B is the formation of typical coronites. They form because olivine + plagioclase becomes an incompatible assemblage [Chap. 5, reaction (15)]. The up to 4-cm-large olivine megacrysts react with matrix plagioclase. The reaction products (orthopyroxene and garnet) form concentric monomineralic shells around the primary olivine with garnet forming the outer shell toward the matrix. This remains the only garnet that ever formed in the LT eclogite facies gabbro. The Opx-Grt coronas were repeatedly modified afterwards by the progress of metamorphism. The most characteristic mineral of this stage is jadeite. It forms from albite component. It is found in matrix plagioclase only as extremely small grains along fractures and cracks that provided access for water to the igneous plagioclase grains. Note that there is no quartz in the rock at any stage. Quartz that is produced from albite breakdown [reaction (70)] dissolved in the aqueous metamorphic fluid. It was transported to the reaction sites in the nearby corona structures where it was used up by olivine-consuming reactions. (C) During the first stage of the eclogite facies, the characteristic assemblage of LT eclogites form; Cr-omphacite pseudomorphs augite, the assemblage jadeite + zoisite + kyanite replaces plagioclase; kyanite + talc + chloritoid pseudo-

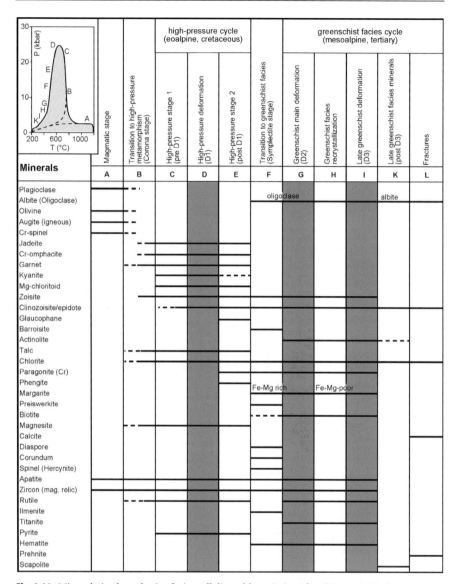

Fig. 9.11. Minerals in the eclogite facies Allalin gabbro, Swiss Alps (Meyer 1983 a)

morph olivine and the coronas. The coarse-grained igneous gabbro structure is often perfectly preserved. Note that Mg-rich chlorite is present in the eclogite-stage assemblage. The 650 °C and 25 kbar LT eclogite facies metagabbro contains a number of hydrate minerals (Chl, Cld, Tlc, Zo) and the high-pressure mineral assemblage owes its very existence to the access of water to the gabbro during the subduction process. (D) Access of water is made possible by deformation D1 (brittle fractures healed with Mg-chloritoid and euhedral omphacite!). Deformation D1 induces, locally, a complete recrystallization of the rock but usually leaves the magmatic gabbro structure intact (microfractures and cracks). (E) The high-pressure stage 2 is characterized by the formation of pure glaucophane, Cr-paragonite and phengite. The assemblage is characteristic of the blueschist facies. Note, however, that the blueschist facies here represents a retrograde overprint on an LT eclogite facies assemblage. Its extent depends, again, on the availability of water and it is irregularly developed. (F) A transitional stage is associated with the return to shallower levels of the lithosphere that is characterized by fine-grained reaction products of local hydration such as margarite, preiswerkite and the Na-Ca-amphibole barroisite; however, spinel, corundum and diaspore are also formed at this stage. The barroisitic amphibole links the blueschist facies glaucophane to the greenschist facies actinolite. Barroisite is typical and widespread in the retro-eclogites of the Zermatt area. The F-stage assemblages are often found in symplectites that pseudomorph and replace eclogite stage minerals. Note also that plagioclase (oligoclase) returned to the rock. (G, H and I) The Tertiary greenschist facies overprint is accompanied by two distinct phases of deformation. The impact of these stages on the Allalin gabbro varies from place to place and ranges from virtually nothing to complete retrogression. The greenschist facies deformation and hydration is often localized along shear zones and deformation is ductile in contrast to the eclogite-stage deformation that has been brittle. The typical assemblage of the greenschist stage is albite + zoisite (epidote) + actinolite + chlorite + biotite. The assemblage is characteristic for conditions around 4–5 kbar and 400–450 °C. (K, L) The minerals of stages of the subgreenschist facies are mainly found in veins (albite veins, calcite, prehnite in veins).

The Allalin gabbro also contains sulfides, oxides and other accessory minerals that formed and were stable at various stages of the P–T history of the rock. This example of a complex reaction history and successive development of minerals and mineral assemblages in the Allalin gabbro shows that:

- Deformation is essential in metamorphic processes. It allows water to enter dry igneous rocks. In the absence of water, igneous and metamorphic assemblages will survive under even extreme P–T conditions (e. g., igneous Pl + Aug + Ol at 25 kbar and 650 °C).
- Mafic rocks that undergo eclogite facies metamorphism do not necessarily become eclogites. The bulk composition of many gabbros may not be favorable for the formation of garnet.
- LT eclogite facies mafic rocks that form by subduction of oceanic lithosphere are characterized by the presence of a wide variety of hydrous minerals at peak-pressure conditions of the eclogite facies (Zo, Chl, Tlc, Cld, Phe).
- LT eclogite facies rocks may pass through the blueschist facies on their return path to the surface. The blueschist facies overprint is then retrograde in nature and adds another selection of hydrous minerals to the high-pressure low-temperature rock (Pg, Gln).
- All stages (A through L) of the complex P–T path followed by the Allalin gabbro and the reaction history involving a large number of minerals can

be preserved in a single rock body of not more than about 2 km in extent. The predominant assemblage that is found in a single piece of rock or within a small area of a given outcrop depends on how much water entered the rock at a given P–T stage.

- The availability of water is, in turn, related to the local extent of deformation. Given a certain stage, if much water is available, the formation of the maximum hydrated assemblage for that stage is possible. Water is consumed by hydration reactions that tend to dry out the rock. Water must be continuously supplied in order to keep the reactions running that produce the maximum hydrated assemblage at that stage. The access for the water to the originally dry rock is provided by micro-fractures and other brittle deformation during high-pressure metamorphism and by pervasive ductile deformation during greenschist facies metamorphism.

References

Abbott RN Jr (1982) A petrogenetic grid for medium and high grade metabasites. Am Mineral 67:865–876

Ahn JH, Peacor DR, Essene EJ (1985) Coexisting paragonite-phengite in blueschist eclogite: a TEM study. Am Mineral 70:1193–1204

Austrheim H, Griffin WL (1985) Shear deformation and eclogite formation within granulite – facies anorthosites of the Bergen Arcs, western Norway. Chem Geol 50:267–281

Banno S (1986) The high pressure metamorphic belts in Japan: a review. Geol Soc Am Mem 164:365–374

Barnicoat AC (1995) High pressure metamorphism and sea-floor alteration: a control on H_2O-release patterns during subduction. Boch Geol Geotechn Arb 44:17–18

Barnicoat AC, Cartwright I (1997) The gabbro-eclogite transformation: an oxygen isotope and petrographic study of west Alpine ophiolites. J Metamorph Geol 15:93–104

Battey MH, McRitchie WD (1975) The petrology of the pyroxene-granulite facies rocks of Jotunheimen, Norway. Norsk Geol Tidsskrift 55:1–49

Bearth P (1963) Chloritoid und Paragonit aus der Ophiolith-Zone von Zermatt-Saas Fee. Schweiz Mineral Petrogr Mitt 43:269–286

Bearth P (1967) Die Ophiolithe der Zone von Zermatt-Saas Fee. Beitr Geol Karte Schweiz NF l32:130 pp

Bégin NJ (1992) Contrasting mineral isograd sequences in metabasites of the Cape Smith Belt, northern Québec, Canada: three new bathograds for mafic rocks. J Metamorph Geol 10:685–704

Bevins RE, Robinson D (1993) Parageneses of Ordovician sub-greenschist to greenschist facies metabasites from Wales, UK. Eur J Mineral 5:925–935

Bohlen SB, Liotta JJ (1986) A barometer for garnet amphibolites and garnet granulites. J Petrol 27:1025–1034

Boles JR, Coombs DS (1975) Mineral reactions in zeolitic Triassic tuff, Hokonui Hills, New Zealand. Geol Soc Am Bull 86:163–173

Brothers RN, Yokoyama K (1982) Comparison of high pressure schists belts of New Caldonia and Sanbagawa, Japan. Contrib Mineral Petrol 79:219–229

Brown EH (1974) Comparison of the mineralogy and phase relations of blueschists from the North Cascades, Washington, and greenschists from Otago, New Zealand. Geol Soc Am Bull 85:333–344

Brown EH (1977) The crossite content of Ca-amphibole as a guide to pressure of metamorphism. J Petrol 18:53–72

Carmichael RS (1989) Practical handbook of physical properties of rocks and minerals. CRC Press, Boca Raton, 834 pp

Chinner GA, Dixon JE (1973) Some high-pressure parageneses of the Allalin gabbro, Valais, Switzerland. J Petrol 14:185–202

Coleman RG, Lee DE, Beatty LB, Brannock WW (1965) Eclogites and eclogites: their differences and similarities. Geol Soc Am Bull 76:483–508

Coombs DS, Ellis AJ, Fyfe WS, Taylor AM (1959) The zeolite facies, with comments on the interpretation of hydrothermal syntheses. Geochim Cosmochim Acta 17:53–107

Cooper AF (1972) Progressive metamorphism of metabasic rocks from the Haast Schist group of southern New Zealand. J Petrol 13:457–492

Cox RA, Indares A (1999) Transformation of Fe-Ti gabbro to coronite, eclogite and amphibolite in the Baie du Nord segment, Maicouagan Imbricate Zone, eastern Grenville Province. J Metamorph Geol 17:537–555

Crawford ML (1966) Composition of plagioclase and associated minerals in some schists from Vermont, USA, and South Westland, New Zealand, with inferences about the peristerite solvus. Contrib Mineral Petrol 13:269–294

Donato MM (1989) Metamorphism of an ophiolitic tectonic melange, northern California Klamath Mountains, USA. J Metamorph Geol 7:515–528

El-Shazly AK (1994) Petrology of lawsonite-, pumpellyite- and sodic amphibole-bearing metabasites from north-east Oman. J Metamorph Geol 12:23–48

Ernst WG (1972) Occurence and mineralogical evolution of blueschist belts with time. Am J Sci 272:657–668

Eskola P (1921) On the eclogites of Norway. Skr Vidensk Selsk Christiania, Mat-Nat Kl I 8:1–118

Evans BW (1990) Phase relations of epidote-blueschists. Lithos 25:3–23

Evans BW, Brown EH (1986) Blueschists and eclogites. Geological Society of America Memoir, The Geological Society of America, Boulder, Colorado, 423 pp

Ferry JM (1984) Phase composition as a measure of reaction progress and an experimental model for the high-temperature metamorphism of mafic igneous rocks. Am Mineral 69:677–691

Fett A, Brey GP (1995) Partitioning of Al, Cr and Fe between garnet, clinopyroxene and rutile in natural eclogites and high pressure experiments. Eur J Mineral 7:64

Forbes RB, Evans BW, Thurston SP (1984) Regional progressive high-pressure metamorphism, Seward peninsula, Alaska. J Metamorph Geol 2:43–54

Frey M, De Capitani C, Liou JG (1991) A new petrogenetic grid for low-grade metabasites. J Metamorph Geol 9:497–509

Frimmel HE, Hartnady CJH (1992) Blue amphiboles and their significance for the history of the Pan-African Gariep belt, Namibia. J Metamorph Geol 10:651–670

Gibson GM (1979) Margarite in kyanite- and corundum-bearing anorthosite, amphibolite, and hornblendite from central Fiordland, New Zealand. Contrib Mineral Petrol 68:171–179

Griffin WL, Austrheim H, Brastad K, Bryhni I, Krill A, Mørk MBE, Qvale H, Tørudbakken B (1983) High-pressure metamorphism in the Scandinavian Caledonides. In: Gee DG, Sturt BA (eds) The caledonide orogen – Scandinavia and related areas. Wiley, Chichester

Harte B, Graham CM (1975) The graphical analysis of greenschist to amphibolite facies mineral assemblages in metabasites. J Petrol 16:347–370

Helms TS, McSween HY Jr, Labotka TC, Jarosewich E (1987) Petrology of a Georgia Blue ridge amphibolite unit with hornblende + gedrite + kyanite + staurolite. Am Mineral 72:1086–1096

Hirajima T, Hiroi Y, Ohta Y (1984) Lawsonite and pumpellyite from the Vestgotabreen Formation in Spitsbergen. Norsk Geol Tidsskrift 4:267–273

Hirajima T, Banno S, Hiroi Y, Ohta Y (1988) Phase petrology of eclogites and related rocks from the Motalafjella high-pressure metamorphic complex in Spitsbergen (Arctic Ocean) and its significance. Lithos 22:75–97

Holland TJB (1979a) Experimental determination of the reaction paragonite=jadeite+kyanite+H_2O, and internally consistent thermodynamic data for part of the system Na_2O–Al_2O_3–SiO_2–H_2O, with applications to eclogites and blueschists. Contrib Mineral Petrol 68:293–301

Holland TJB (1979b) High water activities in the generation of high pressure kyanite eclogites of the Tauern Window, Austria. J Geol 87:1–27

Holland TJB, Richardson SW (1979) Amphibole zonation in metabasites as a guide to the evolution of metamorphic conditions. Contrib Mineral Petrol 70:143–148

Hosotani H, Banno S (1986) Amphibole composition as an indicator of subtle grade variation in epidote-glaucophane schists. J Metamorph Geol 4:23–36

Humphris SE, Thompson G (1978) Hydrothermal alteration of oceanic basalts by seawater. Geochim Cosmochim Acta 42:107–125

James HL (1955) Zones of regional metamorphism in the Precambrian of northern Michigan. Geol Soc Am Bull 66:1455–1488

Johnson CD, Carlson WD (1990) The origin of olivine-plagioclase coronas in metagabbros from the Adirondack Mountains, New York. J Metamorph Geol 8:697–717

Kawachi DE (1975) Pumpellyite-actinolite and contiguous facies metamorphism in the Upper Wakatipu district, southern New Zealand. N Z J Geol Geophys 17:169–208

Klein U, Schumacher JC, Czank M (1996) Mutual exsolution in hornblende and cummingtonit: compositions, lamellar orientations, and exsolution temperatures. Am Mineral 81:928–939

Konzett J, Hoinkes G (1996) Paragonite- hornblende assemblages and their petrological significance: an example from the Austroalpine Schneeberg Complex, Southern Tyrol, Italy. J Metamorph Geol 14:85–101

Leake BE (1964) The chemical distinction between ortho- and para-amphibolite. J Petrol 5:238–254

Liogys VA, Jenkins DM (2000) Hornblende geothermometry of amphibolite layers of the Popple Hill gneiss, north-west Adirondack Lowlands, New York, USA. J Metamorph Geol 18:513–530

Liou JG, Maruyama S, Cho M (1987) Very low-grade metamorphism of volcanic and volcaniclastic rocks – mineral assemblages and mineral facies. In: Frey M (ed) Low temperature metamorphism. Blackie, Glasgow, pp 59–113

López S, Castro A (2001) Determination of the fluid-absent solidus and supersolidus phase relationships of MORB-derived amphibolites in the range of 4–14 kbar. Am Mineral 86:1396–1403

Lucchetti G, Cabella R, Cortesogno L (1990) Pumpellyites and coexisting minerals in different low-grade metamorphic facies of Liguria, Italy. J Metamorph Geol 8:539–550

Markl G, Bucher K (1997) Eclogites from the early Proterozoic Lofoten Island, N Norway. Lithos 42:15–35

Messiga B, Scambelluri M, Piccardo GB (1995) Chloritoid-bearing assemblages in mafic systems and eclogite-facies hydration of alpine Mg-Al metagabbros (Erro-Tobbio Unit, Ligurian Western Alps). Eur J Mineral 7:1149–1167

Messiga B, Kienast JR, Rebay M, Piccardi P, Tribuzio R (1999) Cr-rich magnesiochloritoid eclogites from the Monviso ophiolites (Western Alps, Italy). J Metamorph Geol 17:287–299

Meyer J (1983a) Mineralogie und Petrologie des Allalingabbros. Doctoral dissertation. Doctoral dissertation. Univ Basel, 329 pp

Meyer J (1983b) The development of the high-pressure metamorphism in the Allalin metagabbro (Switzerland). Terra Cognita 3:187

Meyre C, Puschnig AR (1993) High-pressure metamorphism and deformation at Trescolmen, Adula nappe, Central Alps. Schweiz Mineral Petrograph Mitt 73:277–283

Meyre C, De Capitani C, Partsch JH (1997) A ternary solid solution model for omphacite and its application to geothermobarometry of eclogites from the Middle Adula nappe (Central Alps, Switzerland). J Metamorph Geol 15:687–700

Miyashiro A (1979) Metamorphism and metamorphic belts. Allen and Unwin, London, 492 pp

Mongkoltip P, Ashworth JR (1986) Amphibolitization of metagabbros in the Scottish Highlands. J Metamorph Geol 4:261–283

Neuhoff PS, Fridriksson T, Arnórsson S, Bird DK (1999) Porosity evolution and mineral paragenesis during low-grade metamorphism of basaltic lavas at Teigarhorn, eastern Iceland. Am J Sci 299:467–501

Newton RC (1986) Metamorphic temperatures and pressures of group B and C eclogites. Geol Soc Am Memoir 164:17–30

O'Brien PJ (1993) Partially retrograded eclogites of the Münchberg Massif, Germany: records of a multistage Variscan uplift history in the Bohemian Massif. J Metamorph Geol 11:241–260

Okay AI (1994) Sapphirine and Ti-clinohumite in ultra-high-pressure garnet-pyroxenite and eclogite from Dabie Shan, China. Contrib Mineral Petrol 116:145–155

Patrick BE, Day HW (1989) Controls on the first appearance of jadeitic pyroxene, northern Diablo Range, California. J Metamorph Geol 7:629–639

Pattison DR (1991) Infiltration-driven dehydration and anatexis in granulite facies metagabbro, Grenville Province, Ontario, Canada. J Metamorph Geol 9:315–332

Platt JP (1993) Exhumation of high-pressure rocks: a review of concepts and processes. Terra Nova 5:119–133

Pognante U, Kienast J-R (1987) Blueschist and eclogite transformations in Fe-Ti gabbros: a case from the western Alps ophiolites. J Petrol 28:271–292

Powell WG, Carmichael DM, Hodgson CJ (1993) Thermobarometry in a sub-greenschist to greenschist transition in metabasites of the Abitibi greenstone belt, Superior Province, Canada. J Metamorph Geol 11:165–178

Russ-Nabelek C (1989) Isochemical contact metamorphism of mafic schist, Laramie Anorthosite Complex, Wyoming: amphibole compositions and reactions. Am Mineral 74:530–548

Sanders LS (1988) Plagioclase breakdown and regeneration reactions in Grenville kyanite eclogite at Glenelg, NW Scotland. Contrib Mineral Petrol 98:33–39

Seki Y, Oki Y, Matsuda T, Mikami K, Okumura K (1969) Metamorphism in the Tanzawa Mountains, central Japan. J Jpn Assoc Mineral Petrol Econ Geol 61:1–29, 50–75

Selverstone J, Spear FS (1985) Metamorphic P–T paths from pelitic schists and greenstones from the south-west Tauern Window, eastern Alps. J Metamorph Geol 3:439–465

Smelik EA, Veblen DR (1989) A five-amphibole assemblage from blueschists in northern Vermont. Am Mineral 74:960–964

Smith DC, Lappin MA (1989) Coesite in the Straumen kyanite-eclogite pod, Norway. Terra Nova 1:47–56

Soto JI (1993) PTMAFIC: software for thermobarometry and activity calculations with mafic and ultramafic assemblages. Am Mineral 78:840–844

Spear FS (1980) NaSi«CaAl exchange equilibrium between plagioclase and amphibole. An empirical model. Contrib Mineral Petrol 72:33–41

Springer RK, Day HW, Beiersdorfer RE (1992) Prehnite-pumpellyite to greenschist facies transition, Smartville Complex, near Auburn, California. J Metamorph Geol 10:147–170

Starkey RJ, Frost BR (1990) Low-grade metamorphism of the Karmutsen Volcanics, Vancover Island, British Columbia. J Petrol 31:167–195

Terabayshi M (1993) Compositional evolution in Ca-amphibole in the Karmutsen metabasites, Vancouver Island, British Columbia, Canada. J Metamorph Geol 11:677–690

Thieblemont D, Triboulet C, Godard G (1988) Mineralogy, petrology and P-T-t path of Ca-Na amphibole assemblages, Saint-Martin des Noyers formation, Vendee, France. J Metamorph Geol 6:697–716

Thompson AB (1971) P_{CO_2} in low grade metamorphism: zeolite, carbonate, clay mineral, prehnite relations in the system $CaO-Al_2O_3-CO_2-H_2O$. Contrib Mineral Petrol 33:145–161

Thompson JB (1991) Modal space: applications to ultramafic and mafic rocks. Can Mineral 29:615–632

Triboulet C, Thiéblemont D, Audren C (1992) The (Na-Ca) amphibole-albite-chlorite-epidote-quartz geothermobarometer in the system S–A–F–M–C–N–H_2O. 2. Applications to metabasic rocks in different metamorphic settings. J Metamorph Geol 10:557–566

Wenk E, Keller F (1969) Isograde in Amphibolitserien der Zentralalpen. Schweiz Mineral Petrograph Mitt 49:157–198

Wolke C, Truckenbrodt J, Johannes W (1995) Beginning of dehydration-melting in amphibolites at P = 10 kbar. Eur J Mineral 7:273

Zen E-A (1961) The zeolite facies: an interpretation. Am J Sci 259:401–409

CHAPTER 10

Metamorphism of Granitoid Rocks

10.1
Introduction

Granitoid rocks comprise granite, alkali-feldspar granite, granodiorite and tonalite. They constitute the largest portion of the continental crust. Granitoid gneisses, commonly migmatites, are in fact the dominant rock type of the continents. Because the main constituents – alkali-feldspar, plagioclase, quartz, biotite, muscovite, hornblende – are found over a wide range of P–T conditions, this rock group is not a very useful indicator of metamorphic grade and is therefore largely neglected in textbooks on metamorphic petrology. Unlike wet sedimentary rocks, granitoid rocks will enter the metamorphic realm in a predominantly dry state. In order to start metamorphic reactions some hydration is necessary. The access of a water-rich fluid will be facilitated by tectonic activity. Also, in the absence of penetrative deformation, granitoid rocks retain remarkably well their original igneous structures.

In the following, a few selected mineralogical features are mentioned which are useful to determine metamorphic grade, but no attempt is made to treat the progressive metamorphism of granitoid rocks.

10.2
Presence of Prehnite and Pumpellyite

Although prehnite and pumpellyite are commonly known from very low-grade mafic igneous rocks and graywackes, these Ca-Al silicates are sometimes also found in quartzo-feldspathic plutonic rocks (for references, see Tulloch 1979; see also AlDahan 1989). Both minerals typically occur within biotite, secondary chlorite, and hornblende, often forming lenticular or bulbous bodies in them. In the case of the biotite-prehnite association, most authors conclude that the biotite cleavage simply forms a suitable structural site for prehnite crystallizing from a fluid, but prehnite actually replacing its biotite host has also been observed, according to the possible reaction: biotite + anorthite (component in plagioclase) + H_2O = prehnite + chlorite + K-feldspar + titanite + muscovite (Tulloch 1979). The presence of prehnite and pumpellyite in granitoid rocks was earlier attributed to the activity of hydrothermal solutions, but alteration under prehnite-pumpellyite facies conditions is now commonly proposed if signs of low-temperature metamorphism are also observed in nearby other lithologies.

10.3
Presence of Stilpnomelane

Stilpnomelane is a Fe-rich hydrous silicate with low Al and K content which may be mistaken for biotite. The names ferrostilpnomelane and ferristilpnomelane are used respectively for Fe^{2+}- and Fe^{3+}-rich varieties, the latter being formed by secondary oxidation of the former. In metagranitoids, stilpnomelane generally has a sheaf-like habit and is preferentially found within microcline or it is replacing igneous biotite and hornblende. Stilpnomelane-forming reactions in low-grade metagranitoids are poorly understood, but biotite and/or chlorite are considered to be the most important reactants. Stilpnomelane occurs as a mineral of low-grade metamorphism along with chlorite, phengite, epidote, quartz, microcline and albite, but at slightly higher grade also together with green biotite. A possible stilpnomelane-consuming reaction in low-grade metagranitoids is:

$$Stp + Phe = Bt + Chl + Qtz + H_2O$$

This reaction was provisionally investigated by Nitsch (1970). In hydrothermal experiments, oxygen fugacity was controlled by the hematite-magnetite buffer, and apparent equilibrium was obtained at 4 kbar/445 °C and at 7 kbar/460 °C. Summing up, stilpnomelane is a diagnostic index mineral for metagranitoid rocks formed under lower greenschist and blueschist facies conditions.

10.4
The Microcline-Sanidine Transformation Isograd

A discontinuity in the structural state of K-feldspar allows for the mapping of a microcline/sanidine transformation isograd in granitoid rocks of the greenschist facies (Bambauer and Bernotat 1982; Bernotat and Bambauer 1982; Bernotat and Morteani 1982; Bambauer 1984).

To facilitate the understanding of the following sections, a few terms on the mineralogy of K-feldspars are mentioned here. In the stable high-temperature form sanidine, Al and Si are distributed between the two available tetrahedral (T) sites, T_1 and T_2, with Al slightly preferring the T_1 site. In the low-temperature form microcline, there are four T-sites, $T_1 0$, $T_1 m$, $T_2 0$, $T_2 m$, with Al strongly preferring the $T_1 0$ site. Because intermediate structural states exist, the following K-feldspar modifications are distinguished with increasing temperature: low microcline, high microcline, low sanidine, and high sanidine. For low and high microcline, the Al, Si distributions are given as Al site occupancies $t_1 0$, $t_1 m$, $t_2 0$, and $t_2 m$. The Al,Si ordering in T_1 tetrahedral sites is either determined by measuring the optic axial angle, $2V_x$, or by X-ray methods.

The most detailed study on the microcline/sanidine transformation isograd has been performed in the Central Swiss Alps. This area will, therefore, serve as a case study here. The structural state of K-feldspar occurring in granitoid rocks allows three zones to be distinguished, arranged with respect to increasing Alpine metamorphic grade:

Zone 1 comprises unchanged, variable pre-Alpine structural states ranging from low to high microcline, showing a range of $2V_x$ ~55–85°. Metamorphic grade is restricted to the lowest greenschist facies.

Zone 2 comprises distinctly cross-hatched low microcline ($2V_x$ ~75–88°, $t_1 0$–$t_1 m$ ~1.0, 1–4 mol% Ab). Metamorphic grade covers lower greenschist facies conditions.

Zone 3 comprises frequent orthoclase (monoclinic bulk optics predominating, triclinic or monoclinic by X-ray) displaying a variety of high microcline structural states ($2V_x$ ~53–75°, $t_1 0$–$t_1 m = 0.0$–0.4, 2.5–6.5 mol% Ab) together with variable amounts of cross-hatched low microcline ($t_1 0$–$t_1 m$ ~1.0). Metamorphic grade ranges from high greenschist facies to high amphibolite facies conditions.

The microcline/sanidine transformation isograd separating zone 2 from zone 3 is interpreted as a relic of the diffusive transformation sanidine $\Rightarrow$ microcline at the climax of the late-Alpine metamorphism in the following way. K-feldspar from zone 2 was annealed long enough at low temperature during Alpine metamorphism to form low microcline. In zone 3, temperatures surpassed T_{dif} upon slow heating and K-feldspar was transformed to the high-temperature polymorph sanidine. During subsequent cooling, T dropped below the temperature of the phase transformation. Note that during the retrograde sanidine $\Rightarrow$ microcline transition, triclinic domains must form within a monoclinic host; this process has a high kinetic barrier. Because of rapid cooling, diffusion was slowed down and the K-feldspar remained as intermediate to high microcline in the rocks. The low microcline observed at the beginning of zone 3 is interpreted as a relic of zone 2.

The temperature of the microcline-sanidine phase transformation has not been determined experimentally because of the very slow kinetics of Al,Si ordering. From several lines of evidence, discussed in detail by Bambauer and Bernotat (1982), this temperature was estimated to be ~450 °C for an approximate composition $Or_{95-90}Ab_{5-10}$, with a negligible pressure dependence.

In the Central Swiss Alps, the microcline-sanidine transformation isograd was mapped over a distance of more than 140 km. It fits well into the general pattern of metamorphic zonation of this area, and is found between the stilpnomelane-out and staurolite-in mineral zone boundaries. Furthermore, provisional results indicate that the transformation isograd surface is inclined ~15° towards the north.

10.5
Granitoid Rocks in the Eclogite Facies and the Blueschist Facies

A rare example of granite, granodiorite and tonalite metamorphosed under eclogite facies conditions is known from the Sesia Zone of the Western Alps (Compagnoni and Maffeo 1973; Oberhänsli et al. 1985). These quartzo-feldspathic rocks are remnants of the Variscan basement and were transformed during the earliest phase of Alpine metamorphism in mid-Cretaceous time.

The least deformed rocks retain well-preserved igneous textures with relict K-feldspar and biotite. During a first stage, these remarkable rocks were converted to the assemblage: quartz + jadeite + zoisite + garnet + (3T) phengite + K-feldspar. Jadeite, quartz and zoisite have completely replaced plagioclase, whilst biotite has been partly transformed into coronitic garnet, phengite and rutile. During the later decompression stages, jadeite was replaced by omphacite and chloromelanite, followed by glaucophane. The P–T conditions for the climax of metamorphic recrystallization are estimated as > 14 kbar and 500–600 °C.

Orthogneisses of granite, granodiorite and tonalite composition from the Seward Peninsula, Alaska, underwent progressive orogenic blueschist facies metamorphism in the Jurassic, followed by partial re-equilibration during decompression under greenschist facies conditions (Evans and Patrick 1987). The mineral assemblage in granitic orthogneiss is quartz-albite-microcline + (3T) phengite + biotite; almandine-rich garnet occurs in metatonalite in the absence of microcline. No typical high-pressure minerals such as jadeite, omphacite, or glaucophane, or obvious pseudomorphs thereof, were found in the orthogneisses. Diagnostic blueschist facies assemblages occur, however, in adjacent metabasites and metapelites, and P–T conditions were approximately 9–10 kbar/420 °C (Thurston 1985).

The Variscan granitic basement of Corsica was partly involved in early Alpine metamorphism during the middle-to-late Cretaceous (Gibbons and Horak 1984). Metamorphism of this basement, induced by the overthrusting of a blueschist facies nappe, was confined to a major, ca. 1-km-thick ductile shear zone within which deformation increases upwards towards the overlying nappe. Within a mylonitized biotite-hornblende granodiorite, the mafic minerals are replaced by crossite + low-Ti biotite + phengite + titanite ± epidote, recording metamorphic conditions transitional between blueschist and greenschist facies. P–T estimates for the metamorphism at the base of the shear zone are given as 390–490 °C at 6–9 kbar, but these figures are not well constrained.

References

AlDahan AA (1989) The paragenesis of pumpellyite in granitic rocks from the Siljan area, central Sweden. Neues Jahrb Mineral Monatsh 367–383
Ashworth JR (1985) Migmatites. Blackie, Glasgow, 301 pp
Bambauer HU (1984) Das Einfallen der Mikroklin/Sanidin-Isogradenfläche in den Schweizer Zentral-Alpen. Schweiz Mineral Petrogr Mitt 64:288–289
Bambauer HU, Bernotat WH (1982) The microcline/sanidine transformation isograd in metamorphic regions. I. Composition and structural state of alkali feldspars from granitoid rocks of two N–S traverses across the Aar massif and Gotthard "massif", Swiss Alps. Schweiz Mineral Petrogr Mitt 62:185–230
Bea F (1989) A method for modelling mass balance in partial melting and anatectic leucosome segregation. J Metamorph Geol 7:619–628
Bernotat WH, Bambauer HU (1982) The microcline/sanidine transformation isograd in metamorphic regions. II. The region of Lepontine metamorphism, central Swiss Alps. Schweiz Mineral Petrogr Mitt 62:231–244
Bernotat WH, Morteani G (1982) The microcline/sanidine transformation isograd in metamorphic regions: Western Tauern window and Merano-Mules-Anterselva complex (eastern Alps). Am Mineral 67:43–53

Bohlen SR, Boettcher AL, Wall, VJ, Clemens JD (1983) Stability of phlogopite-quartz and sanidine-quartz: a model for melting in the lower crust. Contrib Mineral Petrol 83:270–277

Compagnoni R, Maffeo B (1973) Jadeite-bearing metagranites l.s., related rocks in the Mount Mucrone area (Sesia-Lanzo zone, western Italian Alps). Schweiz Mineral Petrogr Mitt 53:355–378

Engel AEJ, Engel CE (1958) Progressive metamorphism and granitization of the major paragneiss, northwest Adirondack mountains, New York, part 1. Geol Soc Am Bull 69:1369–1414

Evans BW, Patrick BE (1987) Phengite-3T in high-pressure metamorphosed granitic orthogneisses, Seward Peninsula, Alaska. Can Mineral 25:141–158

Ferry JM (1979) Reaction mechanism, physical conditions, and mass transfer during hydrothermal alteration of mica and feldspar in granitic rocks from south-central Maine, USA. Contrib Mineral Petrol 68:125–139

Finger F, Clemens JD (1995) Migmatization and "secondary" granitic magmas: effects of emplacement and crystallization of "primary" granitoids in southern Bohemia, Austria. Contrib Mineral Petrol 120:311–326

Franz L, Harlov DE (1998) High-grade K-feldspar veining in granulites from the Ivrea-Verbano Zone, northern Italy: fluid flow in the lower crust and implications for granulite facies genesis. J Geol 106:455–472

Frey M, Hunziker JC, Jäger E, Stern WB (1983) Regional distribution of white K-mica polymorphs and their phengite content in the Central Alps. Contrib Mineral Petrol 83:185–197

Frost BR, Frost CD, Hulsebosch TP, Swapp SM (2000) Origin of the Charnockites of the Louis Lake Batholith, Wind River Range, Wyoming. J Petrol 41:1759–1776

Früh-Green GL (1994) Interdependence of deformation, fluid infiltration and reaction progress recorded in eclogitic metagranitoids (Sesia Zone, Western Alps). J Metamorph Geol 12:327–343

Ganguly J, Singh RN, Ramana DV (1995) Thermal perturbation during charnockitization and granulite facies metamorphism in southern India. J Metamorph Geol 13:419–430

Gibbons W, Horak J (1984) Alpine metamorphism of Hercynian hornblende granodiorite beneath the blueschist facies schistes lustrés nappe of NE Corsica. J Metamorph Geol 2:95–113

Grant JA (1986) Quartz-phlogopite-liquid equilibria and origins of charnockites. Am Mineral 71:1071–1075

Grapes R, Otsuki M (1983) Peristerite composition in quartzofeldspathic schists, Franz Josef Fox Glacier Area, New Zealand. J Metamorph Geol 1:47–61

Grapes R, Watanabe T (1992) Paragenesis of titanite in metagraywackes of the Franz Josef-Fox Glacier area, Southern Alps, New Zealand. Eur J Mineral 3:547–555

Hirajima T, Compagnoni R (1993) Petrology of a jadeite-quartz/coesite-almandine-phengite fels with retrograde ferro-nyböite from Dora-Maira Massif, Western Alps. Eur J Mineral 5:943–955

Holtz F, Johannes W, Tamic N, Behrens H (2001) Maximum and minimum water contents of granitic melts generated in the crust: a reevaluation and implications. Lithos 56:1–14

Koons PO, Thompson AB (1985) Non-mafic rocks in the greenschist, blueschist and eclogite facies. Chem Geol 50:3–30

Kriegsman LM (2001) Partial melting, partial melt extraction and partial back reaction in anatectic migmatites. Lithos 56:75–96

Le Goff E, Ballèvre M (1990) Geothermobarometry in albite-garnet orthogneisses: a case study from the Gran Paradiso nappe (Western Alps). Lithos 25:261–280

Leake BE (1998) Widespread secondary Ca garnet and other Ca silicates in the Galway Granite and its satellite plutons caused by fluid movements, western Ireland. Mineral Mag 62:381–386

Luth WD, Jahns RH, Tuttle OF (1964) The granite system at pressures of 4 to 10 kilobars. J Geophys Res 69:659–773

Massone HJ, Chopin CP (1989) P–T history of the Gran Paradiso (western Alps) metagranites based on phengite geobarometry. In: Daly JS, Cliff RA, Yardley BWD (eds) Evolution of metamorphic belts. Geological Society Special Publication. Blackwell, Oxford, pp 545–550

Mehnert KR (1968) Migmatites and the origin of granitic rocks. Elsevier, Amsterdam

Morteani G, Raase P (1974) Metamorphic plagioclase crystallization and zones of equal anorthite content in epidote-bearing, amphibole-free rocks of the western Tauernfenster, eastern Alps. Lithos 7:101–111

Mottana A, Carswell DA, Chopin C, Oberhänsli R (1990) Eclogite facies mineral parageneses. In: Carswell DA (ed) Eclogite facies rocks. Blackie, Glasgow, pp 14–52
Nijland TG, Senior A (1991) Sveconorwegian granulite facies metamorphism of polyphase
   migmatites and basic dykes, south Norway. J Geol 99:515–525
Nitsch KH (1970) Experimentelle Bestimmung der oberen Stabilitätsgrenze von Stilpnome-
   lan. Fortschr Mineral 47:48–49
Oberhänsli R, Hunziker JC, Martinotti G, Stern WB (1985) Geochemistry, geochronology
   and petrology of Monte Mucrone: an example of eo-alpine eclogitization of Permian
   granitoids in the Sesia-Lanzo zone, western Alps, Italy. Chem Geol Isotope Geosci Sect
   52:165–184
Olsen SN (1984) Mass-balance and mass-transfer in migmatites from the Colorado Front
   Range. Contrib Mineral Petrol 85:30–44
Osanai Y, Komatsu M, Owada M (1991) Metamorphism and granite genesis in the Hidaka
   Metamorphic Belt, Hokkaido, Japan. J Metamorph Geol 9:111–124
Patiño Douce AE, Humphreys ED, Johnston AD (1990) Anatexis and metamorphism in tec-
   tonically thickened continental crust exemplified by the Sevier hinterland, western North
   America. Earth Planet Sci Lett 97:290–315
Patrick B (1995) High-pressure–low-temperature metamorphism of granitic orthogneiss in
   the Brooks Range, northern Alaska. J Metamorph Geol 13:111–124
Santosh M, Jackson DH, Harris NB, Mattey DP (1991) Carbonic fluid inclusions in South
   Indian granulites: evidence for entrapment during charnockite formation. Contrib
   Mineral Petrol 108:318–330
Schwarcz HP (1966) Chemical and mineralogic variations in an Arkosic Quartzite during
   progressive regional metamorphism. Geol Soc Am Bull 77:509–532
Steck A (1976) Albit-Oligoklas-Mineralgesellschaften der Peristeritlücke aus alpinmetamor-
   phen Granitgneisen des Gotthardmassivs. Schweiz Mineral Petrogr Mitt 56:269–292
Steck A, Burri G (1971) Chemismus und Paragenesen von Granaten aus Granitgneisen der
   Grünschiefer- und Amphibolitfazies der Zentralalpen. Schweiz Mineral Petrogr Mitt
   51:534–538
Thurston SP (1985) Structure, petrology, and metamorphic history of the Nome Group
   blueschist terrane, Salmon Lake area, Seward Peninsula, Alaska. Geol Soc Am Bull
   96:600–617
Török K (1998) Magmatic and high-pressure metamorphic development of orthogneisses in
   the Sopron area, Eastern Alps (W-Hungary). Neues Jahrb Mineral Abh 173:63–91
Tulloch AJ (1979) Secondary Ca-Al silicates as low-grade alteration products of granitoid
   biotite. Contrib Mineral Petrol 69:105–117
Voll G (1976) Recrystallisation of quartz, biotite and feldspars from Erstfeld to the Leventina
   nappe, Swiss Alps, and its geological significance. Schweiz Mineral Petrogr Mitt 56:641–
   647

List of Abbreviations

Ab	albite	Cbz	chabazite	Fa	fayalite
Acm	acmite	Cc	chalcocite	Fac	ferroactinolite
Act	actinolite	Ccl	chrysocolla	Fcp	ferrocarpholite
Adr	andradite	Ccn	cancrinite	Fed	ferroedenite
Agt	aegirine-augite	Ccp	chalcopyrite	Flt	fluorite
Ak	åkermanite	Cel	celadonite	Fo	forsterite
Alm	almandine	Cen	clinoenstatite	Fpa	ferropargasite
Aln	allanite	Cfs	clinoferrosilite	Fs	ferrosilite (ortho)
Als	aluminosilicate	Chl	chlorite	Fst	fassite
Am	amphibole	Chn	chondrodite	Fts	ferrotscher-
An	anorthite	Chr	chromite		makite
And	andalusite	Chu	clinohumite	Gbs	gibbsite
Anh	anhydrite	Cld	chloritoid	Ged	gedrite
Ank	ankerite	Cls	celestite	Gh	gehlenite
Anl	analcite	Cp	carpholite	Gln	glaucophane
Ann	annite	Cpx	Ca clinopyroxene	Glt	glauconite
Ant	anatase	Crd	cordierite	Gn	galena
Ap	apatite	Crn	carnegieite	Gp	gypsum
Apo	apophyllite	Crn	corundum	Gr	graphite
Apy	arsenopyrite	Crs	cristroballite	Grs	grossular
Arf	arfvedsonite	Cs	coesite	Grt	garnet
Arg	aragonite	Cst	cassiterite	Gru	grunerite
Atg	antigorite	Ctl	chrysotile	Gt	goethite
Ath	anthophyllite	Cum	cummingtonite	Hbl	hornblende
Aug	augite	Cv	covellite	Hc	hercynite
Ax	axinite	Czo	clinozoisite	Hd	hedenbergite
Bhm	boehmite	Dg	diginite	Hem	hematite
Bn	bornite	Di	diopside	Hl	halite
Brc	brucite	Dia	diamond	Hs	hastingsite
Brk	brookite	Dol	dolomite	Hu	humite
Brl	beryl	Drv	dravite	Hul	heulandite
Brt	barite	Dsp	diaspore	Hyn	haüyne
Bst	bustamite	Eck	eckermannite	Ill	illite
Bt	biotite	Ed	edenite	Ilm	ilmenite
Cal	calcite	Elb	elbaite	Jd	jadeite
Cam	Ca clinoamphi-	En	enstatite (ortho)	Jh	johannsenite
	bole	Ep	epidote	Kfs	K-feldspar

Kln	kaolinite	Ntr	natrolite	Ser	sericite
Kls	kalsilite	Oam	orthoamphibole	Sil	sillimanite
Krn	kornerupine	Ol	olivine	Sp	sphalerite
Krs	kaersutite	Omp	omphacite	Spd	spodumene
Ktp	kataphorite	Opx	orthopyroxene	Spl	spinel
Ky	kyanite	Or	orthoclase	Spr	sapphirine
Lct	leucite	Osm	osumilite	Sps	spessartine
Lm	limonite	Pct	pectolite	Srl	schorl
Lmt	laumontite	Pen	protoenstatite	Srp	serpentine
Lo	loellingite	Per	periclase	St	staurolite
Lpd	lepidolite	Pg	paragonite	Stb	stilbite
Lws	lawsonite	Pgt	pigeonite	Stp	stilpnomelane
Lz	lizardite	Phe	phengite	Str	strontianite
Mag	magnetite	Phl	phlogopite	Sud	sudoite
Mcp	magnesio-carpholite	Pl	plagioclase	Tlc	talc
		Pmp	pumpellyite	Tmp	thompsonite
Mel	melilite	Pn	pentlandite	Toz	topaz
Mgh	maghemite	Po	pyrrhotite	Tr	tremolite
Mgs	magnesite	Prg	pargasite	Trd	tridymite
Mkt	magnesiokato-phorite	Prh	prehnite	Tro	troilite
		Prl	pyrophyllite	Ts	tschermakite
Mnt	montmorillonite	Prp	pyrope	Ttn	titanite
Mnz	monazite	Prv	perovskite	Tur	tourmaline
Mo	molybdenite	Py	pyrite	Usp	ulvöspinel
Mrb	magnesiorie-beckite	Qtz	quartz	Ves	vesuvianite
		Rbk	riebeckite	Vrm	vermiculite
Mrg	margarite	Rdn	rhodonite	Wa	wairakite
Ms	muscovite	Rds	rhodochrosite	Wo	wollastonite
Mtc	monticellite	Rt	rutile	Wth	witherite
Mul	mullite	Sa	sanidine	Wus	wüstite
Ne	nepheline	Scp	scapolite	Zo	zoisite
Nrb	norbergite	Sd	siderite	Zrn	zircon
Nsn	nosean	Sdl	sodalite		

Subject Index

Printing (Computer to Film): Saladruck Berlin
Binding: Stürtz AG, Würzburg